Political Ecology of Industrial Crops

This book employs a political ecology lens to unravel how industrial crops catalyse ecological, agrarian, socioeconomic, and institutional transformation.

Using the conceptual tools and perspectives of political ecology, namely multi-scalar analysis and attention to marginalisation, social difference, and discourses and narratives, this volume provides a critical and comprehensive assessment of the transformative power of industrial cropping systems. It presents a truly international overview by drawing on a range of case studies from the global South, including soybeans in South America, cashew nuts in Guinea Bissau, cotton in India, maize in China, jatropha in Ghana, sugarcane in Peru and Eswatini, and oil palm in Ghana and Peru. The unique case studies are put into perspective with chapters introducing the key concepts of political ecology and critical dimensions of industrial cropping systems related to large-scale land acquisitions, land grabbing, and marginal land. The individual chapters employ different approaches all rooted in political ecology, thus offering a rich over-view of how the field engages with such cropping systems. Overall, this volume contains valuable propositions for improving current policies and practices in industrial crop settings in both developed and developing countries.

Through its comprehensive and interdisciplinary outlook, this volume will be of great interest to students and scholars of political ecology, agrarian studies, development studies, and ecological economics.

Abubakari Ahmed is a lecturer at the S.D. Dombo University of Business and Integrated Development Studies (SDD-UBIDS), Ghana.

Alexandros Gasparatos is an associate professor at the University of Tokyo and a visiting associate professor at the United Nations University, Japan.

Other books in the Earthscan Food and Agriculture Series

Geographical Indication and Global Agri-Food
Development and Democratization
Edited by Alessandro Bonanno, Kae Sekine and Hart N. Feuer

Multifunctional Land Uses in Africa
Sustainable Food Security Solutions
Elisabeth Simelton and Madelene Ostwald

Food Security Policy, Evaluation and Impact Assessment
Edited by Sheryl L. Hendriks

Transforming Agriculture in Southern Africa
Constraints, Technologies, Policies and Processes
Edited by Richard A. Sikora, Eugene R. Terry, Paul L. G. Vlek and Joyce Chitja

Home Gardens for Improved Food Security
Edited by D. Hashini Gelhena Dissanayake and Karimbhai M. Maredia

The Good Farmer
Culture and Identity in Food and Agriculture
Rob J.F. Burton, Jérémie Forney, Paul Stock and Lee-Ann Sutherland

Deep Agroecology and the Homeric Epics
Global Cultural Reforms for a Natural-Systems Agriculture
John W. Head

Fighting for Farming Justice
Diversity, Food Access and the USDA
Terri R. Jett

Political Ecology of Industrial Crops
Edited by Abubakari Ahmed and Alexandros Gasparatos

For more information about this series, please visit: www.routledge.com/books/series/ECEFA/

Political Ecology of Industrial Crops

Edited by
Abubakari Ahmed and
Alexandros Gasparatos

Routledge
Taylor & Francis Group
LONDON AND NEW YORK

earthscan
from Routledge

First published 2022
by Routledge
2 Park Square, Milton Park, Abingdon, Oxon OX14 4RN

and by Routledge
605 Third Avenue, New York, NY 10158

Routledge is an imprint of the Taylor & Francis Group, an informa business

British Library Cataloguing-in-Publication Data
A catalogue record for this book is available from the British Library

Library of Congress Cataloging-in-Publication Data
Names: Ahmed, Abubakari, editor. | Gasparatos, Alexandros, editor.
Title: Political ecology of industrial crops / edited by Abubakari Ahmed
and Alexandros Gasparatos.
Description: Milton Park, Abingdon, Oxon ; New York, NY : Routledge, 2022. |
Series: Earthscan food and agriculture |
Includes bibliographical references and index.
Identifiers: LCCN 2021009837 (print) | LCCN 2021009838 (ebook)
Subjects: LCSH: Crops–Political aspects. | Political ecology.
Classification: LCC SB91 .P76 2022 (print) |
LCC SB91 (ebook) | DDC 630–dc23
LC record available at https://lccn.loc.gov/2021009837
LC ebook record available at https://lccn.loc.gov/2021009838

ISBN: 978-0-367-35670-5 (hbk)
ISBN: 978-1-032-06213-6 (pbk)
ISBN: 978-0-429-35110-5 (ebk)

Typeset in Bembo
by Newgen Publishing UK

Contents

Acknowledgements

We wish to thank our editorial team at Routledge, Hannah Ferguson, and John Baddeley for their guidance, support, and patience throughout the development of this volume.

This volume would not have been possible without the generous support of the Asahi Glass Foundation through a "*Continuation Grant on Humanities and Social Sciences*" and a "*Research Grant on Humanities and Social Sciences*", the Japan Science and Technology Agency (JST) for the Belmont Forum project FICESSA, and the German Academic Exchange Service (DAAD) through the climapAfrica Postdoctoral Fellowship programme.

Figures

Tables

About the authors

Abubakari Ahmed is a lecturer at the S.D. Dombo University of Business and Integrated Development Studies (SDD-UBIDS), Ghana.

Andrew Flachs is an associate professor at Purdue University, USA.

Alexandros Gasparatos is an associate professor at the University of Tokyo and a visiting associate professor at the United Nations University, Japan.

Richard Helliwell is a researcher (Forsker II) at Ruralis – Institute for Rural and Regional Research, Norway.

Marcin Pawel Jarzebski is a project assistant professor at the University of Tokyo, Tokyo.

Brandon D. Lundy is a professor at Kennesaw State University, USA.

Graham von Maltitz is a research fellow at the University of Stellenbosch, South Africa.

William G. Moseley is the Dewitt Wallace professor of Geography and the director of the Program for Food, Agriculture and Society, at Macalester College, USA.

Shakespear Mudombi is an economist at the Trade and Industrial Policy Strategies (TIPS), South Africa.

Gustavo de L.T. Oliveira is an assistant professor at the University of California, Irvine, USA.

Nikole Roland is an MSc student at the University of Tokyo, Japan.

Frida Segura-Urrunaga is a lawyer and former MSc student at the Pontifical Catholic University of Peru, Peru.

Orla Shortall is an agricultural sociologist at the James Hutton Institute, UK.

Patricia Urteaga-Crovetto is a professor at the Pontifical Catholic University of Peru, Peru.

Li Zhang is a visiting assistant professor at the University of California, Irvine, USA.

Part I
Introductory

1 Industrial crops as agents of transformation

Justifying a political ecology lens

Abubakari Ahmed, Marcin Pawel Jarzebski and Alexandros Gasparatos

Industrial crops: Definition, uses, and modes of production

Industrial crops are crops whose products have important non-food uses such as fibre, bioenergy, and biomaterials (Singh, 2010) (Table 1.1). Some industrial crops such as cotton, tobacco, or jatropha have products with strictly non-food uses, as they cannot be used for direct food consumption or as an input to the food industry. Other industrial crops such as oil palm, sugarcane, and soybean have both products with non-food uses and products that are integral components of the food industry, without being staple food crops (Singh, 2010) (Table 1.1). Finally, a few staple food crops such as maize, wheat, and potatoes have products with important industrial uses (Table 1.1) (Chapter 10).

Some industrial crops are practically mono-functional, as their products are used solely or overwhelmingly for a single end-use such as fibre (e.g. cotton), recreation (e.g. tobacco), energy (e.g. jatropha), or biomaterials (e.g. rubber). Other industrial crops, such as sugarcane, are multi-functional, as their products have multiple uses in the food (i.e. sugar), energy (i.e. ethanol, electricity), and chemical (i.e. bioplastics) industries (Singh, 2010).

In this respect, industrial crops are defined by their functionality and end use (Table 1.1) rather than the mode of their production. In fact, as discussed below, industrial crops are produced in very diverse systems across the world ranging from large-scale intensified plantations to extensive smallholder farms (Singh, 2010).

For this edited volume, we adopt a rather broad definition of industrial crops that encompasses crops whose products have clear industrial uses such as bioenergy and biomaterials and non-food crops whose products are important inputs in the food industry (e.g. cocoa, coffee). It is worth noting that some crops can be considered as industrial in some geographical contexts, and non-industrial in others. For example, maize is clearly an industrial crop in the US, as it is a major feedstock for bioenergy and biomaterials. However, in other geographical contexts, maize has a few industrial uses (e.g. China) (Chapter 10) or practically no such uses (e.g. in sub-Sahara Africa) (Jarzebski et al., 2020).

Table 1.1 Major industrial crops, uses, and production patterns in 2019

Crop	Uses					Production (Mt)	Main producing countries
	Food industry	Bioenergy	Fibre	Biomaterial/biochemials	Recreation		
Cashew nuts	✓					4	Ivory Coast, India, Burundi, Vietnam, Philippines
Cocoa	✓					5.6	Ivory Coast, Ghana, Indonesia, Nigeria, Ecuador
Coffee	✓					10	Brazil, Vietnam, Colombia, Indonesia, Ethiopia
Cotton		✓	✓			82.6	China, India, USA, Brazil, Pakistan
Jatropha		✓	✓	✓		NA	China, Indonesia, Ghana, Tanzania, Mozambique
Jute	✓		✓			3.4	India, Bangladesh, China, Uzbekistan, Nepal
Linseed	✓	✓		✓		3.1	Kazakhstan, Russia, Canada, China, USA
Maize	✓	✓		✓		1148.5	USA, China, Brazil, Argentina, Ukraine
Miscanthus		✓				NA	USA
Mustard seed	✓	✓				0.7	Nepal, Russia, Canada, Myanmar, Ukraine
Oil palm	✓	✓		✓		410.7	Indonesia, Malaysia, Thailand, Nigeria, Colombia
Potatoes	✓	✓		✓		370.4	China, India, Russia, Ukraine, USA
Rapeseed	✓	✓		✓		70.5	Canada, China, India, France, Ukraine
Rubber		✓		✓		14.6	Thailand, Indonesia, Vietnam, India, China
Sisal			✓			0.2	Brazil, Tanzania, Kenya, Madagascar, Haiti
Soybean	✓	✓				333.7	Brazil, USA, Argentina, China, India
Sugar beet	✓	✓				278.5	Russia, France, Germany, USA, Turkey
Sugarcane	✓	✓		✓		1949.3	Brazil, India, Thailand, China, Pakistan
Sunflower seed	✓	✓		✓		56.1	Russia, Ukraine, Argentina, Romania, China
Switchgrass		✓				NA	USA
Tea	✓					6.5	China, India, Kenya, Sri Lanka, Vietnam
Tobacco		✓			✓	6.7	China, India Brazil, Zimbabwe, USA
Wheat	✓	✓				765.8	China, India, Russia, USA, France

Note: There are no reliable global statistics about the production and major producing countries for jatropha, miscanthus, and switchgrass. The major producing countries for these three crops were identified through a literature review.

Source: (FAOSTAT, 2021).

Size

	Smallholder schemes 1s - 10s ha	Large-scale projects 100s - 1000s ha
Local	**Type I projects** Use at the village or farm level (e.g. small-scale biofuel projects for rural electrification)	**Type II projects** Use at plantation level (e.g. large commercial farmers or mines producing biofuel for own use)
National or international	**Type III projects** Outgrowers linked to commercial plantations	**Type IV projects** Large-scale commercial plantations

Market

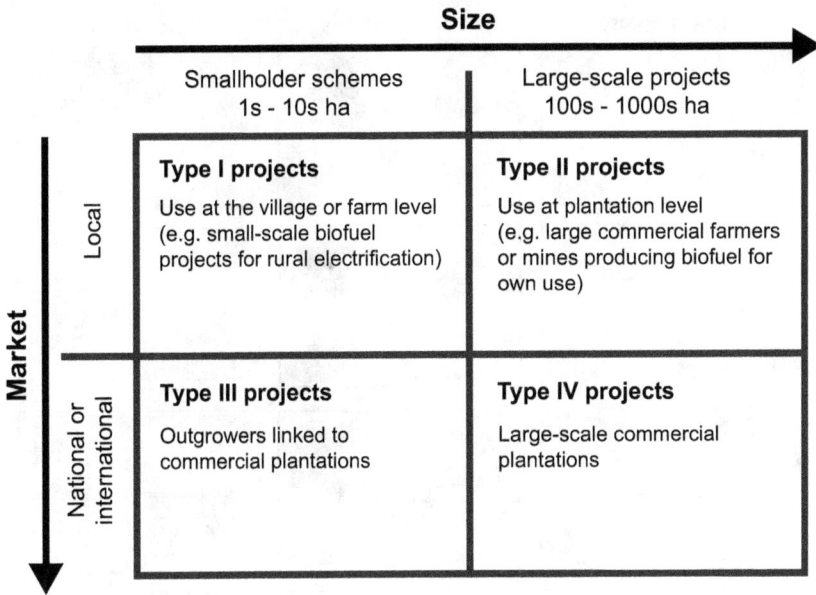

Figure 1.1 Main modes of industrial crop production.
Adapted from (Gasparatos et al., 2015).

There is a large variability in industrial crop production systems depending on the crop, intended market, production location, and various other contextual environmental, socioeconomic, and institutional factors (Figure 1.1). Mindful of the large diversity of industrial crop systems globally, the most common production modes include: (a) large-scale systems (e.g. plantations); (b) smallholder-based schemes; and (c) hybrid systems (Figure 1.2). Such systems are integrated in very diverse landscapes containing different configurations of agricultural land and natural vegetation (Gasparatos et al., 2018).

Large-scale production systems can be sub-divided into large-scale farming and plantation farming depending on ownership and labour practices (Gibbon, 2011). A common characteristic of all these systems is their much larger size compared to family farms (see below), which can extend from a few tens of hectares to several thousands of hectares depending on the crop and the region (Saravia Matus et al., 2013; Smalley, 2013) (Chapter 3). Thus, large-scale production systems tend to convert rather extensive areas, adopt monocultural practices, entail land consolidation processes, and rely on hired labour (Gibbon, 2011; Smalley, 2013). The owners and investors of large-scale production systems are equally diverse, usually including national or multi-national private companies, state agencies, parastatal bodies, or joint partnerships (Chapters 3, 5, 8, 9, and 11).

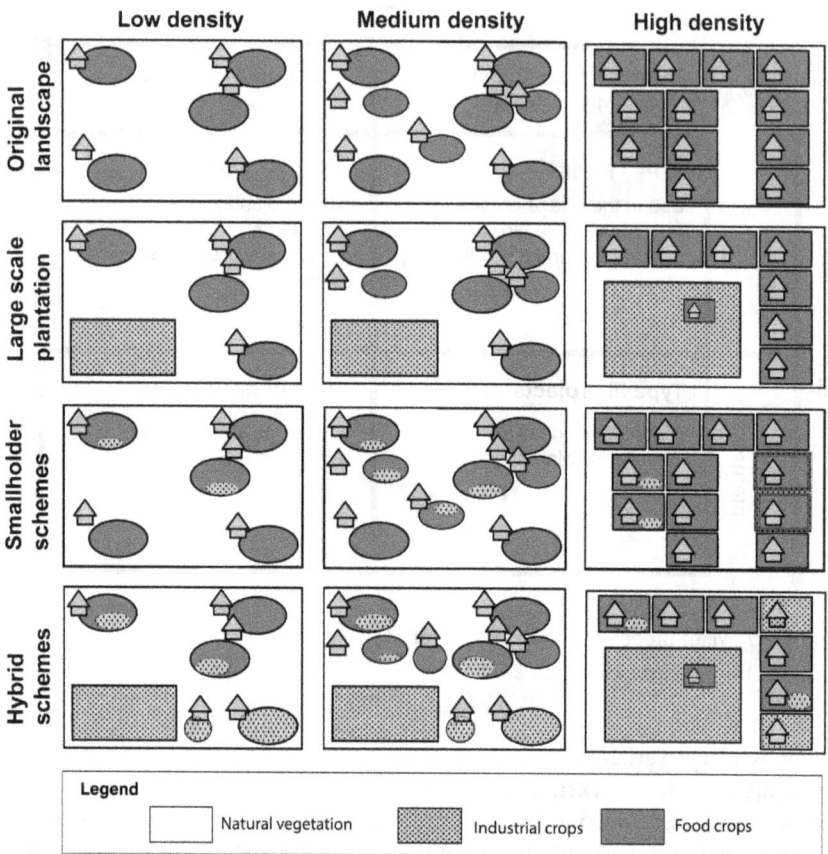

Figure 1.2 Direct land use change from the main modes of industrial crop production. Adapted from (Gasparatos et al., 2018).

Smallholder-based production is mainly undertaken at the level of individual family farms and can take many forms depending on, among others, land allocation, land consolidation, agricultural practices, and integration in industrial crop value chains (Jelsma et al., 2017; Jezeer et al., 2018; Kuivanen et al., 2016; von Maltitz et al., 2019). For example, some smallholders tend to allocate part of their land for industrial crop production, setting aside the remainder for food crop production, while others specialise entirely in industrial crop production, essentially converting all their land into small industrial crop plantations. Some crops such as sugarcane or oil palm require the almost-complete farm conversion, especially if produced under irrigated conditions or outgrower arrangements (von Maltitz et al., 2019; Jelsma et al., 2017). In terms of market integration, industrial crop smallholders can enter in dedicated contractual arrangements with plantations (i.e. sell all their output) receiving in return agricultural inputs and extension services (e.g. outgrower schemes), while others can remain independent selling to different buyers depending on market signals (Ahmed et al., 2019c; von Maltitz et al., 2019) (Chapters 8 and 11).

Hybrid production systems combine large-scale production areas (e.g. core plantations) and smallholder-based schemes (Brüntrup et al., 2018) (Chapter 8). As mentioned above, these smallholders can be contractually linked to single buyers through outgrower schemes, or sell to multiple buyers depending on market signals. Hybrid systems are more common for crops such as sugarcane and oil palm that are perishable and whose production benefits from achieving economies of scale (Ahmed et al., 2019c; von Maltitz et al., 2019) (Chapters 8 and 11).

History and drivers of industrial crop production

In many parts of the world, industrial crop production has traditionally been a major agricultural activity with a long legacy. For example, industrial crops such as cotton, flax, and sugarcane have had multiple centres of domestication across the globe, with their production spanning many thousands of years. Other industrial crops such as cocoa, coffee, tea, and tobacco also have long histories of domestication and use in some parts of the world but have become much more prevalent in the past centuries following the growing trade between Europe, Asia, and Africa, as well as the colonisation of the Americas.

The production of many industrial crops expanded rapidly during the colonial period, particularly in regions colonised by European powers. The booming production of crops such as cotton, rubber, and sugarcane sought to fuel the burgeoning manufacturing sector in Europe during the industrial revolution (Chapter 2). During that period, plantations were the dominant mode of industrial crop production in much of the global South (Gibbon, 2011; Kenney-Lazar and Ishikawa, 2019; Rogers, 2015) (Chapter 3). More often than not, coercion was used to obtain access land and labour, which were essentially seen as the property of the colonial plantation owners (Austin, 2009) (Chapter 3). Coercive practices were also very prevalent when forcing smallholders to produce industrial crops and be integrated into internationalised value chains (Chapter 7). Industrial crops almost invariably transformed profoundly local agrarian systems through multi-dimensional processes, including (a) the actual introduction of new crops and production practices (and the land and labour re-allocation this entailed), (b) the creation of new elites, social classes, and gender roles, and (c) the stronger linkage of farmers and rural areas to international commodity chains (Chapters 2–3). Arguably, such processes coupled with the increasing industrial crop demand gradually precipitated the "cash-crop revolution" of the post-colonial period (Austin, 2009).

During the post-colonial era, many countries in the global South actively attempted to modernise their agricultural sectors and integrate them better into the global economic system through industrial crops (Mellor and Malik, 2017; Timberlake, 1985; van Vliet et al., 2015). For example, some countries in

Southeast Asia became the leading global producers of oil palm and rubber to the extent that these crops dominated their agricultural sectors (FAOSTAT, 2021). Some countries in sub-Saharan Africa went even further to "transform" to cash crop economies as industrial crops became the main exports and drivers of economic growth. Some examples include Malawi with tobacco and sugarcane (Chinangwa et al., 2017), Swaziland and Mauritius with sugarcane (Kwong, 2005; Terry and Ogg, 2017) (Chapter 11), Burkina Faso with cotton (Vitale, 2018), and Guinea Bissau with cashew nuts (Chapter 6).

Although the production trajectories and driving forces have been very different between crops and regions (Chapter 3), it is commonly observed that during the post-colonial era, smallholders became much more prominent in many industrial crop production systems across the global South (Austin, 2009; Byerlee, 2014; Young, 1970). For some crops, this marked a shift from predominately large-scale production approaches to smallholder-based ones (Byerlee, 2014) (see "History and drivers of industrial crop production"). Central to this shift was often the discourse of the comparative advantage in the production of certain industrial crops, and their framing as cash crops ideal for smallholder income generation and poverty alleviation (Austin, 2009; Bryceson, 2019; Hall et al., 2017) (Chapters 2 and 6). As a result, such crops were promoted to subsistence farmers by any means possible, thereby further affecting local agrarian structures (Austin, 2009; Masanjala, 2006) (Chapters 2–3, 7, and 11). This shift towards smallholders came in tandem with the Green Revolution, precipitating major changes in agricultural practices, including intensification through high agrochemical/fertiliser input and/or mechanisation (Chapter 7) (Dawson et al., 2016; FAO, 2015).

Industrial crop production has consistently increased in the past decades, both in terms of area and output, in order to cater to their ever-growing global demand (Chapter 3) (Figure 1.3, Figure 1.4). In the global South, the expansion of some industrial crops such as cocoa, coffee, and cotton has been mainly through smallholders and was assisted through development and poverty alleviation narratives (Chapters 7 and 11). However, for some other industrial crops such as oil palm, sugarcane, and soybean, there has been a resurging interest in their large-scale production (see "History and drivers of industrial crop production"). In the mid-2000s an interconnected set of international circumstances created the pre-conditions for a spike in large-scale land acquisitions (LSLAs), which was assisted through narratives of economic growth and energy security (Chapters 3–5). This period has been branded as the land rush, with many international and national investors actively seeking land for large-scale agricultural production, and especially for bioenergy crops (Cotula, 2012; Dell'Angelo et al., 2017) (Chapters 3–5 and 8).

Indeed, during the recent land rush, the national governments in many countries of the global South started viewing industrial crops (and the foreign direct investments [FDI] they could attract) as engines of national economic growth and rural development (Gasparatos et al., 2015; Schoneveld, 2014) (see

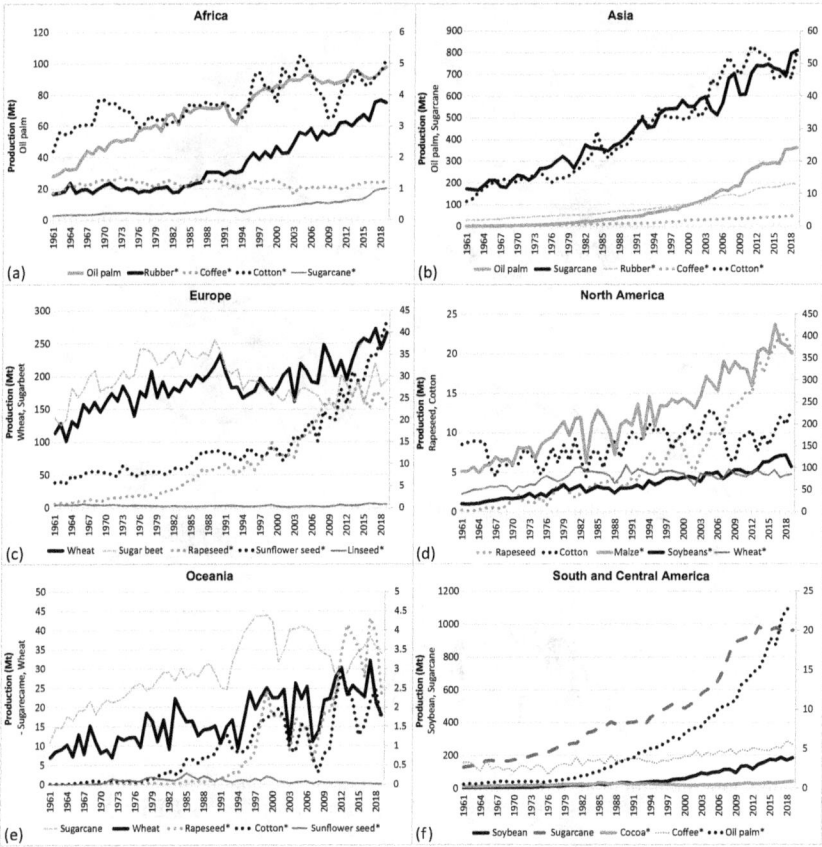

Figure 1.3 Production of major industrial crops in different regions.

Note: The crops included in each panel are not necessarily the five most widely produced in the respective regions in terms of tonnage. The right-hand y-axes apply for crops indicated with asterisks (★).

Source: (FAOSTAT, 2021).

"Socioeconomic transformation"). Many countries opened their rural frontiers for FDIs linked to industrial crops (Giovannetti and Ticci, 2016; Schoneveld, 2014), and especially bioenergy crops that could be exported to the European Union to contribute to its recently enacted biofuel mandate (Chapter 3). Indeed bioenergy crops such as jatropha, oil palm, and sugarcane have accounted for a significant fraction of LSLAs across the global South (Mechiche-Alami et al., 2019; Schoneveld, 2014).

However, the expansion of industrial crops during the recent land rush also entailed very conscious and coordinated efforts to further increase land

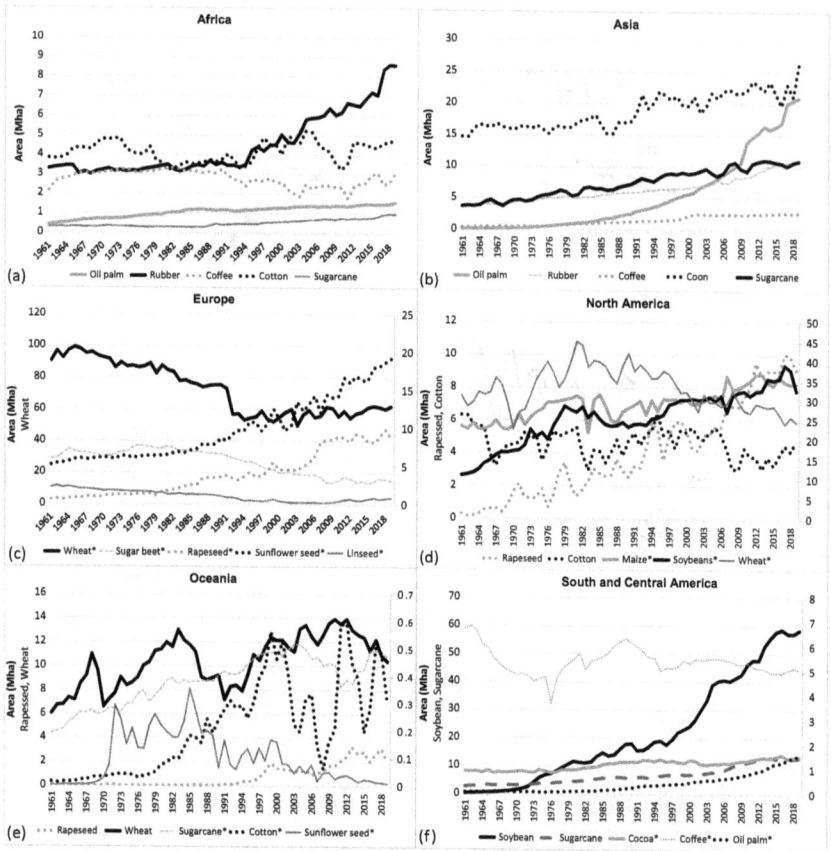

Figure 1.4 Cultivated area for major industrial crops in different regions.

Note: The crops included in each panel are not necessarily the five most widely cultivated in the respective regions in terms of area. The right-hand y-axes apply for crops indicated with asterisks (★).

Source: (FAOSTAT, 2021).

availability for agro-industrial investments, intensify agricultural production practices, and integrate smallholders to global commodity value chains (Chapters 5–7 and 11), all of which have deeply transformed ecosystems, and agrarian and social systems (see "Ecological transformation" through "Socioeconomic transformation"). However, while industrial crop expansion had entailed in the past the rather overt exercise of power from colonial authorities and national governments, it now entails more "subtle" efforts to create (or even manipulate) institutional frameworks that are conducive to the interests of international and national investors, agribusinesses, and other elite actors (Chapters 3–5, and 9–10) (see "Institutional transformation").

Industrial crops as agents of transformation

Ecological transformation

As an agricultural activity, industrial crop production causes direct and indirect land use change. For example, all major types of industrial crop production systems have been associated with the direct conversion of natural vegetation, agricultural land, and other land uses (Gasparatos et al., 2018). Some industrial crop production systems have also been linked to indirect land use change, in that they displace other land uses (especially agricultural land), which in turn requires the conversion of natural vegetation to accommodate the displaced land uses (Khanna and Crago, 2012).[1] Many studies have found that the direct and indirect land use change from the large-scale production of industrial crops causes habitat loss and degradation (including deforestation), biodiversity loss, and carbon stock change. Such examples include the large-scale production of sugarcane (Filoso et al., 2015; Semie et al., 2019), maize (Searchinger et al., 2008; Wright and Wimberly, 2013), oil palm (Savilaakso et al., 2014; Vijay et al., 2016), soybeans (Boerema et al., 2016; Green et al., 2019), and rubber (Ahrends et al., 2015; Warren-Thomas et al., 2018). Similarly, habitat loss and degradation can also be caused through smallholder-based production, e.g. for oil palm (Meijaard et al., 2020). Direct and indirect land use change has also been linked to the degradation and loss of different ecosystem services (Ahmed et al., 2018; Gasparatos et al., 2018).

The commercial production of industrial crops often "demands" substantial yield increases and profit maximisation through the adoption of intensified production practices such as irrigation, mechanisation, and high fertiliser/agrochemical input (Chapter 9). Although such intensification practices are usually associated with large-scale production systems (Chapter 3), it is very common for smallholders to intensify production through such practices (Chapter 7 and 11) (see "Agrarian transformation"). Some of these practices such as irrigation and agrochemical/fertiliser use can affect hydrological cycles and cause pollution, respectively. For example, many studies have identified that the extensive irrigation of sugarcane, oil palm, and soybean can affect hydrological cycles and increase water scarcity (Chiarelli et al., 2020; Hess et al., 2016) (Chapter 3). Similarly, studies have shown that the extensive fertilisers and agrochemical input during sugarcane, rubber, oil palm, maize, and soybean production causes water pollution both adjacent to production areas and further downstream (Donner and Kucharik, 2008; Filoso et al., 2015; Obidzinski et al., 2012) (Chapter 3).

Agrarian transformation

Agrarian transformation is the multidimensional process through which agrarian systems change due to factors as diverse as new production practices, technologies, knowledge, markets, and globalisation, among others (Brockett, 2019; Harriss and Harriss, 1989). Agrarian transformation can manifest in many different ways including, for example, the adoption of new crops and production practices (and the loss/decline of others), the emergence of new social classes, as well as shifts in market orientation, labour dynamics, gender roles, and land tenure rules (Borras, 2009). Mindful of the substantial variation between contexts, below, we outline some of the most important mechanisms and processes underpinning agrarian transformation in the context of industrial crops.

First, the inherent commercialisation of industrial crops creates new roles for farmers and local communities. As most industrial crops are not staple food crops but are produced for profit generation through sales in domestic or international commodity markets (see "Industrial crops: Definition, uses and modes of production"), their commercial production in areas previously under food crop production (and usually subsistence agriculture) marks a radical departure from the pre-existing agricultural systems. This precipitates major changes in the agrarian structure of host rural communities (Mwanika et al., 2020; Woods, 2020), including through the (a) rapid conversion and consolidation of subsistence farmlands and common land into monoculture plantations (Nghiem et al., 2020; Woods, 2020), (b) shifts from smallholder-based food crop production (often subsistence) to paid labour in large-scale industrial crop plantations (Bolt, 2017; Nghiem et al., 2020), and (c) emergence of contract farming (e.g. outgrower schemes) (Hall et al., 2017; Yaro et al., 2017) (see "Industrial crops: Definition, uses and modes of production"). Such mechanisms have profound implications for capital formation and the allocation of land and labour.

Second, again due to their inherent commercialisation, concerns over productivity and profitability have traditionally permeated industrial crop systems (Gibbon, 2011; Smalley, 2013). Thus, there have been conscious efforts to intensify industrial crop production systems, whether large-scale, smallholder-based, or hybrid (see "Industrial crops: Definition, uses and modes of production" and "Ecological transformation"), again having implications for agrarian transformation (Leguizamón, 2014; Rambo, 2017). Intensification is usually approached through the increased mechanisation and use of farm inputs, the adoption of improved crop varieties, and the improvement of access to knowledge, credit, and extension services, all of which have been found to transform agrarian systems (Dawson et al., 2019; Greaves, 2015; Kansanga et al., 2019; Taylor, 2020) (Chapters 7, 9, and 11).

Third, the shifts in agrarian roles and production practices outlined above dictate also changes in rural livelihoods (Dao, 2018; Nyambura, 2015). For example, contract farming, direct employment in plantations, and engagement in other off-farm income-generating activities (within and around large-scale production areas), all shift labour from traditional livelihoods to off-farm or contract-based livelihoods (Bolt, 2017; Borras, 2009; Patel and Mishra, 2019). Such changes in livelihood activities often come in tandem with changes in gender roles and dynamics (Elmhirst et al., 2017; Gyapong, 2019; Hall et al., 2017). Over time, the changes outlined above create or deepen social differentiation, and even catalyse the emergence of distinct social classes, thus cementing agrarian transformation processes (Cousins, 2010) (Chapters 8 and 11).

Socioeconomic transformation

Economic narratives have dominated the discourses related to industrial crops even before the global land rush of the mid-to-late 2000s. On the one hand, industrial crop production has been perceived as a capitalistic means of continuous exploitation during and after the colonial rule (Darkoh and Ould-Mey, 1992; Timberlake, 1985). On the other hand, industrial crop production has been viewed as one of the few comparative economic advantages in some developing countries, and an avenue to modernise their agricultural sectors and integrate them better into the global economic system (Mellor and Malik, 2017; Timberlake, 1985; van Vliet et al., 2015) (Chapters 2 and 6).

This optimistic viewpoint that industrial crop production can spur economic growth, rural development, and poverty alleviation has made them very appealing to many national governments in developing countries, influencing coordinated efforts both to attract land-based investments in rural areas and promote smallholder-based schemes. Conversely, the pessimistic viewpoint that industrial crop production systems are inherently unfair has fuelled many of the discontents about their production and further expansion. These very different viewpoints often converge in areas of industrial crop production, considering that it gives rise to various socioeconomic impacts, which are often unevenly distributed between different stakeholders.

In terms of economic change, industrial crops can have various macro- and micro-level economic impacts, which depend between crops, countries, and production systems (Chapter 3). Some of the more common macro-level economic impacts include (a) national economic growth (e.g. by attracting FDIs and generating taxes, employment, and foreign exchange) and energy security (e.g. by reducing dependence on imported fuels through biofuel production and use) (Arndt et al., 2012) (Chapter 3). However, it has often been pointed out that national economies overlying on industrial crops are quite vulnerable to both short-term price fluctuations and long-term commodity crop drops (e.g. Vitale, 2018) (Chapter 6).

Micro-level economic impacts commonly include (a) income generation through plantation employment and contract farming (Romijn et al., 2014; von Maltitz et al., 2019); (b) generation of secondary employment and off-farm income opportunities (Ahmed et al., 2019c; Hill and Vigneri, 2014); and (c) infrastructure development (Hall et al., 2017; Mudombi et al., 2018). However, despite these rather positive economic impacts, many studies have argued that these benefits are sometimes small, reach few people, and/or unequally distributed (Ahmed et al., 2019b; Ariza-Montobbio et al., 2010; Dao, 2018; Dietz et al., 2014) (Chapters 3, 8, and 11).

Food security has been one of the most commonly discussed social impacts of industrial crop production, especially for bioenergy crops (Chapter 4).

However, the intersection of industrial crop production and food security is multi-faceted with numerous mechanism at play including (Hervas, 2019; Jarzebski et al., 2020) (Chapter 3), including, among others, the (a) loss/decline of food crop production due to cropland conversion and labour diversion; (b) improved ability to purchase food due to plantation wages or smallholder income; (c) reduced ability to buy food due to food price increases; (d) changes in nutritional quality due to time diversion (especially of women) from food preparation to waged employment or smallholder-based production.

Some of the modalities of land acquisition and engagement in industrial crop value chains have been considered as unethical or a source of major inequalities, and have sometimes precipitated social conflicts. For example, many scholars have tracked how the land dispossession and inappropriate compensation following industrial crop LSLAs has escalated to contestations and social conflicts in many rural contexts of the global South (Ahmed et al., 2019a; Dell'Angelo et al., 2017; Mingorría et al., 2014). Other studies have reported that coercion and unethical corporate practices have been employed to gain cheap land and labour, again leading to agitation and social conflict (Brun, 1991; Byerlee, 2014; Cotula, 2012) (Chapter 8). Some of the exploitative labour practices have even been framed as a new form of slavery and colonialisation (Austin, 2009). Very importantly, many of the aspects discussed above are gender-differentiated. Although women can theoretically benefit from engagement in industrial crop production (Arndt et al., 2011), in reality, they often engage differently (and often unfairly) compared to men, and face disproportionally the negative impacts (Behrman et al., 2012; Elmhirst et al., 2017; Fonjong and Gyapong, 2021; Li, 2015).

Institutional transformation

Institutional transformation can be both a facilitator and an outcome of industrial crop expansion. Regarding the former, the change or manipulation (often through illicit means) of policy frameworks was a pre-condition to obtain access to large tracts of land or to allow certain intensified production practices (Chapter 5), while the use of informal practices during LSLAs has led to many instances of land grabbing (Costantino, 2014; White and Dasgupta, 2010). Another example of institutional transformation facilitating industrial crop production has been the reclassification of maize and soybean in China from staple food crops to industrial crops, which is expected to alter production trajectories in the country and other parts of the world (Chapter 10).

Regarding the latter, there have been many examples of how the actual large-scale production of industrial crops transformed the institutions governing access to natural resources, and essentially the distribution of the accruing benefits (Chapter 11). For example, LSLAs and land grabbing catalysed diverse changes in land rights through coerced transfers of lands (Borras and Franco, 2010; Devine, 2016; Kaag and Zoomers, 2014) and the privatisation of common-pool resource, thereby changing land tenure arrangements in rural communities (Kansanga et al., 2019).

More often than not, elites and powerful actors within industrial crop value chains drive these institutional transformations and/or benefit from them. Apart from governments, corporations, and investors, other elite actors include local chiefs, intermediaries in LSLAs, and powerful associations (Borras et al., 2012) (Chapters 3, 5, and 11). These actors often use their privileged position and power to manipulate institutions and draw in the process personal benefits at the expense of communal interest (Ahmed et al., 2019a; Li, 2018; Ndi and Batterbury, 2017; Woods, 2020).

Towards a political ecology of industrial crops

Sections "History and drivers of industrial crop production" through "Industrial crops as agents of transformation" clearly illustrate that industrial crop expansion has various drivers, trajectories, and outcomes, which can be positive or negative depending on the context. Furthermore, such phenomena tend to be tightly interlinked and operate at different temporal and spatial scales. Thus, it is quite difficult to delineate the ecological, agrarian, socioeconomic, and institutional transformations associated with industrial crop expansion in the global South, especially when they unfold in parallel or in close succession (Chapter 2). Arguably, many of these transformative processes (and their underlying mechanisms) are highly political, as they are embedded with the interests of diverse actors, with different powers, using diverse narratives and discourses to achieve their goals (Chapters 2–4).

In this complex and highly value-laden context, political ecology offers an ideal analytical lens to approach and study industrial crop systems in the global South. In a nutshell, political ecology is an interdisciplinary field that mobilises very diverse theoretical, conceptual, and methodological tools to study the change in social-ecological systems (Perreault et al., 2015). Political ecology focusses especially on how political processes and power underlie the uneven distribution of the costs and benefits of environmental change, acknowledging that these are mediated by factors as diverse as class, race, ethnicity, or gender (Robbins, 2012).

Approaches rooted in Political Ecology are increasingly mobilised to study industrial crop systems in very diverse agrarian contexts of the global South (Ahmed et al., 2019a; Bennett et al., 2018; Dietz et al., 2014; Elmhirst et al., 2017; Kenney-Lazar and Ishikawa, 2019). However, despite the proliferation of this literature, it still remains rather fragmented, as studies tend to focus on individual crops (or classes of crops), geographies, or themes (e.g. gender dynamics). This edited volume aims to provide a comprehensive overview of how approaches rooted in Political Ecology can be mobilised to critically analyse the ecological, agrarian, socioeconomic, and institutional transformation caused by industrial crops. The individual chapters focus on different themes, crops, and production systems across the global South as a means of providing a critical, yet comprehensive, outlook of the challenges and promises posed by industrial crop expansion. Furthermore, the chapters employ very different approaches rooted in Political Ecology, thus providing a rich overview of how this interdisciplinary field engages with these diverse cropping systems.

Part I is essentially the introductory part of the edited volume, containing four chapters that collectively frame the subsequent chapters. Chapter 1 introduces key concepts and aspects of industrial crop production, as well as the overall structure of the edited volume (Ahmed et al., 2021). Chapter 2 introduces some of the main dimensions of the field of Political Ecology, and how they can be collectively mobilised to critically study industrial crop systems (Moseley, 2021). Chapter 3 tracks the drivers, trajectories, processes, and outcomes of industrial crop-based LSLAs and land grabbing, especially focussing on jatropha in sub-Saharan Africa, rubber in Southeast Asia, and oil palm in Latin America (Ahmed and Gasparatos, 2021a). Chapter 4 delineates the main elements of the marginal land discourse and how it has been mobilised to justify the expansion of bioenergy crops (Shortall and Helliwell, 2021).

Parts II–IV form the main body of the edited volume. They consist of empirical chapters exploring how different types of industrial crops catalyse ecological, agrarian, socioeconomic, and institutional transformation in different areas, as well as the underlying mechanisms, outcomes, and responses to these transformative processes. Despite significant overlap between the different types of transformation, each part tries to focus more on a distinct type of transformation.

Part II focusses on processes and phenomena associated with ecological transformation. Chapter 5 discusses how institutional reforms and manipulation enabled cocoa, oil palm, and sugarcane LSLAs in Peru, transforming highly biodiverse forests and local communities (Urteaga-Crovetto and Segura-Urrunaga, 2021). Chapter 6 elucidates how cashews have come to dominate the agrarian systems of Guinea-Bissau, invariably transforming natural vegetation and agricultural land and having major ramifications for the national economy and local communities (Lundy, 2021).

Part III focusses on processes and phenomena associated with agrarian transformation. Chapter 7 outlines how organic and genetically modified cotton have come to dominate the Indian cotton smallholder sector, affecting the process farming practices, market connections, and livelihoods (Flachs, 2021). Chapter 8 unravels how oil palm and jatropha LSLAs have catalysed social differentiation and shaped gender roles in the agrarian contexts of Ghana (Ahmed and Gasparatos, 2021b).

Part IV focusses on processes and phenomena associated with socioeconomic and institutional transformation. Chapter 9 provides a comprehensive overview of the history of soybean expansion in Latin America, how political and economic priorities have shaped it, and how it has affected landscapes, agrarian systems, and local communities throughout the continent (Oliveira, 2021). Chapter 10 describes the institutional transformation surrounding the re-classification of maize from being a staple to industrial crop in China and identifies who is expected to benefit and lose from this process (Zhang, 2021). Chapter 11 outlines the history, institutions, and impacts of sugarcane production in the northern Lowveld of Eswatini, and how it has catalysed major changes for land and water access at different levels (Gasparatos et al., 2021).

Part V synthesises and systematises the main insights generated across the individual chapters. Chapter 12 employs the main conceptual tools of Political Ecology, namely multi-scalar analysis, and attention to marginalisation, social difference, and discourses and narratives (Gasparatos and Ahmed, 2021). Through this critical synthesis, Chapter 12 identifies some priority areas for conceptual and empirical research to help understand better the critical aspects of industrial crop production in the global South.

Note

1 Indirect land use change has been mostly discussed in the context of bioenergy crops, but has been extremely difficult to be quantified in an accurate and non-controversial manner (Khanna and Crago, 2012).

References

Ahmed, A., Abubakari, Z., Gasparatos, A., 2019a. Labelling large-scale land acquisitions as land grabs: Procedural and distributional considerations from two cases in Ghana. Geoforum 105, 1–15. doi.org/10.1016/j.geoforum.2019.05.022

Ahmed, A., Campion, B.B., Gasparatos, A., 2019b. Towards a classification of the drivers of jatropha collapse in Ghana elicited from the perceptions of multiple stakeholders. Sustain. Sci. 14, 315–339. doi.org/10.1007/s11625-018-0568-z

Ahmed, A., Dompreh, E., Gasparatos, A., 2019c. Human wellbeing outcomes of involvement in industrial crop production: Evidence from sugarcane, oil palm and jatropha sites in Ghana. PLoS One 14. doi.org/10.1371/journal.pone.0215433

Ahmed, A., Gasparatos, A., 2021a. Political ecology of large-scale land acquisitions and land grabbing for industrial crops, in: Ahmed, A., Gasparatos, A. (Eds.), Political Ecology of Industrial Crops. Routledge, London.

Ahmed, A., Gasparatos, A., 2021b. Changing agrarian dynamics in oil palm and jatropha production areas of Ghana: A feminist political ecology perspective, in: Ahmed, A., Gasparatos, A. (Eds.), Political Ecology of Industrial Crops. Routledge, London.

Ahmed, A., Jarzebski, M.P., Gasparatos, A., 2018. Using the ecosystem service approach to determine whether jatropha projects were located in marginal lands in Ghana: Implications for site selection. Biomass and Bioenergy 114, 112–124. doi.org/10.1016/j.biombioe.2017.07.020

Ahmed, A., Jarzebski, M., Gasparatos, A., 2021. Industrial crops as agents of transformation: Justifying a Political Ecology lens, in: Ahmed, A., Gasparatos, A. (Eds.), Political Ecology of Industrial Crops. Routledge, London.

Ahrends, A., Hollingsworth, P.M., Ziegler, A.D., Fox, J.M., Chen, H., Su, Y., Xu, J., 2015. Current trends of rubber plantation expansion may threaten biodiversity and livelihoods. Glob. Environ. Chang. 34, 48–58. doi.org/10.1016/j.gloenvcha.2015.06.002

Ariza-Montobbio, P., Lele, S., Kallis, G., Martinez-Alier, J., 2010. The political ecology of *Jatropha* plantations for biodiesel in Tamil Nadu, India. J. Peasant Stud. 37, 875–897. doi.org/10.1080/03066150.2010.512462

Arndt, C., Benfica, R., Thurlow, J., 2011. Gender implications of biofuels expansion in Africa: The case of Mozambique. World Dev. 39, 1649–1662. doi.org/10.1016/j.worlddev.2011.02.012

Arndt, C., Pauw, K., Thurlow, J., 2012. Biofuels and economic development: A computable general equilibrium analysis for Tanzania. Energy Econ. 34, 1922–1930. doi.org/10.1016/j.eneco.2012.07.020

Austin, G., 2009. Cash crops and freedom: Export agriculture and the decline of slavery in colonial West Africa. Int. Rev. Soc. Hist. 54, 1–37. doi.org/10.1017/S0020859009000017

Behrman, J., Meinzen-Dick, R., Quisumbing, A., 2012. The gender implications of large-scale land deals. J. Peasant Stud. 39, 49–79. doi.org/10.1080/03066150.2011.652621

Bennett, A., Ravikumar, A., Paltán, H., 2018. The Political Ecology of Oil Palm Company-Community partnerships in the Peruvian Amazon: Deforestation consequences of the privatization of rural development. World Dev. 109, 29–41. doi.org/10.1016/J.WORLDDEV.2018.04.001

Boerema, A., Peeters, A., Swolfs, S., Vandevenne, F., Jacobs, S., Staes, J., Meire, P., 2016. Soybean trade: Balancing environmental and socio-economic impacts of an intercontinental market. PLoS One 11, e0155222. doi.org/10.1371/journal.pone.0155222

Bolt, M., 2017. Becoming and unbecoming farm workers in Southern Africa. Anthropol. South Africa 40, 241–247. doi.org/10.1080/23323256.2017.1406313

Borras, S.M., 2009. Agrarian change and peasant studies: Changes, continuities and challenges – an introduction. J. Peasant Stud. 36, 5–31. doi.org/10.1080/03066150902820297

Borras, S., Franco, J., 2010. From threat to opportunity? Problems with the idea of a "Code of Conduct" for land-grabbing. Yale Hum. Rights Dev. Law J. 13, 507–523.

Borras, S.M., Franco, J.C., Gómez, S., Kay, C., Spoor, M., 2012. Land grabbing in Latin America and the Caribbean. J. Peasant Stud. 39, 845–872. doi.org/10.1080/03066150.2012.679931

Brockett, C.D., 2019. Land, power, and poverty: Agrarian transformation and political conflict in Central America, land, power, and poverty: Agrarian transformation and political conflict in Central America. Routledge, New York. doi.org/10.4324/9780429034060

Brun, T.A., 1991. The nutrition and health impact of cash cropping in west Africa: A historical perspective. World Rev. Nutr. Diet. 65, 124–162.

Brüntrup, M., Schwarz, F., Absmayr, T., Dylla, J., Eckhard, F., Remke, K., Sternisko, K., 2018. Nucleus-outgrower schemes as an alternative to traditional smallholder agriculture in Tanzania – strengths, weaknesses and policy requirements. Food Secur. 10, 807–826. doi.org/10.1007/s12571-018-0797-0

Bryceson, D.F., 2019. Gender and generational patterns of African deagrarianization: Evolving labour and land allocation in smallholder peasant household farming, 1980–2015. World Dev. 113, 60–72. doi.org/10.1016/j.worlddev.2018.08.021

Byerlee, D., 2014. The fall and rise again of plantations in tropical Asia: History repeated? Land 3, 574–597. doi.org/10.3390/land3030574

Chiarelli, D.D., Passera, C., Rulli, M.C., Rosa, L., Ciraolo, G., D'Odorico, P., 2020. Hydrological consequences of natural rubber plantations in Southeast Asia. L. Degrad. Dev. 31, 2060–2073. doi.org/10.1002/ldr.3591

Chinangwa, L., Gasparatos, A., Saito, O., 2017. Forest conservation and the private sector: Stakeholder perceptions towards payment for ecosystem service schemes in the tobacco and sugarcane sectors in Malawi. Sustain. Sci. 12, 727–746. doi.org/10.1007/s11625-017-0469-6

Costantino, A., 2014. Land grabbing in Latin America: Another natural resource curse? Agrar. South J. Polit. Econ. 3, 17–43. doi.org/10.1177/2277976014530217

Cotula, L., 2012. The international political economy of the global land rush: A critical appraisal of trends, scale, geography and drivers. J. Peasant Stud. 39, 649–680. doi.org/10.1080/03066150.2012.674940

Cousins, B., 2010. What is a "smallholder"?: Class-analytic perspectives on small-scale farming and agrarian reform in South Africa, in: Hebinck, P., Shackleton, C. (Eds.), Reforming Land and Resource Use in South Africa: Impact on Livelihoods. Routledge, London, pp. 86–109. doi.org/10.4324/9780203839645

Dao, N., 2018. Rubber plantations and their implications on gender roles and relations in northern uplands Vietnam. Gender, Place Cult. 25, 1579–1600. doi.org/10.1080/0966369X.2018.1553851

Darkoh, M.B.K., Ould-Mey, M., 1992. Cash crops versus food crops in Africa: A conflict between dependency and autonomy. Transafrican J. Hist. 21, 36–50. doi.org/10.2307/24520419

Dawson, N., Martin, A., Camfield, L., 2019. Can agricultural intensification help attain Sustainable Development Goals? Evidence from Africa and Asia. Third World Q. 40, 926–946. doi.org/10.1080/01436597.2019.1568190

Dawson, N., Martin, A., Sikor, T., 2016. Green revolution in sub-Saharan Africa: Implications of imposed innovation for the wellbeing of rural smallholders. World Dev. 78, 204–218. doi.org/10.1016/j.worlddev.2015.10.008

Dell'Angelo, J., D'Odorico, P., Rulli, M.C., Marchand, P., 2017. The tragedy of the grabbed commons: Coercion and dispossession in the global land rush. World Dev. 92, 1–12. doi.org/10.1016/J.WORLDDEV.2016.11.005

Devine, J.A., 2016. Community forest concessionaires: Resisting green grabs and producing political subjects in Guatemala. J. Peasant Stud. 1–20. doi.org/10.1080/03066150.2016.1215305

Dietz, K., Engels, B., Pye, O., Brunnengräber, A., 2014. The Political Ecology of Agrofuels, Routledge ISS Studies in Rural Livelihoods. Routledge, London.

Donner, S.D., Kucharik, C.J., 2008. Corn-based ethanol production compromises goal of reducing nitrogen export by the Mississippi River. Proc. Natl. Acad. Sci. 105, 4513–4518.

Elmhirst, R., Siscawati, M., Sijapati Basnett, B., Ekowati, D., 2017. Gender and generation in engagements with oil palm in East Kalimantan, Indonesia: Insights from feminist political ecology. J. Peasant Stud. 44, 1135–1157. doi.org/10.1080/03066150.2017.1337002

FAO, 2015. The Economic Lives of Smallholder Farmers: An Analysis Based on Household Data from Nine Countries. Food and Agriculture Organisation (FAO), Rome.

FAOSTAT, 2021. Crop Production [WWW Document]. URL www.fao.org/faostat/en/#faq (accessed 2.15.21).

Filoso, S., Do Carmo, J.B., Mardegan, S.F., Lins, S.R.M., Gomes, T.F., Martinelli, L.A., 2015. Reassessing the environmental impacts of sugarcane ethanol production in Brazil to help meet sustainability goals. Renew. Sustain. Energy Rev. doi.org/10.1016/j.rser.2015.08.012

Flachs, A., 2021. The political ecology of genetically modified and organic cotton in India as agents of Agrarian transformation, in: Ahmed, A., Gasparatos, A. (Eds.), Political Ecology of Industrial Crops. Routledge, London.

Fonjong, L.N., Gyapong, A.Y., 2021. Plantations, women, and food security in Africa: Interrogating the investment pathway towards zero hunger in Cameroon and Ghana. World Dev. 138, 105293. doi.org/10.1016/j.worlddev.2020.105293

Gasparatos, A., Ahmed, A., 2021. Political ecology of industrial crops: Towards a synthesis and systematization, in: Ahmed, A., Gasparatos, A. (Eds.), Political Ecology of Industrial Crops. Routledge, London.

Gasparatos, A., Romeu-Dalmau, C., von Maltitz, G.P., Johnson, F.X., Shackleton, C., Jarzebski, M.P., Jumbe, C., Ochieng, C., Mudombi, S., Nyambane, A., Willis, K.J., 2018. Mechanisms and indicators for assessing the impact of biofuel feedstock production on ecosystem services. Biomass Bioenerg. 114, 157–173.

Gasparatos, A., von Maltitz, G.P., Johnson, F.X., Lee, L., Mathai, M., Puppim de Oliveira, J.A., Willis, K.J., 2015. Biofuels in sub-Sahara Africa: Drivers, impacts and priority policy areas. Renew. Sustain. Energy Rev. 45, 879–901. doi.org/10.1016/j.rser.2015.02.006

Gasparatos, A., von Maltitz, G., Roland, N., Ahmed, A., Mudombi, S., Jarzebski, M., 2021. Institutional and socioeconomic transformation from sugarcane expansion in northern Eswatini, in: Ahmed, A., Gasparatos, A. (Eds.), Political Ecology of Industrial Crops. Routledge, London.

Gibbon, P., 2011. Experiences of Plantation and Large-scale Farming in 20th Century Africa (No. 20), DIIS Working Paper 2011. Copenhagen.

Giovannetti, G., Ticci, E., 2016. Determinants of biofuel-oriented land acquisitions in sub-Saharan Africa. Renew. Sustain. Energy Rev. 54, 678–687. doi.org/10.1016/j.rser.2015.10.008

Greaves, M., 2015. The rethinking of technology in class struggle: Communicative affirmation and foreclosure politics. Rethink. Marx. 27, 195–211. doi.org/10.1080/08935696.2015.1007792

Green, J.M.H., Croft, S.A., Durán, A.P., Balmford, A.P., Burgess, N.D., Fick, S., Gardner, T.A., Godar, J., Suavet, C., Virah-Sawmy, M., Young, L.E., West, C.D., 2019. Linking global drivers of agricultural trade to on-the-ground impacts on biodiversity. Proc. Natl. Acad. Sci. U. S. A. 116, 23202–23208. doi.org/10.1073/pnas.1905618116

Gyapong, A.Y., 2019. Land deals, wage labour, and everyday politics. Land 8, 94. doi.org/10.3390/land8060094

Hall, R., Scoones, I., Tsikata, D., 2017. Plantations, outgrowers and commercial farming in Africa: Agricultural commercialisation and implications for agrarian change. J. Peasant Stud. 44, 515–537. doi.org/10.1080/03066150.2016.1263187

Harriss, J., Harriss, B., 1989. Agrarian transformation in the third world, in: Gregory, D., Walford, R. (Eds.), Horizons in Human Geography. Palgrave, London, pp. 258–278. doi.org/10.1007/978-1-349-19839-9_14

Hervas, A., 2019. Land, development and contract farming on the Guatemalan oil palm frontier. J. Peasant Stud. 46, 115–141. doi.org/10.1080/03066150.2017.1351435

Hess, T.M., Sumberg, J., Biggs, T., Georgescu, M., Haro-Monteagudo, D., Jewitt, G., Ozdogan, M., Marshall, M., Thenkabail, P., Daccache, A., Marin, F., Knox, J.W., 2016. A sweet deal? Sugarcane, water and agricultural transformation in sub-Saharan Africa. Glob. Environ. Chang. 39, 181–194. doi.org/10.1016/j.gloenvcha.2016.05.003

Hill, R.V., Vigneri, M., 2014. Mainstreaming gender sensitivity in cash crop market supply chains, in: Gender in Agriculture: Closing the Knowledge Gap. Springer, Netherlands, pp. 315–342. doi.org/10.1007/978-94-017-8616-4_13

Jarzebski, M.P., Ahmed, A., Boafo, Y.A., Balde, B.S., Chinangwa, L., Saito, O., von Maltitz, G., Gasparatos, A., 2020. Food security impacts of industrial crop production in sub-Saharan Africa: A systematic review of the impact mechanisms. Food Secur. 12, 105–135. doi.org/10.1007/s12571-019-00988-x

Jelsma, I., Schoneveld, G.C., Zoomers, A., van Westen, A.C.M., 2017. Unpacking Indonesia's independent oil palm smallholders: An actor-disaggregated approach to identifying environmental and social performance challenges. Land use policy 69, 281–297. doi.org/10.1016/j.landusepol.2017.08.012

Jezeer, R.E., Santos, M.J., Boot, R.G.A., Junginger, M., Verweij, P.A., 2018. Effects of shade and input management on economic performance of small-scale Peruvian coffee systems. Agric. Syst. 162, 179–190. doi.org/10.1016/j.agsy.2018.01.014

Kaag, M.M.A., Zoomers, E.B., 2014. The Global Land Grab: Beyond the Hype. Zed Books, London.

Kansanga, M., Andersen, P., Kpienbaareh, D., Mason-Renton, S., Atuoye, K., Sano, Y., Antabe, R., Luginaah, I., 2019. Traditional agriculture in transition: Examining the impacts of agricultural modernization on smallholder farming in Ghana under the new Green Revolution. Int. J. Sustain. Dev. World Ecol. 26, 11–24. doi.org/10.1080/13504509.2018.1491429

Kenney-Lazar, M., Ishikawa, N., 2019. Mega-plantations in southeast asia landscapes of displacement. Environ. Soc. Adv. Res. 10, 63–82. doi.org/10.3167/ares.2019.100105

Khanna, M., Crago, C.L., 2012. Measuring Indirect land use change with biofuels: Implications for policy. Annu. Rev. Resour. Econ. 4, 161–184. doi.org/10.1146/annurev-resource-110811-114523

Kuivanen, K.S., Alvarez, S., Michalscheck, M., Adjei-Nsiah, S., Descheemaeker, K., Mellon-Bedi, S., Groot, J.C.J., 2016. Characterising the diversity of smallholder farming systems and their constraints and opportunities for innovation: A case study from the Northern Region, Ghana. NJAS – Wageningen J. Life Sci. 78, 153–166. doi.org/10.1016/j.njas.2016.04.003

Kwong, R.N.G.K., 2005. Status of sugar industry in Mauritius: Constraints and future research strategies. Sugar Tech 7, 5–10. doi.org/10.1007/BF02942411

Leguizamón, A., 2014. Modifying Argentina: GM soy and socio-environmental change. Geoforum 53, 149–160. doi.org/10.1016/j.geoforum.2013.04.001

Li, T.M., 2015. Social impacts of oil palm in Indonesia: A gendered perspective from West Kalimantan (No. 124), Social impacts of oil palm in Indonesia: A gendered perspective from West Kalimantan, Occasional Paper. Center for International Forestry Research (CIFOR), Bogor. doi.org/10.17528/cifor/005579

Li, T.M., 2018. After the land grab: Infrastructural violence and the "Mafia System" in Indonesia's oil palm plantation zones. Geoforum 96, 328–337. doi.org/10.1016/j.geoforum.2017.10.012

Lundy, B.D., 2021. The political ecology of cashew pomiculture in Guinea-Bissau, in: Ahmed, A., Gasparatos, A. (Eds.), Political Ecology of Industrial Crops. Routledge, London.

Masanjala, W.H., 2006. Cash crop liberalization and poverty alleviation in Africa: Evidence from Malawi. Agric. Econ. 35, 231–240. doi.org/10.1111/j.1574-0862.2006.00156.x

Mechiche-Alami, A., Piccardi, C., Nicholas, K.A., Seaquist, J.W., 2019. Transnational land acquisitions beyond the food and financial crises. Environ. Res. Lett. 14, 084021. doi.org/10.1088/1748-9326/ab2e4b

Meijaard, E., Brooks, T.M., Carlson, K.M., Slade, E.M., Garcia-Ulloa, J., Gaveau, D.L.A., Lee, J.S.H., Santika, T., Juffe-Bignoli, D., Struebig, M.J., Wich, S.A., Ancrenaz, M., Koh, L.P., Zamira, N., Abrams, J.F., Prins, H.H.T., Sendashonga, C.N., Murdiyarso, D., Furumo, P.R., Macfarlane, N., Hoffmann, R., Persio, M., Descals, A., Szantoi, Z., Sheil, D., 2020. The environmental impacts of palm oil in context. Nat. Plants. doi.org/10.1038/s41477-020-00813-w

Mellor, J.W., Malik, S.J., 2017. The impact of growth in small commercial farm productivity on rural poverty reduction. World Dev. 91, 1–10. doi.org/10.1016/j.worlddev.2016.09.004

Mingorría, S., Gamboa, G., Martín-López, B., Corbera, E., 2014. The oil palm boom: Socio-economic implications for Q'eqchi' households in the Polochic valley, Guatemala. Environ. Dev. Sustain. 16, 841–871. doi.org/10.1007/s10668-014-9530-0

Moseley, W.G., 2021. Political Agronomy 101: An introduction to the political ecology of industrial cropping systems, in: Ahmed, A., Gasparatos, A. (Eds.), Political Ecology of Industrial Crops. Routledge, London.

Mudombi, S., Von Maltitz, G.P., Gasparatos, A., Romeu-Dalmau, C., Johnson, F.X., Jumbe, C., Ochieng, C., Luhanga, D., Lopes, P., Balde, B.S., Willis, K.J., 2018. Multi-dimensional poverty effects around operational biofuel projects in Malawi, Mozambique and Swaziland. Biomass and Bioenergy 114, 41–54. doi.org/10.1016/j.biombioe.2016.09.003

Mwanika, K., State, A.E., Atekyereza, P., Österberg, T., 2020. Colonial legacies and contemporary commercial farming outcomes: sugarcane in Eastern Uganda. Third World Q. 1–19. doi.org/10.1080/01436597.2020.1783999

Ndi, F.A., Batterbury, S., 2017. Land grabbing and the axis of political conflicts: Insights from Southwest Cameroon. Africa Spectr. 52, 33–63. doi.org/10.1177/0002039717 05200102

Nghiem, T., Kono, Y., Leisz, S.J., 2020. Crop boom as a trigger of smallholder livelihood and land use transformations: The case of coffee production in the Northern mountain region of Vietnam. Land 9, 56. doi.org/10.3390/land9020056

Nyambura, R., 2015. Agrarian transformation(s) in Africa: What's in it for women in rural Africa? Dev. 58, 306–313. doi.org/10.1057/s41301-016-0034-0

Obidzinski, K., Andriani, R., Komarudin, H., Andrianto, A., 2012. Environmental and social impacts of oil palm plantations and their implications for biofuel production in Indonesia. Ecol. Soc. 17, 25. doi.org/10.5751/ES-04775-170125

Oliveira, G.D.L.T., 2021. Political ecology of soybeans in Southern America, in: Ahmed, A., Gasparatos, A. (Eds.), Political Ecology of Industrial Crops. Routledge, London.

Patel, R.R., Mishra, D.K., 2019. Agrarian transformation and changing labour relations in Kalahandi, Odisha. J. South Asian Dev. 14, 314–337. doi.org/10.1177/0973174119889831

Perreault, T.A., Bridge, G., McCarthy, J.P., 2015. The Routledge Handbook of Political Ecology, Routledge international handbooks. Routledge, New York.

Rambo, A.T., 2017. The agrarian transformation in Northeastern Thailand: A review of recent research. Southeast Asian Stud. 6, 211–245. doi.org/10.20495/seas.6.2_211

Robbins, P., 2012. Political Ecology: A Critical Introduction. J. Wiley & Sons, Sussex.

Rogers, T.D., 2015. Agricultural Transformations in Sugarcane and Labor in Brazil, Oxford Research Encyclopedia of Latin American History. Oxford University Press, Oxford. doi.org/10.1093/acrefore/9780199366439.013.55

Romijn, H., Heijnen, S., Colthoff, J., de Jong, B., van Eijck, J., 2014. Economic and social sustainability performance of Jatropha Projects: Results from field surveys in Mozambique, Tanzania and Mali. Sustainability 6, 6203–6235. doi.org/10.3390/su6096203

Saravia Matus, S.L., Cimpoies, D., Ronzon, T., 2013. Panorama of Typologies of Agricultural Holdings. Food and Agriculture Organisation (FAO), Rome.

Savilaakso, S., Garcia, C., Garcia-Ulloa, J., Ghazoul, J., Groom, M., Guariguata, M.R., Laumonier, Y., Nasi, R., Petrokofsky, G., Snaddon, J., Zrust, M., 2014. Systematic review of effects on biodiversity from oil palm production. Environ. Evid. 3, 1–20. doi.org/10.1186/2047-2382-3-4

Schoneveld, G., 2014. The geographic and sectoral patterns of large-scale farmland investments in sub-Saharan Africa. Food Policy 48, 34–50. doi.org/10.1016/j.foodpol.2014.03.007

Searchinger, T., Heimlich, R., Houghton, R.A., Dong, F., Elobeid, A., Fabiosa, J., Tokgoz, S., Hayes, D., Yu, T.H., 2008. Use of U.S. croplands for biofuels increases greenhouse gases through emissions from land-use change. Science 319, 1238–1240. doi.org/10.1126/science.1151861

Semie, T.K., Silalertruksa, T., Gheewala, S.H., 2019. The impact of sugarcane production on biodiversity related to land use change in Ethiopia. Glob. Ecol. Conserv. 18, e00650. doi.org/10.1016/j.gecco.2019.e00650

Shortall, O., Helliwell, R., 2021. Marginal land for bioenergy crop production: Ambiguities, contradictions and cultural significance in policy and farmer discourses, in: Ahmed, A., Gasparatos, A. (Eds.), Political Ecology of Industrial Crops. Routledge, London.

Singh, B.P., 2010. Overview of industrial crops, in: Singh, B.P. (Ed.), Industrial Crops and Uses. CABI, Wallingford, pp. 1–20. doi.org/10.1079/9781845936167.0001

Smalley, R., 2013. Plantations, Contract Farming and Commercial Farming Areas in Africa: A Comparative Review (No. 055), e Land and Agricultural Commercialisation in Africa. Bellville.

Taylor, M., 2020. Hybrid Realities: making a new Green Revolution for rice in south India. J. Peasant Stud. 47, 483–502. doi.org/10.1080/03066150.2019.1568246

Terry, A., Ogg, M., 2017. Restructuring the Swazi sugar industry: The changing role and political significance of smallholders. J. South. Afr. Stud. 43, 585–603. doi.org/10.1080/03057070.2016.1190520

Timberlake, L., 1985. Africa in Crisis: The Causes, the Cures of Environmental Bankruptcy. Earthscan, Washington DC.

Urteaga-Crovetto, P., Segura-Urrunaga, F., 2021. Transforming nature, crafting irrelevance: The commodification of marginal land for sugarcane and cocoa agroindustry in Peru, in: Ahmed, A., Gasparatos, A. (Eds.), Political Ecology of Industrial Crops. Routledge, London.

van Vliet, J.A., Schut, A.G.T., Reidsma, P., Descheemaeker, K., Slingerland, M., Van De Ven, G.W.J., Giller, K.E., 2015. De-mystifying family farming: Features, diversity and trends across the globe. Glob. Food Sec. 5, 11–18. doi.org/10.1016/j.gfs.2015.03.001

Vijay, V., Pimm, S.L., Jenkins, C.N., Smith, S.J., 2016. The impacts of oil palm on recent deforestation and biodiversity loss. PLoS One 11. doi.org/10.1371/journal.pone.0159668

Vitale, J., 2018. Economic importance of cotton in Burkina Faso: Background paper to the UNCTAD-FAO Commodities and Development Report 2017 Commodity markets, economic growth and development. Rome.

von Maltitz, G.P., Henley, G., Ogg, M., Samboko, P.C., Gasparatos, A., Read, M., Engelbrecht, F., Ahmed, A., 2019. Institutional arrangements of outgrower sugarcane production in Southern Africa. Dev. South. Afr. 36, 175–197. doi.org/10.1080/0376835X.2018.1527215

Warren-Thomas, E.M., Edwards, D.P., Bebber, D.P., Chhang, P., Diment, A.N., Evans, T.D., Lambrick, F.H., Maxwell, J.F., Nut, M., O'Kelly, H.J., Theilade, I., Dolman, P.M., 2018. Protecting tropical forests from the rapid expansion of rubber using carbon payments. Nat. Commun. 9, 1–12. doi.org/10.1038/s41467-018-03287-9

White, B., Dasgupta, A., 2010. Agrofuels capitalism: a view from political economy. J. Peasant Stud. 37, 593–607. doi.org/10.1080/03066150.2010.512449

Woods, K.M., 2020. Smaller-scale land grabs and accumulation from below: Violence, coercion and consent in spatially uneven agrarian change in Shan State, Myanmar. World Dev. 127, 104780. doi.org/10.1016/j.worlddev.2019.104780

Wright, C.K., Wimberly, M.C., 2013. Recent land use change in the Western Corn Belt threatens grasslands and wetlands. Proc. Natl. Acad. Sci. U.S.A. 110, 4134–4139. doi. org/10.1073/pnas.1215404110

Yaro, J.A., Teye, J.K., Torvikey, G.D., 2017. Agricultural commercialisation models, agrarian dynamics and local development in Ghana. J. Peasant Stud. 44, 538–554. doi.org/10.1080/03066150.2016.1259222

Young, R.C., 1970. The plantation economy and industrial development in Latin America. Econ. Dev. Cult. Change 18, 342–361. doi.org/10.1086/450437

Zhang, L., 2021. The political ecology of maize in China: National Food Security and the Reclassification of Maize from Staple to Industrial Crop, in: Ahmed, A., Gasparatos, A. (Eds.), Political Ecology of Industrial Crops. Routledge, London.

2 Political Agronomy 101

An introduction to the political ecology of industrial cropping systems

William G. Moseley

Introduction

In 1987 I stepped off a plane onto the tarmac and the warm night air at the international airport in Bamako, Mali. It was the aftermath of the African Sahelian drought of the mid-1980s and I, a young American Peace Corps volunteer, was there as part of a US government initiative known as the African Food Systems Initiative. As an agriculture volunteer, I was to promote gardening and locally appropriate food crops, while my fellow volunteers in other sectors would work on forest regeneration, access to water, and improved primary health care. After several months of training, I moved to a small community of 200 people in southern Mali, where I worked for the next two years. There I met my government counterpart, the local agricultural extension agent from the provincial agricultural authority. While he found my interest in vegetable production quaint, he was focused on getting local farmers to produce more cotton. When he and I met the other extension agents at our monthly meetings, I could not figure out at first why the only crop we ever discussed was cotton, including endless discussions about hitting cotton quotas. What I thought was the Ministry of Agriculture was apparently the Ministry of Cotton. I would subsequently come to understand that Mali had signed agreements with the World Bank to pursue structural adjustment reforms in exchange for loans. Part of these commitments involved a doubling down on cotton production, a crop for which Mali was deemed to have a comparative advantage (see Chapters 6 and 11). In fact, as a result of these efforts, Mali would become the leading producer of cotton in all of Africa by the 1990s (Moseley 2001; Keeley and Scoones 2003).

About a year into my time in Mali, my volunteers and I had a chance to meet the director of the US Agency for International Development (USAID) in Mali. I remember asking the director why, in the aftermath of a major drought and famine, was the US government supporting cotton production, both directly via bilateral assistance and indirectly via the World Bank. This was an industrial crop that could not be eaten, degraded the soil, and was causing farmers to become indebted (Moseley 2005; Moseley and Gray 2008). His response was illuminating. He made three points that I would subsequently hear repeated by Malian civil servants: (a) that Mali needed to produce cotton to meet its debt obligations; (b) that cotton production was key to poverty alleviation in a country like Mali; and (c) that cotton production and food crop production were complementary or symbiotic. Cotton, he argued, solved a number of problems across different scales, including boosting food production.

Returning to Mali in the 1990s for my graduate thesis and dissertation research, I interviewed a large number of civil servants and agricultural extension agents. Many shared in French that "Grace à la CMDT, nos funtionnaires sont payées" (because of the government cotton company, our civil servants are paid). Another refrain I heard over and over again in the local language Bambara, when agricultural extension agents met with farmers, was that "kori tigi ye nyo tigi ye" (big cotton producers are big sorghum producers). It was true that cotton met a lot of the financial responsibilities for the government in Mali, but it was also, unfortunately, leading many farmers to indebtedness rather than poverty alleviation (Lacy 2008). Furthermore, the synergies between cotton farming and food crop production were only true for the wealthiest of farmers, with many becoming increasingly food insecure as they tried to produce more cotton (Moseley 2001).

How is it that civil servants, largely trained in crop science or agronomy, were dispensing advice to farmers that (while arguably good for the state and International Financial Institutions) was not good for smallholders? The answer is that the science of agronomy, and related extension, is not apolitical, but rather inflected with power and politics (Sumberg 2017). In other words, it can be argued that agronomists have focused on developing specific types of crops, with certain characteristics, to be cultivated in particular places, for the benefit of certain entities and groups of people. All of these breeding decisions have inherently involved choices, with these choices having been affected by those who have power, money, and influence.

This chapter takes the basic conceptual tools of political ecology and demonstrates how these may be productively employed to better understand industrial cropping systems in diverse geographies around the world. In particular, the chapter examines how attention to power, multi-scalar analysis, processes of marginalization, social difference, and discourse allows for a more incisive diagnosis of problems and development of alternatives. The chapter begins with an exploration of tropical agronomy (see "Political agronomy: The political economy of crop science"). It then details the various contours of political ecology, a field sometimes broadly defined as the political economy of human-environment interactions (Blaikie 1985; Robbins 2012) (see "Political ecology: Origins and major elements"). "Towards a political ecology of industrial crops" then describes how this analytical framework could be applied to industrial crops.[1]

Political agronomy: The political economy of crop science

Agronomy is a crop science and arguably the field of study most closely associated with the rise, spread, and maintenance of industrial cropping systems. Like most natural sciences, agronomy has been presented as apolitical and objective. However, as noted above, a major argument of this chapter is that agronomy is not apolitical, but inflected power and politics. As such, the moniker "political agronomy" is simply an attempt to be more explicit about these politics. Sections "Industrial crop development in the global South" through "Tropical and development agronomy" briefly review the development of industrial cropping systems in the Global South alongside the associated science of tropical agronomy or development agronomy.

Industrial crop development in the global South

Attention to power is critical for understanding the rise and spread of industrial cropping systems. The replacement of subsistence cropping systems by industrial cropping systems was not easy in many areas of the Global South, often requiring extraordinary levels of organized violence and coercion (Davis 2002) (see Chapters 3, 5, 9 and 11). The term "industrial" by definition implies manufacturing and mass production (Chapter 1). Thus, industrial cropping systems are inherently linked to manufacturing and emerged shortly after the rise of mass production in 18th century Europe (although arguably much earlier if accounting for the 17th-century sugarcane plantation economy in the Caribbean) (Mintz 1989). Mass production, also known as Fordism, not only drove down prices and fueled demand, but also thrived on cheap labor as feudalism disintegrated and landless peasants flooded into European cities seeking employment in the emerging manufacturing sector (Lohmann 2003). This thriving and expanding manufacturing sector also needed a stable, abundant, and cheap supply of raw materials. As such, there was a historically strong connection between European colonialism and the need for raw materials to feed industrialization domestically. In this sense, this emergence of different industrial cropping systems in the tropics and subtropics under colonial and post-colonial regimes often arose in tandem with different manufacturing phases in Europe.

Cotton was one of the first industrial cropping systems, as the textile industry was part of the first wave of industrialization in late 18th century Britain (Beckert 2015). The subtropical southern United States was the pre-eminent supplier of cotton to European textile mills up until the American Civil War (1861–1865). Following a blockade of southern US ports, the disruption of cotton supplies to Europe would create the so-called "cotton famine," a period when textile mills shut down and laid off workers, leading to hunger and social unrest (Tripathi 1967). This created a need to develop new cotton production zones outside of the United States, in Africa, Brazil, and South Asia especially (Chapter 7). Here Europeans used forced labor or coercive taxation policies to obligate local farmers to produce new cotton varieties for European markets. This process has been well documented for the British in India and Brazil (Davis, 2002) (Chapter 7), for the British in northern Nigeria (Watts, 1983a), and for the French in Côte d'Ivoire (Bassett, 2001). However, in India,

Brazil, and West Africa, the loss of local food production and the reduction of surplus food storage led to more frequent periods of hunger and famine, something colonial authorities blamed on drought, overpopulation, and indigenous agricultural practices, rather than extractive, industrial crop production (Davis 2002) (Chapter 7). Davis (2002) compares the plight of these areas to China, which did comparatively better in the face of poor rainfall until the British became more influential in the country after the mid-19th century Opium Wars.

As the industrial revolution evolved and new manufacturing industries emerged, there was a need for new raw materials, such as rubber for tire manufacturers, or tobacco, tea, cocoa, and coffee to supply recreational consumption. For example, Firetone's development of rubber plantations in Liberia has been well documented (Mitman 2017), with an entire country becoming associated with a single crop and one multinational firm playing a disproportional role in national politics. The West African cocoa boom in the late 19th and early 20th centuries (mainly in what is now contemporary Ghana and Côte d'Ivoire) is another well-studied case, where the extraordinary growth in cocoa production drew on the past experience of cocoa estates in other parts of the world (namely the Caribbean) (Lewis 1996; Ross 2014). Interestingly, in spite of the competitive success of the extensive agricultural practices used by African farmers during this period, Europeans retained a preference for intensive production techniques under centralized management (Ross 2014) (Chapter 1).

This "hijacking" of subsistence cropping systems for industrial purposes, however, is not limited to the colonial period; it continues today as novel uses for crops are found and new industries emerge (see Chapters 3–4, 6, and 10). While power is at play in the development of new industrial cropping systems, it tends to take the form of a more subtle power of market incorporation, rather than the heavy-handed forced labor or coercion of the colonial period. A good example from the 21st century has been shea butter derived from the kernel of the fruit of the shea tree (*Butyrospermum Parkii*) grown across the savanna belt of West and Central Africa where it is often interspersed in farm fields as an agroforestry species. In many places, shea nuts were collected and processed by women who used the produced shea butter as the staple cooking oil, and then sold some fraction of their output to local traders who in turn supplied urban areas in the region (Rousseau et al. 2019).

However, this situation has been changing, as the international trade in shea nuts has experienced enormous growth since the early 2000s, driven by new uses for shea butter, including as a cocoa butter equivalent (CBE) and an ingredient in cosmetics. In Burkina Faso for example, shea nut exports increased three-fold between 2000 and 2005, and seven-fold between 2005 and 2012 (Rousseau et al. 2015). Today, about half of shea nut production is traditionally self-consumed in Ghana, Burkina Faso, Benin, Côte d'Ivoire, Nigeria, Mali, and Togo, while the rest is exported. While this has arguably had a positive effect on women's incomes, it also has led to some other not-so-positive changes. First, with the increasing shea butter exports, there is less shea butter available for local consumption in several countries where malnutrition is very prevalent. While many families may replace shea butter with purchased cooking oil, the poorest of the poor may be entirely unable to do so as a cost-saving measure (Morgan and Moseley 2020). Furthermore, purchased oil produced from local peanut varieties may contain aflatoxins, which are a known carcinogen. Second, while shea is a traditional tree crop for women, in most producing countries, there is some evidence that men are beginning to engage in shea production, as it becomes a more lucrative crop (Rousseau et al. 2015). Lastly, tree tenure is important for determining who has access to shea trees and their products. When the market for shea nuts was less globalized, most women in rural communities had access to shea nuts if they were willing to invest the time for fruit collection and processing. Today, it is the wealthier women from founding families in rural communities that control access to most trees and are less willing to share the wealth. This has caused an increasing differentiation between poor and wealthy families in rural communities (Rousseau et al. 2019), with markets (often conceived and promoted in the development literature as value chains) as the power driving this transition (Gengenbach et al. 2018).

Tropical and development agronomy

Tropical agronomy, now known as development agronomy, emerged during the colonial era as European powers were expanding into the tropics. The primary objective was to increase the production of certain commodity crops to supply European consumers and industries, namely sugar, tea, coffee, cocoa, tobacco, cotton, rubber, sisal, and bananas, among others (Sumberg et al. 2012). If colonialism was ultimately about resource extraction (a process that fed industrialization in the global North), then tropical agronomists were the ground-level engineers who attempted to design and implement a system that produced industrial crops for the core countries. This was no easy task because it was not just about growing different crops, but often about reorganizing entire agrarian and socioeconomic systems (Chapters 7–11). Their limited understanding of tropical ecosystems and the logics of local production systems essentially made this task more difficult.

Most farm fields are simplified ecosystems (i.e. agro-ecosystems), which humans actively manage for certain outcomes. Given abundant land, relatively scarce human labor resources, and the challenges posed by crop insect predation in many areas of the Tropics, most traditional farming practices were aimed at nurturing agroecological synergies, or positive relationships between different cultivars and insects, in order to reduce labor demand and manage pests (Richards 1985). European farmers and agronomists from the temperate zone were accustomed to an entirely different set of challenges, namely scarce land, relatively abundant labor, and more limited pest problems because of the cold winters. As such, Europeans were often accustomed to less intensive intercropping (or the mixing of crops in farm fields) and more simplified farming systems. Perhaps not surprisingly, many European crop scientists during that period "imported" certain temperate zone assumptions when trying to establish industrial cropping systems in the Global South (Chapter 9). In many cases, the farms they established (often large monocultures) failed quite spectacularly at first as they succumbed to disease and insect predation (Ross 2014).

With colonialism on the ropes following WWII, tropical agronomy morphed into development agronomy in the 1950s and 1960s. It became closely associated with the Green Revolution, a concerted attempt to bring industrial farming practices to the Global South in the form of improved seeds, pesticides, and inorganic fertilizers (Chapters 7 and 9). While the increasing production of key food crops has often been associated with the Green Revolution (namely wheat and irrigated rice varieties), a number of industrial, commodity crops also benefitted. In the short run, this more energy-intensive approach seemed to solve many of the problems that had bedeviled tropical agronomists in the colonial period, namely the pest problems previously discussed, as well as tropical soils that lost their fertility after a few years of continuous cultivation (e.g. oxisols, ultisols) (Chapter 5). However, pesticide resistance and soil acidification would become problems associated with the use of chemical pesticides and inorganic fertilizers, respectively, not to mention social stratification and the depletion of surface and groundwater resources (Patel 2013). Despite these problems caused by industrial approaches to agriculture in the Tropics, the first Green Revolution would be heralded as a great success for its yield increases, thereby solidifying the position of development agronomy in academia and development practice.

Certain elements of development agronomy went into something of a lull in the 1980s, 1990s, and 2000s, as overall donor support and public spending on agriculture declined. This decline was associated with a period of neoliberal economic reform known as structural adjustment, which called for the rollback of the state and the reduced support for domestic food production, especially public spending on agricultural extension and agricultural subsidies, as well as tariff barriers that often protected local food producers (Carney 2008) (Chapter 6). However, such reforms also emphasized free trade and a doubling down on commodity crop exports by tropical countries because they were deemed (albeit dubiously) as having a comparative advantage in this area (Chapters 6 and 11). While this was framed in the language of neoliberal economics and was promoted as necessary for growth, it was essentially a return to a colonial world system where the poorest countries exported largely unprocessed minerals and commodity crops to wealthy countries in exchange for manufactured goods (Moseley et al. 2010).

The implications of these changes for academic and professional agronomists were profound. Those working on food crops saw much of their funding and jobs dry up, whereas those working on industrial commodity crops continued to find support and employment (Dowswell and Borlaug 1995). However, while public funding support for industrial crops was also in decline, the private sector stepped in to support agronomic research and agricultural extension in many cases. All of this came at a price in many contexts, both in the Global North and

Global South. It essentially meant that agricultural schools, and their research output, were often shaped by the corporate interests that funded them. As private companies took over agricultural extension functions, this also meant that farmers were receiving advice from a particular perspective and designed to support the use of certain products (Morgan et al. 2009).

Other aspects of development agronomy, namely those focused on food crop production, have rebounded sharply since the 2007–2008 global food crisis (Giller et al. 2017). This food crisis (and the related social unrest) led to a resurgence of donor interest in food production. A key initiative has been the New Green Revolution for Africa (GR4A). By arguing that the first Green Revolution had little success in improving food production in Africa, the GR4A focused on African food crops and pushed a similar package of improved seeds, pesticides, and fertilizers (Moseley et al. 2016). However, a key difference from the first Green Revolution has been the increased involvement of agri-business, which may have been a legacy of the neoliberal period discussed earlier when the private sector became a major source of funding for agronomic research and extension. Another important aspect of the GR4A has been the idea of a value chain, namely a linear model of linkages between input providers, producers, processors, and markets. This approach has received attention in different contexts around the world and suggests that smallholders need to be better integrated into the global system via value chains, clearly delineating the important role of private sector actors in this process (Gengenbach et al. 2018). While the GR4A is largely focused on food crops, value chain thinking has influenced how development scholars and practitioners think about industrial crops. In contrast to the decades of prevailing neoliberalism, development thinkers are once again discussing the possibility of the on-shoring of agroprocessing activities related to industrial crop production, such as textile manufacturing in cotton-producing countries or chocolate production in cocoa-growing areas (Moseley 2015) (Chapter 9). This trend of on-shoring has only been accelerated by the COVID-19 pandemic and the associated supply chain disruptions (Moseley and Battersby 2020).

Political ecology: Origins and major elements

From cultural ecology to political ecology

The post-colonial period, or the decades following WWII, was a good time to reassess the colonial interpretations of indigenous agricultural practices in the tropics. For starters, political decolonization in Asia and Africa in the 1940s to 1960s (Latin American decolonization happened earlier), was accompanied by intellectual decolonization, or the rethinking of many Eurocentric paradigms. At the same time, citizens in the Global North were beginning to question the uncritical acceptance of modern agricultural practices, and perhaps more broadly the triumphalist narrative of technologically driven progress. Emblematic of this period is Rachel Carson's *Silent Spring* (1962), which was a ground-breaking text raising important health and environmental concerns related to the heavy use of pesticides in industrial agriculture.

It is out of this environment (i.e. the questioning of modernization in an era of decolonization) that cultural ecology emerged in the 1950s and 1960s. Cultural ecologists, many hailing from anthropology and geography, had a strong interest in small-scale, subsistence agricultural production systems in the tropics (Moseley et al. 2013). By spending extended periods of time in these communities, cultural ecologists sought to carefully document indigenous agricultural practices and to understand their ecological and social rationale. Unlike colonial interpretations, which depicted these farming systems as primitive, backward, wasteful, and inefficient, cultural ecologists were able to demonstrate their underlying rationality. Such systems were often extraordinarily efficient, relying, for example, on intercropping and agroforestry to manage insect predation, maintain soil quality, and minimize soil erosion (e.g. Richards 1985). Cultural ecologists were also able to effectively demonstrate that indigenous farming systems were not static but able to evolve and adapt to changing environmental and social conditions (e.g. Denevan 1983). In this respect, cultural ecology scholarship was an important step in intellectual decolonization processes, as it legitimized approaches to agriculture in the Global South on their own terms.

However, the criticisms of cultural ecology started in the early 1980s, with the most searing of these attacks coming from Michael Watts (1983b). His essential concern was that cultural ecologists often explored small-scale agricultural systems in isolation without considering how historical and contemporary policies and phenomena at broader scales might affect the local circumstances. He argued that this myopia could lead to a complete misinterpretation of the situation (Watts, 1983a). Furthermore, small-scale subsistence agricultural systems were quickly becoming a thing of the past, with most farmers connected to larger markets in one way or another. Drawing from evidence in northern Nigeria, Watts (1983a) suggested that the causes of famine were not overpopulation[2] or drought, but British policies that encouraged cash cropping over food production and the dismantling of local grain storage practices. Historically, surplus grain was stored at the household, village, and kingdom levels in good years, so that farmers could tap into these resources when the rains inevitably failed in this highly variable rainfall environment. The aggressive taxation (in the form of a head tax) and the elimination of African kingdoms effectively destroyed surplus grain storage as an important safety net.

In an attempt to conceptualize the broader economic system that was influencing local human-environment interactions, Watts (1983b) and other scholars such as Blaikie (1985) effectively drew on contemporary structuralist thinking. The essential insight of structuralism was that countries are not entirely independent of one another, but operate in a system where their margin of maneuverability is constrained by historical trading relationships (Frank, 1979; Wallerstein, 1979). Frank (1979), by using his dependency theory, showed how European colonizers effectively underdeveloped tropical colonies by restructuring their economies to produce goods for their benefit and not for the local economy. Wallerstein (1979) expanded on this with his world systems theory wherein he divided the world into different spheres: periphery, semi-periphery, and core. Countries in the periphery supplied raw materials to semi-peripheral and core countries, with the semi-periphery engaging in low-end, heavy manufacturing, and the core producing high-end goods and financial services. Central to this structuralist paradigm was the notion that the roles of countries were more or less fixed unless they took deliberate steps to break out and change their development, such as erecting tariff barriers or engaging in import substitution. Watts (1983b) and Blaikie (1985) married these ideas with cultural ecology, using structuralist thinking to explain how the degradation related to resource extraction and commodity crop production was often linked to broader scale political economy, be it national policies, demand in other countries, or international institutional arrangements. This new analytical formulation would come to be known as political ecology (Bassett 1988; Robbins 2012).

Major elements and dimensions of political ecology

The major conceptual and analytical perspectives of political ecology revolve around the elements of multi-scalar analysis, marginalization, attention to social difference, and environmental narratives or discourse (Robbins 2012; Moseley et al. 2013).

Multi-scalar analysis has long been central to political ecology. The main idea is that local human-environment interactions are rarely just about local conditions, but are often affected by political-economic conditions at other scales, ranging from the global, down to the regional, national, and local. This can include, for example, the influence of policies at the national scale, trading relationships between countries at the regional scale, or policies promulgated by multilateral agencies and/or financial institutions at the global scale. Political ecologists sometimes establish connections across scales via chains of explanation, a term coined by British academic Piers Blaikie (Blaikie 1985). A chain of explanation often shows how a certain policy, program, or set of international relationships influences the practices of individual farmers or companies, encourages the growing of particular crops, or the organization of producers. A chain of explanation also allows moving from the proximate or immediate cause of an issue to the ultimate cause. While the majority of development and environmental organizations focus on proximate causes of ecosystem degradation such as the cutting of trees, hunting, or poor agricultural management, political ecologists suggest that these will not be resolved until ultimate causes such as business practices, poverty, or international demand for certain products are addressed (Moseley et al. 2013).

We can use these ideas of thinking across scales, chains of explanation, and proximate versus ultimate causation to better understand and systematize the example of the Malian cotton smallholders outlined in the "Introduction." As presented in Figure 2.1, at the local level in Mali, wealthy farmers grow most of

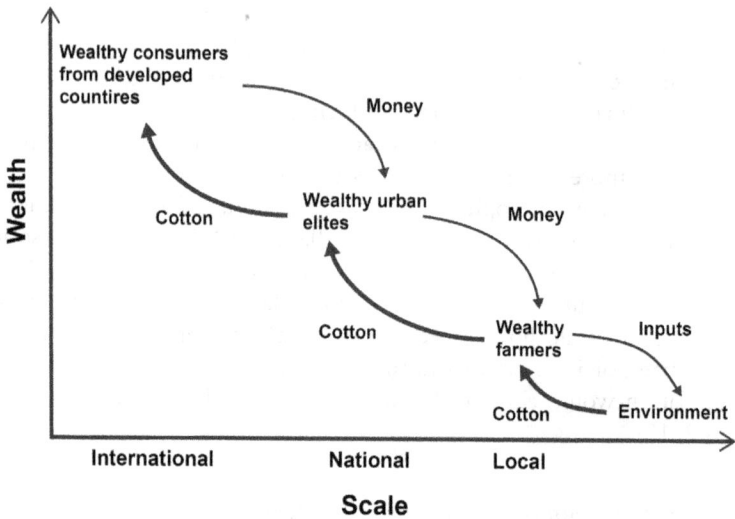

Figure 2.1 Multi-scalar linkages of cotton production in Mali.
Source: Adapted from (Moseley 2001).

the cotton, applying large quantities of inorganic fertilizers that cause soil acid-ification. While the management practices of these farmers are the proximate cause of soil degradation, they are incentivized by government policies that encourage cotton cultivation in terms of a guaranteed floor price and access to credit to purchase inputs and equipment. At the national level, bureaucrats in Mali are keen to support policies that boost cotton production because they know that its sales account for a large share of government revenue and effect-ively pay their salaries. However, when moving further up the chain of explan-ation to the international scale, other important drivers of cotton-induced environmental degradation become visible. First, the World Bank and the International Monetary Fund (IMF) have encouraged countries like Mali to grow cotton to pay off their multilateral loans. Second, wealthy consumers in the Global North demand cotton clothing, as they perceive it to be natural fiber and good for the environment. This perception has been bolstered by the cotton industry advertising campaigns, even though cotton production entails some of the most chemical-intensive cultivation practices (Chapter 7). While many of the same ideas about cotton cultivation were shared in the "Introduction," here, it becomes possible to appreciate the driving factors organized in terms of a political ecology lens (Moseley 2001, 2005, 2008).

Another key concept in political ecology is that of marginalization. Blaikie (1985) suggested that there were at least three types of marginalization, namely social, environmental, and economic, that often work in concert and are mutu-ally reinforcing. According to Blaikie (1985), socially marginal groups are more likely to be pushed into environmentally marginal areas that are less productive agriculturally. Farmers in more environmentally marginal lands would then be less productive and increase the risks of becoming economically marginal. Drawing on work in Nepal, Blaikie and Brookfield (1987) showed how less powerful groups often got pushed out of productive valleys into more environmentally marginal and precarious hillside farming areas. Not only were these farmers less productive on hillsides, but also their farming practices were often (unfairly) blamed for the ensuing erosion. In a similar way, when the British colonized Zimbabwe (then known as Rhodesia), they undertook an extensive land survey, zoning and categorizing land all over the country in terms of its agricultural potential, and gave the best land (typically in cooler upland plateau areas known as the Highveld) to white European farmers for tobacco production. Local black Africans were moved out of these areas into lower-lying, warmer, and drier areas known as the Lowveld that had lower agricultural potential (Thomas 2003). These areas also contained more wildlife, and subsequently national parks, which precipitated extensive human–wildlife conflicts and loss of crops to animals (mainly elephants) (Logan and Moseley 2002). This history of social, environmental, and economic marginalization was one of the major rationales for land reform more recently in the country. Another classic historical example is the marginalization that occurred in South Africa, as it became a white settler colony, a process that started with the Dutch in 1652 and then transitioned to the British. Under the apartheid regime in South Africa, the architects of racialized capitalism created an agrarian system where black smallholders held so little land that they had to work on white, commercial farms in order to make ends meet. Under this system, white farmers benefitted from the cheap and consistent supply of black labor, not to mention the diminished competition from African smallholders. Ironically, the apartheid regime also discursively contrasted the struggling black smallholders with the large white-owned plantations as an example of white superiority (Bundy 1988; McCusker et al. 2015).

However, processes of marginalization are not just historical, as has been recently been observed through the phenomenon of large-scale land leases, also known as "land grabs," which have accelerated since the 2007–2008 global food crisis as international investors and sovereign wealth funds have sought to secure land in poor countries for export-oriented agriculture (Edelman et al. 2013) (Chapters 3 and 5). Ethiopia under the previous regime has been such an example, where large tracks of land were leased to outside investors and countries. This land was often occupied by poor smallholders or pastoralists and was declared vacant, as the government did not recognize formal tenure to local communities. Marginalized to the periphery of these new industrial crop plantations, they often ended up working as laborers for low salaries.[3]

Attention to social difference is another critical element to political ecology analysis, be it in terms of gender, race, class, or ethnicity, among multiple other factors. While most political ecologists understand that these differences are socially constructed, they also understand that these constructed differences have real impacts in terms of how people interact with the environment. For example, in places where women are tasked to collect water or firewood, they may be the first to notice that the water table is dropping or forest resources are declining. Consequently, they may need to travel further to collect water or harvest firewood, meaning that such environmental changes directly impact their time allocation and labor more significantly than other members of their local community with different roles (Awumbila and Momsen 1995). Similarly, herders, because of their occupation, may be the first to notice when pasture resources are degrading, and they are also more directly impacted by these changes (Oba and Kaitira 2006).

Some of the defining work in this realm has been conducted under the banner of feminist political ecology, a subfield of political ecology that was established by scholars such as Carney (1993) and Rocheleau et al. (1996). Carney (1993), for example, documented how Green Revolution rice projects in The Gambia disproportionately impacted women in terms of labor demands, with women eventually rebelling when pushed to work twice as hard growing two, instead of one, rice crops per year with little to no compensation.

It is also important to note that sometimes social differences are not as important as conventional wisdom would suggest. A common trope in some conventional scholarships is that farmer–herder conflicts are due to ethnic animosity, overpopulation, and/or resource scarcity. In contrast, Bassett (1988) reveals that ethnic tensions between Fulani herders and Senufo farmers in northern Côte d'Ivoire were largely the result of broader political-economic factors. In this case, Ivoirian policies favoring food self-sufficiency encouraged Fulani herders to enter the country from its Sahelian neighbors so that local beef production would increase, a situation also fostered by droughts in Mali and Burkina Faso (Bassett, 1988). However, increasing cattle numbers led to land use conflicts with Senufo farmers who were also being squeezed by low international prices for the cotton they were producing. In sum, it was the national policy encouraging beef production, as well as the low profit margins garnered by cotton producers, that set-up a situation for conflict when cattle did occasionally wander into farmers' fields (Bassett, 1988). This lens offered a very different understanding of the problem than the traditional ethnic-driven, over-population, and resource-scarcity discourse.

The last important dimension of political ecology is the attention to discursive power, particularly in the form of environmental narratives, as well as the ability of social movements to generate counter-narratives. Like many of the social sciences, political ecology was influenced by post-structuralism. Post-structuralists accept the basic problems highlighted by structuralists (discussed earlier in "From cultural ecology to political ecology"), namely that we need to acknowledge a world system in which resource flows between countries are often unequal, leading to underdevelopment, over-extraction, and degradation. A critique of structuralism, however, is the dismal prognosis that one country or segment of society is essentially stuck in a bad position with little chance of improving it. Post-structuralists did not accept this dismal prognosis, acknowledging the power of "agency," or the ability of individuals and groups to change their situation (Lawson 2007). Post-structuralist scholars were particularly interested in the agency of individuals working together in the form of social movements, as well as in discursive power and narrative. In this context, a narrative is the dominant story or conventional wisdom that has been internalized by society, often resulting in a so-called hegemonic discourse, an explanation that the majority of the population unquestionably accepts (Foucault 1972). The essential insight of post-structuralists was that these discourses or narratives were often shaped and maintained by powerful groups, even if they were not accurate representations of reality.

Beginning in the 1990s, political ecologists took on board the insights of post-structuralism and increasingly used these ideas to understand environmental narratives regarding degradation, resource management, and agriculture (Robbins 2012; Moseley et al. 2013). For example, up to that point, desertification and deforestation had long been the dominant understanding of environmental change in the grasslands of West Africa. In Guinea, Fairhead and Leach (1995) examined the phenomenon of forest islands, namely patches of forest in the savanna grassland zone. The conventional wisdom since the colonial period was that these grasslands were relics of a previously forested landscape, a landscape that had largely been deforested by human activity. However, Fairhead and Leach (1995) revealed through ethnographic interviews and time-series analysis of aerial photographs that forest islands were in fact products of afforestation and human activity, rather than relics of a previously forested landscape. Seeking to understand why inaccurate environmental narratives persisted for so long, Leach and Mearns (1996) described how colonial agronomists and foresters, as well as contemporary land managers, NGOs, and expatriate scientists also played role in maintaining these notions because they benefitted from them in some way. Other scholars have described similar phenomena in South Asia (Robbins 1998) and Latin America (Bebbington and Bebbington 2011). Of course, social movements, or participatory research or development approaches, may disrupt these dominant narratives, often proposing counter-narratives that may describe reality better (Robbins 2012). For example, the Frierian-inspired, bottom-up participatory development approaches may enable local communities and farmers, with a deep understanding of local knowledge, to articulate a different narrative of environmental change (Friere 1982; Moseley and Laris 2008).

Toward a political ecology of industrial crops

Considering the arguments made throughout this chapter, it is proposed that a political ecology toolkit may be applied to explain critical aspects pertaining to agronomic practices and industrial crop production. Political ecology can contribute to a better and more multi-dimensional understanding of industrial cropping systems, using conceptual tools related to multi-scalar analysis, marginalization, attention to social difference, and discourse, as outlined below (Table 2.1).

Table 2.1 Conceptual tools for explaining the political ecology of industrial crops

Conceptual tools	Major insights
Multi-scalar analysis	– Strong connections between aspects of industrial crop production, processing, trading, and consumption at different spatial and temporal scales; – Industrial crops (and their production and processing systems) are inherently connected to other parts of the world via markets; – International and national policies influence farmer behavior at the local scale; – Industrial crops foster relationships of dependency within and between countries.
Marginalization	– Environmental: industrial crop production and related market forces often push poor and disenfranchised farmers onto more marginal land; – Economic: industrial or food crop production on marginal land leads to low yields and economic output, deepening poverty; – Social: poverty reinforces many negative social impacts (e.g. food insecurity) and deepens social marginalization creating a vicious cycle.
Attention to social difference	– Industrial cropping systems engage different groups, which often have differentiated characteristics and roles, e.g. in terms of gender, ethnicity, race, age, or affluence; – Socially constructed roles and differences have implications for who accumulates wealth or experiences certain impacts (e.g. occupational hazards) within the context of industrial cropping systems.
Discourse	– Different discourses operate at different levels, e.g. policy discourses justifying industrial crop production, academic discourses criticizing or supporting major aspects of industrial cropping systems; – Dominant discourses and narratives surrounding industrial crop production, processing, trade, and consumption are often shaped by the powerful entities that benefit most from these systems.

In terms of multi-scalar analysis, unlike subsistence food crop production, industrial cropping systems are inherently connected to other parts of the world via markets. This happens across all major stages of these systems, starting with the actual industrial crop cultivation, to processing, trade, and consumption. Thus, it is virtually impossible to understand the local dynamics of industrial crop production without also considering how these are connected to other parts of the world, and the respective dynamics in those faraway places. Such a multi-scalar analysis allows, for example, to obtain a better understanding of how the surging demand for palm oil in the global food products and bioenergy industries is leading to land dispossession, plantation expansion, and deforestation in Indonesia (Pichler 2015). This multi-scalar lens also allows for scaling-down and pushing scholars to think about relationships of dependency within (and between) countries. This could be, for example, policies set at the national level that drive farmer behavior locally, such as subsidy structures encouraging farmers to cultivate certain crops following a particular production model. It could also mean paying attention to exploitative relationships between different actors in the rural sector, or the weak positions of local farmers in global value chains. A classic manifestation of the former is agricultural dualism, wherein plantation-based agricultural systems draw on cheap labor from subsistence farming, which may exist in parallel with commercial agriculture in the same areas (Chapters 8 and 11). For example, tea plantations in southern India regularly draw on cheap labor from the surrounding smallholders, which are often female tea pickers who work on a piece-rate basis (Hayami and Damodaran 2004). These strong linkages between aspects of industrial crop production occurring at different scales is very visible in practically all of the empirical chapters in this edited volume (Chapters 5–11).

In terms of marginalization, while it varies by the type of the industrial crop and the mode of production (e.g. some industrial crops are grown solely by smallholders, others in plantations and others in hybrid systems, Chapter 1), industrial cropping systems often demand large amounts of land that is managed in a more centralized manner (Chapters 3, 5, 8–11). This often results in less powerful farmers or actors becoming marginalized through different mechanisms such as being squeezed out in the production systems (e.g. Chapter 9), losing their land and livelihoods during land grabs (Chapters 3 and 5), working for low salaries in large-scale plantations due to lack of other options (Chapter 8), or being pushed onto more environmentally marginal land or having disproportionate access to resources (Chapter 11), which then creates or deepens their economic marginality.

In terms of social difference, it is critical to pay attention to socially differentiated actors, across the entire chain of industrial cropping systems. Often gender and norms (both local and outside) have a significant effect on who grows industrial crops or performs certain tasks (e.g. Chapters 6 and 8). These socially constructed gender roles have major implications for who accumulates wealth or experiences certain occupational hazards. For example, even though cotton was traditionally grown in West Africa by women in small gardens for sale to local weavers, it was essentially framed as a male crop when the French and British introduced it as an industrial crop during the colonial era (with new seed varieties, inputs, and tilling practices). The result was that the male heads of household ended up accumulating most of the wealth from export-oriented cotton production, a process that skewed male–female power dynamics within households (Bassett 2001, Moseley 2008). This edited volume contains many examples of social differentiation in the context of industrial crops, including the lower access of young people and women on agricultural land for cashew production in Guinea-Bissau (Chapter 6), gender-differentiated impacts from engagement industrial crop plantations in Ghana (Chapter 8), and disproportionate benefits of higher-caste farmers in cotton cooperatives in India (Chapter 7).

The dominant discourses and narratives surrounding industrial crop production are often shaped by the powerful entities that benefit most from these systems. For example, while considerable social disruption occurs in some contexts when plantations are established, including land grabs and loss of sub-sistence production (Chapter 3), these systems are often portrayed as engines of economic development that generate employment and bolster economic devel-opment. In South Africa, for example, white farmers (and industrial approaches to farming more broadly) were and continue to be heralded as a key engine of national development, and a model that should be exported to neighboring countries (Moseley 2007). This narrative occludes the dispossession and violence that occurred to establish the system. Even though successful social movements led to important political changes in South Africa, the dominant positive narrative about industrial agriculture has been amazingly resilient and robust because certain dominant actors inside and outside the country (including international financial institutions) continue to support this hegemonic dis-course (Peet 2002). Many other discourses have been used in the context of industrial crop production to justify their production or achieve their expan-sion such as comparative advantage (Chapter 6), marginal lands (Chapters 4–5), or agricultural modernization and diversification (Chapters 9–10).

Finally, it is worth mentioning that the chapters in this edited volume use a variety of discourses to critique the justification, promotion, expansion, and impacts of industrial crops including land grabbing (Chapters 3 and 5), food vs. fuel (Chapter 4), value crafting (Chapter 5), flexing (Chapter 9), neo-natures (Chapter 9), and theory of access (Chapter 11), among others. Such discourses offer very useful perspectives to discuss aspects of industrial cropping systems when using a political ecology lens.

Conclusions

Like many fields of knowledge, agronomy is inflected with power and politics. Once we acknowledge how power dynamics have shaped the dominant forms of agricultural knowledge and cropping systems around the world, it is possible to begin imagining more equitable and sustainable alternatives. This chapter took the basic conceptual tools of political ecology and demonstrated how these may be productively employed to better understand a range of indus-trial cropping systems in diverse geographies around the world. In particular, the chapter focused on how paying attention to power, multi-scalar analysis, processes of marginalization, social difference, and discourse can allow for a more incisive diagnosis of problems and development of alternatives in contexts of industrial crop expansion.

Notes

1 Industrial crops are crops that can be used for non-food uses, such as fiber, bioenergy, and other industrial products (Singh 2010). Refer to Chapter 1 for a more compre-hensive description.

2 It is worth noting that many early political ecologists are deeply distrustful of Malthusian explanations for degradation and hunger (Robbins 2012).

3 It is worth noting that even though the Ethiopian government framed these "land grabs" as employment generation opportunities, and a showcase for agricultural modernization, the fact remains that they also contributed to national food insecurity as most of their output was not food crops and/or was exported (Moseley 2012).

References

Awumbila, M. and J.H. Momsen. 1995. "Gender and the Environment: Women's Time Use as a Measure of Environmental Change." *Global Environmental Change*. 5(4): 337–346.

Bassett, T.J. 1988. "The Political Ecology of Peasant-Herder Conflicts in the Northern Ivory Coast." *Annals of the Association of American Geographers*. 78(3): 453–472.

Bassett, T.J. 2001. *The Peasant Cotton Revolution in West Africa. Côte d'Ivoire, 1880–1995*. London: Cambridge University Press.

Bebbington, D.H. and A. Bebbington. 2011. "Post What? Extractive Industries, Narratives of Development and Socio-Environmental Disputes across the (ostensibly) Changing Andean Region." In: *New Political Space in Latin American Natural Resource Governance*. Edited by H. Haarstad. New York: Palgrave.

Beckert, S. 2015. *Empire of Cotton: A Global History*. New York: Vintage.

Blaikie, P.M. 1985. *The Political Economy of Soil Erosion in Developing Countries*. London: Longman.

Blaikie, P.M. and H. Brookfield. 1987. *Land Degradation and Society*. New York: Methuen & Co.

Bundy, C. 1988. *The Rise and Fall of the South African Peasantry*. Cape Town: David Philip, James Currey.

Carney, J. 1993. "Converting the Wetlands, Engendering the Environment: The Intersection of Gender with Agrarian Change in the Gambia." *Economic Geography*. 69(4): 329–348.

Carney, J. 2008. "The Bitter Harvest of Gambian Rice Policies." *Globalizations*. 5(2): 129–142, DOI: 10.1080/14747730802057456

Carson, R. 1962. *Silent Spring*. New York: Houghton Mifflin Company.

Davis, M. 2002. *Late Victorian Holocausts: El Niño Famines and the Making of the Third World*. London: Verso Press.

Denevan, W.M. 1983. "Adaptation, Variation, and Cultural Geography." *The Professional Geographer*. 35(4): 399–406.

Dowswell, C. and N. Borlaug. 1995. "Mobilising Science and Technology to Get Agriculture Moving in Africa." *Development Policy Review*. 13(2):115–129.

Edelman, M., C. Oya and S.M. Borras. 2013. "Global Land Grabs: Historical Processes, Theoretical and Methodological Implications and Current Trajectories." *Third World Quarterly*. 34(9): 1517–1531.

Fairhead, J. and M. Leach. 1995. "False Forest History, Complicit Social Analysis – Rethinking Some West African Environmental Narratives." *World Development*. 23(6): 1023–1035.

Frank, A.G. 1979. *Dependent Accumulation and Underdevelopment*. London: MacMillan Press.

Friere, P. 1982. *Pedagogy of the Oppressed*. New York: Continuum.

Foucault, M. 1972. *The Archeology of Knowledge*. New York: Pantheon.

Gengenbach, H., R. Schurman, T. Bassett, W. Munro and W. Moseley. 2018. "Limits of the New Green Revolution for Africa: Reconceptualising Gendered Agricultural Value Chains." *The Geographical Journal*. 184(2): 208–214.

Giller, K., J.A. Andersson, J. Sumberg and J. Thompson. 2017. "A golden Age for Agronomy?" In: Sumberg, J. (ed). *Agronomy for Development. The Politics of Knowledge in Agricultural Research*. London: Earthscan. pp. 150–160.

Hayami, Y. and A. Damodaran. 2004. "Towards an Alternative Agrarian Reform: Tea Plantations in South India." *Economic and Political Weekly*. 39(36): 3992–3997.

Keeley, J. and I. Scoones. 2003. *Understanding Environmental Policy Processes*. London: Earthscan.

Lacy, S. 2008. "Cotton Casualties and Cooperatives: Reinventing Farmer Collectives at the Expense of Rural Malian Communities?" In: Moseley, W.G. and L.C. Gray (eds). *Hanging by a Thread: Cotton, Globalization and Poverty in Africa*. Athens, OH: Ohio University Press. pp. 207–226.

Lawson, V. 2007. *Making Development Geography*. London: Hodder Arnold Publication.

Leach, M. and Mearns, R. 1996. "Environmental Change and Policy: Challenging Received Wisdom in Africa." In Leach, M. and Mearns, R. (eds.). *The Lie of the Land: Challenging Received Wisdom on the African Environment*. Oxford: James Curry.

Lewis, K.P. 1996. "The Trinidad Cocoa Industry and the Struggle for Crown Land during the Nineteenth Century." In: Clarence-Smith, W.G. (ed.) *Cocoa Pioneer Fronts since 1800*. London: Palgrave Macmillan.

Logan, B.I. and W.G. Moseley. 2002. "The Political Ecology of Poverty Alleviation in Zimbabwe's Communal Areas Management Programme for Indigenous Resources (CAMPFIRE)." *Geoforum*. 33(1): 1–14.

Lohmann, Larry. 2003. "Re-imagining the population debate." *Corner House Briefing 28*. London: The Corner House.

McCusker, B., W.G. Moseley and M. Ramutsindela. 2015. *Land Reform in South Africa: An Uneven Transformation*. Lanham, MD: Rowman and Littlefield Publishers, Inc.

Mintz, S. 1989. *Caribbean Transformations*. New York: Columbia University Press.

Mitman, G. 2017. "Forgotten Paths of Empire: Ecology, Disease, and Commerce in the Making of Liberia's Plantation Economy: President's Address." *Environmental History*. 22(1): 1–22.

Morgan, J. and W.G. Moseley. 2020. "The Secret is in the Sauce: Foraged Food and Dietary Diversity Among Female Farmers in Southwestern Burkina Faso." *Canadian Journal of Development Studies*. DOI: 10.1080/02255189.2020.1781600.

Morgan, K., T. Marsden, J. Murdoch. 2009. *Worlds of Food: Place, Power, and Provenance in the Food Chain*. New York: Oxford University Press.

Moseley, W.G. 2001. *Sahelian "White Gold" and Rural Poverty-Environment Interactions: The Political Ecology of Cotton Production, Environmental Change, and Household Food Economy in Mali*. Athens, GA: Department of Geography, University of Georgia.

Moseley, W.G. 2005. "Global Cotton and Local Environmental Management: The Political Ecology of Rich and Poor Small-Hold Farmers in Southern Mali." *Geographical Journal*. 171(1): 36–55.

Moseley, W.G. 2007. "Neoliberal Agricultural Policy versus Agrarian Justice: Farm Workers and Land Redistribution in South Africa's Western Cape Province." *South African Geographical Journal*. 89(1): 4–13.

Moseley, W.G. 2008. "Mali's Cotton Conundrum: Commodity Production and Development on the Periphery." In: Moseley, W.G. and L.C. Gray (eds). *Hanging by a*

Thread: Cotton, Globalization and Poverty in Africa. Athens, OH: Ohio University Press. pp. 83–102.

Moseley, W.G. 2015. "Regional Value Chains and Productivity Enhancement in Africa." In: Hanson, K (ed). *Contemporary Regional Development in Africa.* London: Ashgate.

Moseley, W.G. and J. Battersby. 2020. "The Vulnerability and Resilience of African Food Systems, Food Security and Nutrition in the Context of the COVID-19 Pandemic." *African Studies Review.* 63(3). DOI: 10.1017/asr.2020.72

Moseley, W.G., J. Carney and L. Becker. 2010. "Neoliberal Policy, Rural Livelihoods and Urban Food Security in West Africa: A Comparative Study of The Gambia, Côte d'Ivoire and Mali." *Proceedings of the National Academy of Sciences of the United States of America.* 107(13): 5774–5779.

Moseley, W.G. and L.C. Gray (eds). 2008. *Hanging by a Thread: Cotton, Globalization and Poverty in Africa.* Athens, OH: Ohio University Press and Nordic Africa Press.

Moseley, W.G. and P. Laris. 2008. "West African Environmental Narratives and Development-Volunteer Praxis." *The Geographical Review.* 98(1): 59–81.

Moseley, W.G., E. Perramond and H. Hapke and P. Laris. 2013. *An Introduction to Human-Environment Geography: Local Dynamics and Global Processes.* Hoboken, NJ: Wiley/Blackwell.

Moseley, W.G., M. Schnurr and R. Bezner Kerr. 2016. *Africa's Green Revolution: Critical Perspectives on New Agricultural Technologies and Systems.* Oxford, UK: Taylor & Francis.

Oba, G. and L.M. Kaitira. 2006. "Herder Knowledge of Landscape Assessments in Arid Rangelands in Northern Tanzania." *Journal of Arid Environments.* 66(1): 168–186.

Patel, R. 2013. "The Long Green Revolution." *Journal of Peasant Studies.* 40(1): 1–63.

Peet, R. 2002. "Neoliberalism in South Africa." In: Logan, B.I. (ed.) *Globalization, the Third World State and Poverty-Alleviation in the Twenty-First Century.* Hampshire, UK: Ashgate Publishing Ltd.

Pichler, M. 2015. "Legal Dispossession: State Strategies and Selectivities in the Expansion of Indonesian Palm Oil and Agrofuel Production." *Development and Change.* 46: 508–533.

Richards, P. 1985. *Indigenous Agricultural Revolution.* Boulder: Westview Press.

Robbins, P. 1998. "Paper Forests: Imagining and Deploying Exogenous Ecologies in Arid India" *Geoforum* 29(1): 69–89.

Robbins, P. 2012. *Political Ecology: A Critical Introduction.* 2nd edition. Malden, MA: Wiley-Blackwell.

Rocheleau, D.E., B. Thomas-Slayter and E. Wangari (eds.). 1996. *Feminist Political Ecology: Global Perspectives and Local Experience.* New York: Routledge.

Ross, C. 2014. The Plantation Paradigm: Colonial Agronomy, African Farmers, and the Global Cocoa Boom, 1870s–1940s. *Journal of Global History.* 9(1): 49–71.

Rousseau, K., D. Gautier and D.A. Wardell. 2015. "Coping with the Upheavals of Globalization in the Shea Value Chain: The Maintenance and Relevance of Upstream Shea Nut Supply Chain Organization in Western Burkina Faso." *World Development.* 66: 413–427.

Rousseau, K., D. Gautier and D.A. Wardell. 2019. "Socio-Economic Differentiation and Shea Globalization in Western Burkina Faso: Integrating Gender Politics and Agrarian Change." *The Journal of Peasant Studies.* 46(4): 747–766.

Singh, B. 2010. *Industrial Crops and Uses.* Fort Valley, GA: Fort Valley State University.

Sumberg, J. 2017. *Agronomy for Development. The Politics of Knowledge in Agricultural Research.* London: Earthscan.

Sumberg, J., J. Thomson, and P. Woodhouse. 2012. "Why Agronomy in the Developing World Has Become Contentious." *Agriculture and Human Values*. 30(1): 71–83.

Thomas, N.H. 2003. "Land Reform in Zimbabwe." *Third World Quarterly*. 24(4): 691–712.

Tripathi, D. 1967. Opportunism of Free Trade: Lancashire Cotton Famine and Indian Cotton Cultivation. *The Indian Economic & Social History Review*. 4(3): 255–263.

Wallerstein, I.M. 1979. *The Capitalist World Economy: Essays*. London: Cambridge University Press.

Watts, M., 1983a. *Silent Violence: Food, Famine and Peasantry in Northern Nigeria*. Berkeley: University of California Press.

Watts, M. 1983b. "On the Poverty of Theory: Natural Hazards Research in Context. In K. Hewitt (ed.) *Interpretations of Calamity*. Boston: Allen and Unwin. pp. 231–262.

3 Political ecology of large-scale land acquisitions and land grabs for industrial crops

Abubakari Ahmed and Alexandros Gasparatos

Introduction

The large-scale production of industrial crops has shaped the political, socio-economic, and environmental trajectories of many developing countries in the global South (Chapters 1–2). For example, the extensive consolidation and intensification of large-scale industrial crop production models have been observed for sugarcane in Brazil (Rogers, 2015), oil palm in Indonesia and Malaysia (CIFOR, 2009), and soybeans in South America (Oliveira and Hecht, 2016), among many other examples.

The large-scale production of industrial crops is usually undertaken for purely commercial purposes in intensified monocultures spanning several hundred to tens of thousands of hectares (depending on the crop and the geographical context) (Matus et al., 2013; Smalley, 2013) (Chapter 1). Diverse actors such as private companies (both domestic and international), national governments, para-statal agencies, or public-private partnerships establish and operate such large-scale production systems (Smalley, 2013). Large-scale production models often integrate smallholders in different configurations (e.g. outgrowers, independent growers, see Chapters 1 and 8), particularly for some crops such as sugarcane (von Maltitz et al., 2019), jatropha (Romijn et al., 2014), and oil palm (CIFOR, 2015), among others.

In many geographical contexts, large-scale production systems have been perceived to be efficient production models that could maximise industrial crop yields, but also have broader development benefits by modernising and industrialising agrarian systems and boosting economic growth and rural development (Gibbon, 2011; White et al., 2012; Smalley, 2013). Indeed, due to their inherent approach towards land consolidation (Chapter 1), large-scale production models can facilitate economies of scale, especially for perishable crops such as sugarcane and oil palm (Chapters 8 and 11). Furthermore, their centralised management can facilitate their easier integration in vertical corporate structures, which facilitates access to financial capital, agricultural inputs, and markets, further increasing profitability (Pramudya et al., 2017; Oliveira and Hecht, 2016) (Chapter 9). However, this concentration of land and capital has underpinned many of the criticisms and negative impacts of large-scale industrial crop production (Smalley, 2013).

First, in many countries, the processes and modalities of gaining access to the large tracks of uninterrupted land necessary for plantations and ancillary infrastructure have been criticised. Large-scale land acquisitions (LSLAs), which are "*broadly defined as acquisitions (whether purchases, leases or other) of land areas over 1,000 ha*" are one of the most common mechanisms for accessing such extensive land areas (Cotula et al., 2009: 3). Depending on the national and geographical context, LSLAs can entail the direct purchase/lease of public land, direct purchase/lease from large individual landowners, consolidation, and purchase of smallholder lands, or combinations of the above (Cotula et al., 2009; Behrman et al., 2012; German et al., 2013; Notess et al., 2020; Dell'Angelo et al., 2017).

However, again depending on the context, LSLAs for industrial crop production have been associated with the (a) manipulation of existing (or the development of new and more amenable) institutional frameworks, (b) the inappropriate compensation of previous landowners, (c) the lack of respect towards customary rights, or (d) predatory, intimidating, coercive, and misinforming practices from investors and national governments (German et al., 2013; Notess et al., 2020; Cotula, 2012; Mechiche-Alami et al., 2019) (Chapters 4–5, 8–9, and 11). Furthermore, extensive landscape conversion for large-scale industrial crop production can have serious environmental and socioeconomic ramifications, which may, among others, strip the access of local communities or customary landowners to natural resources (White et al., 2012). Collectively, such phenomena have marred the reputation of many industrial crop-based LSLAs, leading to widespread allegations of land grabbing (Adams et al., 2019; Borras et al., 2012; Hall, 2011a, 2011b; Kenney-Lazar, 2012).[1]

Considering that much of the recent industrial crop expansion globally has occurred through LSLAs (Chapter 1), the aim of this chapter is to (a) synthesise the knowledge about the drivers, processes, actors, and impacts of industrial crop LSLAs and land grabs and (b) rationalise this interface through a political ecology lens. For the purpose of this chapter, the term large-scale production bundles diverse production models such as plantations and commercial farming areas (Smalley, 2013) but does not include outgrower schemes unless explicitly stated. Furthermore, the chapter focusses on industrial crops as defined in Chapter 1, and thus does not cover tree-based production models.[2]

The section "Critical aspects of LSLAs and land grabbing for industrial crops" identifies the main drivers, expansion trajectories, actors, and impacts of industrial crop LSLAs as they unfolded during the contemporary land rush in the global South. Mindful of the large global variability between crops, national policies, and local contexts, the section "Unpacking the interface of industrial crops, LSLAs and land-grabbing" focusses on experiences from the recent expansion of jatropha in Sub-Saharan Africa (SSA), rubber in Southeast Asia, and oil palm in Latin America. The section "Political ecology of LSLAs for industrial crop and land grabbing" systematises the main insights through a political ecology lens seeking to rationalise how human–nature relationships are shaped at the interface of industrial crops, LSLAs, and land grabbing.

Critical aspects of LSLAs and land grabbing for industrial crops

Drivers and extent of LSLAs

A rich literature has critically discussed the long colonial history and legacy of plantations-based models for industrial crop production (Gibbon, 2011; Kenney-Lazar and Ishikawa, 2019; Pietilainen and Otero, 2019). This literature points to how plantations were instrumental in fuelling with raw materials the industrial revolution in Europe, causing much harm in the process by permeating the slavery model, changing fundamentally local agrarian systems, creating local elites, causing food insecurity, and triggering famines, among many others (Gibbon, 2011; Kenney-Lazar and Ishikawa, 2019) (Chapters 2 and 7).

For some industrial crops, plantations remained very prevalent in many developing countries during the post-colonial period (Gibbon, 2011; Kenney-Lazar and Ishikawa, 2019; Pietilainen and Otero, 2019). On the contrary, their prevalence declined for other crops due to the rising interest in smallholder agriculture (Byerlee, 2014). Yet, a series of circumstances since the 2000s has created the preconditions for a resurging interest in large-scale industrial crop production through LSLAs (Cotula, 2012; Mechiche-Alami et al., 2019). Many scholars have dubbed this period the global land rush, as it has been characterised by rapid changes in global land use patterns. One of the most notable has been the unprecedented transfer of land rights through LSLAs in many developing countries by (or on behalf of) domestic and foreign private investors and corporations, or state-affiliated institutions (e.g. Cotula, 2012; White et al., 2012).

White et al. (2012) identify six aspects underpinning the rapid expansion of LSLAs since the 2000s, namely the:

(a) concerns over food insecurity (partly connected to changing diets), which have triggered huge corporate investments in food and feed value chains;
(b) concerns over energy insecurity, which have driven the demand for land-based energy options (e.g. biofuels);
(c) promotion of new environmental imperatives and tools, which are associated with green-grabbing (e.g. eco-tourism, REDD+);
(d) establishment of extensive infrastructure corridors and Special Economic Zones, which link areas of natural resource extraction to distant markets and consumption centres;
(e) creation of new financial instruments, which are intended to reduce market risk and generate large profits to investors;

(f)   development of rules, regulations, and incentives by the international community, which provide an amenable environment for the engagement of investors in LSLAs.

When it comes to the demand side, the rapid increase of industrial crop-based LSLAs is strongly linked to the wider increasing global demand for agricultural land over the last six decades. While it is not possible to pinpoint a single driver of this demand, it is widely accepted that this has been the combined effect of population growth, changing diets (and consumption practices more widely), and increasing living standards, among others (Alexander et al., 2015; De Schutter, 2009). For example, dietary changes, and especially the increased prevalence of meat-based diets, have driven the re-configuration of the feed industry, causing in the process the large expansion of some industrial crops such as soybean (White et al., 2012). Similarly, fears over energy insecurity have fuelled the expansion of bioenergy-related LSLAs, especially through coordinated energy policies in some developed countries (Cotula, 2012) (Box 3.1). Arguably, some of the main actors have a strong economic motive to engage in such LSLAs (see "Unpacking the interface of industrial crops, LSLAs and land-grabbing"). However, it is also fair to argue that the underlying concerns over resource access security have invariably influenced many of these LSLAs, especially those related to bioenergy as discussed below (Cotula, 2012; White et al., 2012).

When it comes to the supply side, it has been argued that certain countries were targeted for LSLAs for reasons beyond their amenable agro-ecological conditions for some industrial crops. These reasons are context-specific, but have almost invariably included the (a) relative abundance and low cost of land, (b) weak institutional frameworks to regulate LSLAs and/or international investments, and (c) expectations of national governments and other actors for economic growth, rural development, and/or energy security (Lay and Nolte, 2018).

Box 3.1 Biofuel policies as a driver of LSLAs for bioenergy crops

Many scholars have pointed that biofuel demand (and related policies) have been perhaps the main driver of LSLAs for industrial crops in the last couple of decades, especially for oil palm, jatropha, sugarcane, and soybean (Cotula, 2012) (Chapter 4). The global biofuel boom during the 2000s was the manifestation of a set of interlocked global circumstances, namely the (a) high fuel prices that sparked fears over energy insecurity, (b) globalisation and trade liberalisation that facilitated global trade, and (c) assortment of national policies that sought to provide alternative markets for some agricultural commodities that could be used as biofuel feedstock (White et al., 2012). Although the actual drivers of biofuel expansion vary considerably among countries, arguably the most common are energy security, economic development, and climate change mitigation (Gasparatos et al., 2013).

The early adoption of biofuel mandates and targets in Brazil and the US certainly propelled biofuels into the international spotlight. However, it is unclear whether they ultimately had a large *direct* effect on LSLAs, as the respective biofuel mandates relied more on domestic recourses such as maize (in the US), sugarcane (in Brazil), and soybean (in both countries) (REN21, 2020). Direct LSLAs for biofuel feedstock tend to be more closely associated with biofuel developments in the European Union (EU), and especially with the EU Renewable Energy Directive (EU-RED, Directive 2009/28/EC) (EU, 2009). The EU-RED not only mandated member states to derive 20% of their transport fuel from renewable energy sources such as biofuels, but also allowed imports if these targets could not be met through national production (EU, 2009). This made LSLAs in developing countries a possible avenue of sourcing feedstock and/or biofuels for the EU market, especially jatropha, oil palm, and soybean (Borras et al., 2011).

The biofuel market established through these mandates certainly provided a strong impetus for biofuel-related LSLAs in developing countries, but other factors facilitated them further. First, discursive frames such as marginal land and rural development were mobilised to bypass some of the emerging criticisms of biofuel production (Murnaghan, 2017) (Chapter 4). Second, the large availability of financial capital before the 2008 economic crisis and the gradual consolidation of an international land market (Chapter 5) made Foreign Direct investments (FDIs) in biofuel LSLAs quite attractive to many investors (Borras et al., 2020).

However, despite the spike in biofuel-related LSLAs, their subsequent number declined (Mechiche-Alami et al., 2019), possibly due to amendments in the EU-RED, the collapse of the jatropha sector, low fuel prices, and the 2008 economic crisis. However, with 70 countries around the world enacting biofuel mandates, it is possible that biofuel-related LSLAs might take up again (REN21, 2020).

Since 2000, the above set of circumstances has catalysed the sharp increase in the number of LSLAs in developing countries (Mechiche-Alami et al., 2019). However, despite their high visibility in international policy debates, it is still not possible to establish with certainty the number, type, functionality, status, and distribution of LSLAs (Nolte et al., 2016). The Land Matrix and other sources have reported that approximately 1000–1217 LSLAs occulted in developed countries between 2000 and 2010, amounting to 83.2 Mha (Anseeuw et al., 2012). However, following criticisms of overestimating the number and extent of LSLAs in developing countries (Edelman, 2013; Edelman et al., 2013), the

Table 3.1 Number, status, and coverage of LSLAs

Negotiation status	Number of LSLAs	Actual contract size (Mha)
Concluded	1204	42.4
Intended deals	212	–
Failed deals	97	0.9
No information on status	36	0.2
Total	**1549**	**43.6**

Source: (Nolte et al., 2016).

Land Matrix updated its database through an in-country verification data campaign (Nolte et al., 2016). Following this update, 2016 estimates indicated 1204 concluded land deals (42.4 Mha), 212 intended deals (>20 Mha), and 97 failed deals (7.2 Mha) (Nolte et al., 2016) (see Table 3.1). Out of the 903 concluded deals with geographic information, most are located in Africa (44.1%), followed by Asia (23.8%) and the Americas (21.7%) (Nolte et al., 2016).

Linking LSLAs to land grabbing

As will be discussed in sections "Actors in LSLAs and land grabbing: Roles and vested interests" through "Impacts of LSLAs and land grabbing" many LSLAs followed questionable practices and had negative impacts. Such LSLAs have been branded as land grabs and have received a fair share of criticism. However, it is important to emphasise that not all LSLAs are land grabs and that indeed misrepresentations can occur and have negative ramifications both for the individual LSLAs, but also more widely (Ahmed et al., 2019a).

There is no single all-encompassing criterion to judge which LSLAs constitute land grabs (EcoRuralis, 2016). Yet, there is a broad consensus over the need to ascertain whether the investors (a) followed the legal procedures, (b) conducted adequate consultation with appropriate stakeholders (including local communities), and (c) paid fair compensation that was shared adequately between affected stakeholders (German et al., 2013; Hansen et al., 2016; Schoneveld and German, 2014). In any case, when establishing the possibility of land grabbing, most scholars tend to explore the procedural aspects and distributional effects mediating LSLAs (Ahmed et al., 2019a).

Procedural lenses tend to focus on whether LSLAs adhere to the prevailing formal rules and regulations during the initial land acquisition and the subsequent operation. Such lenses seek to track any deliberate or accidental illicit activities from the different actors engaged in LSLAs, such as investors, governments, traditional authorities, and other elite actors (see "Actors in LSLAs and land grabbing: Roles and vested interests"). The focus is on practices that are not consistent with domestic laws such as the acquisition and conversion of land not intended for agricultural production, the lack of prior informed consent, the limited participation or consultation of local landowners, and the non-payment of adequate and timely compensation (De Schutter, 2011; Wolford et al., 2013) (Chapter 5). When exploring the procedural aspects of LSLAs, many studies have alluded that land grabbing is essentially "control grabbing", as it often dwells in the effective control of land (Huggins, 2014). However, scholars have pointed to the limited viewpoints that procedural lenses offer when linking LSLAs and land grabbing, not the least due to the inherently unfair, incomplete, or easily manipulated policy frameworks governing LSLAs in many of the targeted countries (Ahmed et al., 2019a) (Chapter 5).

Distributional lenses tend to focus on the distribution of the positive and negative socioeconomic and environmental outcomes of LSLAs (Ahmed et al., 2019a) (see "Impacts of LSLAs and land grabbing"). Central to these exercises is the effort to identify whether disproportionate benefits and costs occur during LSLAs, and how they are distributed among the involved actors (see "Actors in LSLAs and land grabbing: Roles and vested interests"). Within this purview, some of the studies employing distributional lenses have branded land grabbing as "benefit grabbing" considering the unequal distribution of social, economic, and environmental cost and benefits among the different actors (Ahmed et al., 2019a).

It is worth pointing that contrary to LSLA (see "Drivers and extent of LSLAs"), there are no readily available estimates of land grabbing globally. This is due to the inherent difficulty and deep scrutiny needed to establish linkages between LSLAs and land grabbing. Thus, land grabbing estimates vary widely between sources, with some organisations claiming that the overall extent might be in the order of millions of hectares in the last few decades (GRAIN, 2018; GRAIN, 2016; Oxfam et al., 2016).

Actors in LSLAs and land grabbing: Roles and vested interests

Many different types of national and international stakeholders have engaged in LSLAs during the recent global land rush (Chapters 2, 5, 9, and 11). This has included actors with radically different roles and interests, coming from the private sector, national government agencies, international financial institutions, civil society, and local elites (EcoRuralis, 2016; Behrman et al., 2012; Mechiche-Alami et al., 2019). Mindful of the huge variability of the actors involved in LSLAs across industrial crops and geographical contexts, below we provide a very brief outline of their roles and interests, as well as how some of their actions can create conditions amenable for land grabbing and negative impacts on ecosystems and local communities (see "Actors in LSLAs and land grabbing: Roles and vested interests" and "Unpacking the interface of industrial crops, LSLAs and land-grabbing").

National and international private companies have undertaken the large majority of LSLAs during the recent land rush. Companies essentially purchase or lease large tracts of land to produce, process, and trade industrial crops in national and/or international markets. Profit generation is the main motive of actors from the private sector, usually following market signals (e.g. increases in biofuel triggered by biofuel mandates) and incentives from national governments (e.g. tax breaks, subsidies, friendly institutional framework, low regulation). There have been multiple accounts of both appropriate and inappropriate engagement of private sector actors in LSLAs (Cotula, 2012), including predatory and unethical corporate practices, manipulation of local rules and regulations, and bribery that have led to land grabbing (see "Actors in LSLAs and land grabbing: Roles and vested interests"). However, private sector actors often perceive land grabbing and lack of community consent for accessing land to be "technical issues". The former is sometimes perceived as an offshoot of uncertainties in land rights and land ownership regulations, and the latter an offshoot of poor due diligence (EcoRuralis, 2016).

Financial institutions are a rather diverse type of stakeholders that engage with LSLA in rather different ways. For example, governmental financial institutions such as the International Monetary Fund (IMF) and the World Bank have been credited for creating international circumstances favouring LSLAs, through, for example, their strong push towards agricultural liberalisation in the 1980s (see "Drivers and extent of LSLAs") (Chapter 2). More recently, LSLAs have flourished partly due to the Multilateral Investment Guarantee Agency that provides guarantees against risks for investors and funds for infrastructural projects (Borras et al., 2020; Nally, 2015). Such organisations tend to adopt narratives branding LSLAs as huge economic and development opportunities that can boost the national economies of developing countries (Borras et al., 2020; Nally, 2015). Consequently, many of these international organisations (and especially the World Bank) have started instituting an international code of conduct for LSLAs, as a step towards achieving win–win outcomes (Borras and Franco, 2010). Conversely, private financial institutions such as banks and investment funds tend to see land (and thus LSLAs) as a commodity. More importantly, they view it as a commodity that can be traded in international markets (and is thus amenable to speculation), holding land until prices increase and then selling it for profits (EcoRuralis, 2016) (Chapter 5). In this sense, similar to private companies, their main motive is economic gain and profit.

National governments of developing countries have often facilitated LSLAs within their national borders (Van Der Wulp, 2013; Wolford et al., 2013). On many occasions, government agencies have acted as intermediaries between other national governments, companies, investors, and local authorities (Gonda, 2019; Rutten et al., 2017), or even acquire land on behalf of investors (Maloa Nnoko-Mewanu, 2016; Schlimmer, 2019). Governments often act under discursive frameworks that point to the opportunities that LSLAs offer for national economic growth, rural development, and/or energy security. However, many scholars have pointed that in their desperate hunt for FDIs, many national governments have failed to scrutinise the business models of investors or put in place appropriate regulatory mechanisms for LSLAs (Nkansah-Dwamena and Bonnie Raschke, 2020; Wolford et al., 2013).

Civil society organisations have played a rather uneven role in the context of LSLAs and land grabbing. Some NGOs (usually international) have advocated certain types of industrial crops, LSLAs, or outgrower schemes around new or existing LSLAs as avenues for rural development and poverty alleviation (e.g. Hunsberger, 2010). Others have vehemently opposed LSLAs and/or specific industrial crops, especially related to bioenergy, due to their possible negative outcomes on ecosystems, rural livelihoods, and food security (see "Impacts of LSLAs and land grabbing") (e.g. Oxfam, 2016; Greenpeace, 2016). On many occasions, such organisations have assisted local communities during social conflicts with agribusinesses during LSLAs or the operation of such projects (e.g. Ahmed et al., 2017).

Traditional authorities and local elites have been major players for LSLAs in some geographical contexts. For example, such stakeholders have often acted as development brokers, seeking to attract FDIs and financial capital to their areas (Ahmed et al., 2018b; Boamah, 2014), facilitating LSLAs through formal and informal processes (Anseeuw et al., 2012; Hall, 2011a). Depending on the context, such actors have assumed different roles as land acquirers, negotiators, land allocators, and signatories to land deals (Campion and Acheampong, 2014). There are very divergent accounts of their motives, ranging from purely altruistic by viewing LSLAs as real development opportunities for their areas, to purely egotistical by catering to their self-interest (Ahmed et al., 2018b).

Impacts of LSLAs and land grabbing

Chapters throughout this edited volume discuss the highly variable impacts of LSLAs and land grabbing related to industrial crops. Evidence suggests that the types and magnitude of the impacts, as well as the affected stakeholders, depend on several factors including the industrial crop itself (and its production practices), the previous land use, the local ecological and socioeconomic context of the LSLA, and the policies guiding the LSLA implementation and operation (Gasparatos et al., 2018; White et al., 2012). These positive and negative environmental, economic, and social impacts emerge through various context-specific mechanisms (Table 3.2). Mindful of this large variability, below we briefly outline only some of the main impacts and associated mechanisms, but the interested reader is diverted to more comprehensive publications (e.g. Gasparatos et al., 2015; Filoso et al., 2015; Byerlee, 2014; Gyapong, 2019; Ragsdale et al., 2018) and "Unpacking the interface of industrial crops, LSLAs and land-grabbing" focusses on some specific contexts of industrial crop expansion.

The main environmental impacts of industrial crop-based LSLAs are underpinned by the conversion of large tracts of land to accommodate agricultural activities and ancillary infrastructure, as well as the pollution associated with intensive crop production. According to Messerli et al. (2014) a large fraction of the LSLAs they studied were implemented in remote forestland with low population density (34% of LSLAs) and moderately populated and moderately accessible shrub or grasslands (26% of LSLAs). This direct and indirect land use change has significant ramifications for habitat and biodiversity loss and

Table 3.2 Impacts of LSLAs and land grabbing, and their underlying mechanisms

Dimension	Impact	Mechanism	References
Environmental	Biodiversity loss	Natural vegetation conversion for large-scale industrial crop production and ancillary infrastructure can cause habitat loss, degradation, and/or fragmentation, having negative biodiversity outcomes	Gasparatos et al., 2018; He and Martin, 2015; Savilaakso et al. 2014; Vijay et al., 2014; Green et al., 2019; Boerema et al., 2016; Semie et al., 2019; Filoso et al., 2015; Ahrends et al., 2015; Warren-Thomas et al., 2015
	Carbon stock change	Natural and/or agricultural land conversion for large-scale industrial crop production and ancillary infrastructure can increase or decrease the carbon stocks in above/below ground vegetation and soil	Achten et al., 2011; Romeu-Dalmau et al., 2018; Liu et al, 2017; Fargione et al., 2008; Blagodatsky et al., 2016
	Water depletion	Large-scale, dense, and intensified industrial crop production under irrigated conditions, can affect water balances, especially in water-scarce environments	Kgathi et al., 2017; Chiarelli et al., 2020; Hess et al., 2016
	Water pollution	Large-scale and intensified industrial crop production through the extensive use of fertilizers and agrochemicals can degrade water resources, causing water pollution	Chiarelli et al., 2020; Filoso et al., 2015; Obidzinski et al., 2012; Sari et al., 2019
Economic	Economic growth (macro-level)	Large-scale production of industrial crops can boost national economic growth through employment generation, attraction of FDIs, taxation, and foreign exchange earnings	Boccanfuso et al., 2018; CIFOR, 2015; Gasparatos et al., 2015
	Energy security (macro-level)	Large-scale production and use of biofuels can reduce reliance on imported foreign fuel, offering a renewable energy option	Gasparatos et al., 2015
	Employment (micro-level)	Large-scale industrial crop production projects generate employment (of variable reach and quality) in rural areas with few employment opportunities	Ahmed et al., 2019b; Dib et al., 2018; Matenga and Hichaambwa 2017
	Income (micro-level)	Employment in large-scale industrial crop projects provides formal and stable income (of variable levels and payment modalities)	Ahmed et al., 2019b; Dib et al., 2018; Schoneveld et al., 2011; Matenga and Hichaambwa, 2017
	Livelihoods (micro-level)	Landscape conversion for large-scale industrial crop production and ancillary infrastructure can degrade the provision of multiple ecosystem services, affecting the livelihoods of local communities	Gasparatos et al., 2018; Ahmed et al., 2018a

Social	Food security	Large-scale industrial crop projects invest in infrastructure (e.g. roads, clinics), improving the livelihoods of local communities (e.g. by facilitating market access)	von Maltitz et al., 2016; Mudombi et al., 2018
		Large-scale conversion of agricultural and natural land can reduce food crop production or the wild food availability (reduced food availability)	Jarzebski et al., 2020; Li and Fox, 2012; Hervas and Isakson, 2020
		Engagement in paid plantation employment can divert labour from (and eventually reduce) food crop production (reduced food availability)	Ahmed et al., 2019b; Mingorría et al., 2014
		Engagement in waged plantation employment can provide income that can be invested to purchase food (improved access to food)	Jarzebski et al., 2020; Hervas and Isakson, 2020
		Industrial crop expansion can increase food prices, reducing food affordability particularly to urban poor and rural landless households (reduced access to food)	Jarzebski et al., 2020; Gasparatos et al., 2015
		Engagement in paid plantation employment can divert the time of women from child care, nutrition, and unpaid care work, affecting food preparation and child nutrition (reduced food utilisation)	Jarzebski et al., 2020; Mingorría et al., 2014
		Large-scale industrial crop production can affect local environmental conditions through land use/cover change, soil quality degradation, and water depletion/pollution, collectively reducing the capacity of local agro-ecosystems to produce food in a stable manner (reduced food stability)	Jarzebski et al., 2020
	Land tenure and dispossession	LSLAs can alter land ownership and access regimes through different processes (e.g. land consolidation, zoning), causing in some cases land dispossession and the loss of private and communal land rights for local/indigenous communities and/or individual landowners	Kirst, 2020; Bennett et al., 2018; Ahmed et al., 2019a
	Social conflicts	Failed promises, unmet expectations, negative impacts, and/or elite capture of benefits from LSLAs can create social conflicts between commercial operators, local communities, and/or other stakeholders	Kusakabe and Myae, 2019; Sabogal, 2013; Ahmed et al., 2019c; Campion and Acheampong, 2014
	Gender inequality	Women can engage unfairly or face disproportionately the negative impacts of large-scale industrial crop production	Ahmed et al., 2018a; Potter, 2020; Fonjong, 2017; Lamb et al., 2017; Dao, 2018

ecosystem services degradation, as has been discussed both in conceptual studies (Gasparatos et al., 2018) and studies for individual industrial crops such as oil palm (Savilaakso et al., 2014; Vijay et al., 2016), soybean (Green et al., 2019; Boerema et al., 2016), sugarcane (Semie et al., 2019; Filoso et al., 2015), and rubber (Ahrends et al., 2015; Warren-Thomas et al., 2015).

This extensive land use change also has important ramifications for carbon stock change, creating extensive carbon stock losses if converting natural vegetation (Achten and Verchot, 2011) or gains if converting degraded woodlands or former agricultural land (Romeu-Dalmau et al., 2018). Carbon stock change can create (a) high carbon debts for some bioenergy crops such as oil palm, sugarcane, and soybean that might take decades or even centuries to repay (pointing to the negative climate change mitigation effect of some biofuel options) (Fargione et al., 2008), or (b) high carbon footprints for some industrial crop products such as rubber (pointing to the unsustainability of some current production and consumption practices) (Blagodatsky et al., 2016).

Furthermore, intensified production methods such as the extensive use of irrigation water (whether surface water or groundwater) and agrochemicals such as fertilisers and pesticides to improve yields have been linked to many negative environmental impacts such as water scarcity (Hess et al., 2016; Chiarelli et al., 2020; Chapter 11) or water pollution (Filoso et al., 2015; Obidzinski et al., 2012; Sari et al., 2019; Chapter 7).

In terms of economic impacts, at the macro-level, the large-scale production of industrial crops has been linked to national economic growth (e.g. by attracting FDIs and generating taxes, employment, and foreign exchange) and energy security (e.g. by reducing dependence on imported fuels through biofuels) (Gasparatos et al., 2015; McCarthy et al., 2012). Numerous national-level modelling studies have explored such macro-economic effects suggesting that the actual magnitude of the impacts depends on different production and penetration scenarios (Arndt et al., 2011). Although such macro-economic impacts vary substantially between industrial crops and geographical contexts, they are usually linked to discursive constructs such as the competitive advantage, rural poverty alleviation, and energy security (e.g. McCarthy et al., 2012) (Chapters 2, 4, and 6).

At the micro-level, the large-scale production of industrial crops can generate employment and rural incomes, both directly (e.g. through waged employment in plantations) and indirectly through broader development effects (e.g. secondary employment generation in services or shops around plantations) (Chapter 8). Although plantation wages are often higher and more stable than most other income opportunities in the rural areas of most developing countries (including when compared to outgrower schemes) (Matenga and Hichaambwa, 2017; Obidzinski et al., 2012; Dib et al., 2018), this is not always the case (Ahmed et al., 2019b). Furthermore, these employment opportunities tend to reach relatively fewer people, especially when compared with the wide reach of smallholder and outgrower schemes (Castellanos-Navarrete et al., 2021; Hall et al., 2017) (Chapter 1). Furthermore, many of these jobs are low-skill, seasonal, precarious (especially in contexts of rapid mechanisation or poor labour standards), and/or target or exclude disproportionally specific social groups (Smalley, 2013; Baccarin et al., 2020; Dib et al., 2018) (Chapter 8). The above are often blamed on flawed policy frameworks, poor corporate practices, or entrenched social inequalities (Li, 2011; Castellanos-Navarrete et al., 2021; Gyapong, 2020). Furthermore, large-scale industrial crop production has sometimes been linked to negative livelihood outcomes for surrounding communities, for example, by compromising access to ecosystem services or causing the loss of other livelihood options, especially if compensation was inappropriate (Ahmed et al., 2018a; Obidzinski et al., 2012). However, some industrial crop plantations have improved infrastructure, thus improving the quality of life for some rural communities (von Maltitz et al., 2016; Mudombi et al., 2018).

Social impacts are perhaps the most hotly contested impacts of LSLAs and land grabs for industrial crops. Like the other impact categories discussed above, the magnitude and mechanisms of social impacts are highly variable, depending on the actual context of industrial crop production (Table 3.2). Some of the most commonly studied social impacts include food security, land dispossession, and social conflicts, which are often gender-differentiated.

Food security is one of the most debated impacts of industrial crop expansion, especially for bioenergy crops (Chapter 4). However, there are multiple mechanisms through which industrial crop productions intersect with food security, considering that it is a multi-dimensional concept.[3] Some of the most commonly identified mechanisms include: (a) reduced availability of food crops and wild food due to the extensive conversion of cropland and/or natural vegetation; (b) reduced food crop production due to labour diversion; (c) improved ability to purchase food due to high and/or stable salaries; (d) increased food prices locally and nationally; (e) diverted time, especially of women, from household activities related to food preparation and nutrition; and (f) reduced ability of agro-ecosystems to produce food due to environmental degradation (Jarzebski et al., 2020; Hervas and Isakson, 2020) (Table 3.2).

In many areas of industrial crop expansion, LSLAs and land grabs have been linked to land dispossession and the loss of land rights (Chapters 5, 8, and 11). A series of underlying processes may lead to land dispossession, including the (a) unclear and conflicting land tenure systems in many developing countries, (b) outright manipulation of land policies from powerful actors, (c) coercion, intimidation, and misinformation from investors and local elites, and/or (d) outright "land stealing" during periods of civil unrest (Potter, 2020; Bennett et al., 2018; Dell'Angelo et al., 2017) (see "Linking LSLAs to land-grabbing").

Many of the contestations emerging from the different impacts and vested interests outlined above can cause social conflicts. Depending on the context, social conflicts can manifest in different forms and involve different actors such as traditional authorities, local communities, investors, and/or civil society (see "Linking LSLAs to land-grabbing"). Social conflicts are often an outcome of the (in)action of these actors, as well as illicit processes for land acquisitions, failed investor promises, land access restrictions/losses, contradictory regulations, and low or lack of compensation (Ahmed et al., 2018b; McCarthy, 2010). Some scholars have identified violent conflict as a convenient vehicle to push for the expansion of industrial crops (Potter, 2020). It is worth noting that many social conflicts emerge across lines of gender (Fonjong, 2017), ethnicity (Krieger and Meierrieks, 2016), or age (Ahmed et al., 2019c).

Finally, it should be emphasised that occasionally the impacts of LSLAs and land grabs are gender-differentiated. Many studies have noted that women engage differently (and often unfairly) in industrial crop production compared to men and face disproportionally the negative impacts (Behrman et al., 2012; Elmhirst et al., 2017; Li, 2015). Similar to other impacts, such differentiated effects are very context-specific but tend to manifest in (a) lower salaries and access to lower-skill or seasonal employment (Ahmed et al., 2019b); (b) worse working conditions (e.g. disproportionate exposure to agrochemicals) (Li, 2015); (c) more precarious land tenure rights and higher risk of eviction (Fonjong, 2017); (d) loss of livelihood opportunities due to the degradation of (or restricted access to) ecosystem services (Ahmed et al., 2018a); (e) higher time investment to access ecosystem services following ecosystem services loss following landscape conversion (Ahmed et al., 2018a; Fonjong, 2017).

Unpacking the interface of industrial crops, LSLAs, and land grabbing

Jatropha in Sub-Saharan Africa

Jatropha was promoted in the mid-to-late 2000s in SSA as a promising, yet untested, biofuel feedstock. Jatropha-based biofuels were primarily destined for the transport sector as a substitute for diesel fuel and secondarily for rural electrification in off-grid generators (Gasparatos et al., 2015). Depending on the country, jatropha was produced for domestic fuel use (e.g. in Malawi) or exports mainly to the EU following the Renewable Energy Directive (e.g. Mozambique) (von Maltitz et al., 2014) (Box 3.1). In some countries such as Ghana, jatropha was initially promoted as an alternative transport fuel for the domestic market, but due to domestic and international circumstances, it was eventually exported to the EU (Ahmed et al., 2017). The above reflects that different drivers affected jatropha expansion across different SSA contexts, including energy security, economic growth, foreign exchange generation, and rural development (Gasparatos et al., 2015).

Due to jatropha's low perishability compared to other biofuel crops such as oil palm and sugarcane, jatropha was produced in very different systems, including large-scale plantations, smallholder-based schemes, and hybrids of the two (Gasparatos et al., 2015; Romijn et al., 2014; von Maltitz and Setzkorn, 2013) (Chapter 1).

Jatropha was the biggest source of LSLAs in SSA during the recent land rush, with international investors acquiring large tracts of land in countries, such as Ghana, Tanzania, and Mozambique (Schoneveld, 2014; Gasparatos et al., 2015).

Whereas the extent of jatropha-related land acquisitions varies considerably between sources (both in terms of extent and quality), 2014 estimates suggest at least 92 concluded land deals covering 2.4 Mha (Nolte et al., 2016). In reality, the number of LSLAs might have been much larger considering the rather inadequate reporting systems in most targeted countries, while the actual extent of land conversion might have been much lower considering the collapse of most jatropha projects before reaching their intended size (Locke and Henley, 2013). Indeed, despite its rapid initial increase, the number of jatropha LSLAs slowed down markedly in the early 2010s and practically ceased by the mid-2010s following the widespread collapse of the sector due to various interrelated factors such as low productivity, lack of markets, and local contestations (von Maltitz et al., 2014; Ahmed et al., 2019c).

Jatropha production was often justified, often on grounds that (a) it is a non-food crop that does not directly compete with food crops, avoiding "food vs. fuel" conflicts; and (b) it can be grown under rainfed conditions in marginal areas and with few agricultural inputs (von Maltitz et al., 2014) (Chapter 4). However, it is not possible to reach high yields under such conditions (Gasparatos et al., 2015). Furthermore, despite early studies pointing to the extensive semi-arid areas supposedly appropriate for jatropha production throughout the contingent (Wicke et al., 2011; Watson et al., 2011), much of the expansion happened in productive cropland and (often degraded) woodlands (Romeu-Dalmau et al., 2018; Degerickx et al., 2016).

Many of the jatropha-related LSLAs in SSA have been considered to be land grabs due to the questionable practices employed when gaining access to land (Cotula, 2012; Hall, 2011b; Vermeulen and Cotula, 2010). Some studies have also shown that many of the jatropha projects supposedly located on marginal lands benefited from good water availability, thereby implying coupled land-water grabs (Adams et al., 2019). Many studies have alluded that investors and intermediaries used illicit practices when acquiring land for large-scale projects (Van Der Wulp, 2013; Wolford et al., 2013) and that the (in)actions of chiefs and traditional authorities led to the dispossession of local community members (Kirst, 2020). In many cases, compensation was not properly determined or distributed, with many examples of procedural or distributional irregularities (Ahmed et al., 2019a). Indeed, many actors have "benefited" from the weak institutional frameworks governing LSLAs, which were in some cases designed and implemented after the major wave of LSLAs. This has been observed in countries such as Ghana and Mozambique, which were major targets for jatropha LSLAs during the land rush (Ahmed et al., 2017; Schut et al., 2010).

Land use change is the main mechanism underpinning the environmental impacts of jatropha LSLAs in SSA. In particular, jatropha LSLAs have mainly converted agricultural land and natural vegetation (e.g. woodland, grassland), and have thus been associated with carbon stock change, including large losses when converting woodland (Romeu-Dalmau et al., 2018; Achten and Verchot, 2011; Degerickx et al., 2016). Other possible environmental impacts such as biodiversity loss due to land use change or invasiveness have not been established concretely (Gasparatos et al., 2015). Similarly, as most jatropha plantations were not irrigated and used rather low levels of agricultural inputs, it is also likely that water depletion and pollution effects were moderate (Gasparatos et al., 2015).

In terms of economic impacts, many jatropha-related LSLAs have generated employment in rural areas with few such formal opportunities (von Maltitz et al., 2016; Romijn et al., 2014; Schoneveld et al., 2011). However, the actual levels of employment generation have most likely been rather modest, considering the low employment intensity especially after the plantation establishment phase (Bosch and Zeller, 2019), while the income was not always high (Schoneveld et al., 2011; Ahmed et al., 2019b; Romijn et al., 2014). However, plantation workers have usually valued the stability of this income, as it facilitates better household planning (Schoneveld et al., 2011). Nevertheless, the collapse of the jatropha sector has practically nullified any possible positive impact on rural development or national economic growth (Gasparatos et al., 2015; Ahmed et al., 2017; von Maltitz et al., 2014). However, it is possible that many of the jatropha LSLAs have had negative livelihood outcomes for local communities in expansion areas when considering that many were located on land that provided multiple ecosystem services (even when branded as marginal) (Ahmed et al., 2018a) and that compensation was not always adequate or properly distributed (Ahmed et al., 2018b).

When it comes to social impacts, studies have identified rather different food security outcomes for jatropha LSLAs. For example, some studies have identified negative outcomes through reduced food availability, while others benefit through improved access to food (Jarzebski et al., 2020; Bosch and Zeller, 2019). Furthermore, in many settings, jatropha LSLAs have been linked to negative social impacts, such as (a) land dispossession and loss of land rights (Tufa et al., 2018; Sulle and Nelson, 2009), (b) social conflicts between local communities, traditional authorities, civil society, and investors (Ahmed et al., 2019c; Hunsberger, 2010), and (c) gender inequality (Ahmed et al., 2019a) (Chapter 8).

Rubber in Southeast Asia

The global production and demand for rubber have increased markedly in the past decades from 5.8 Mt in 1994 to 14.6 Mt in 2019 (FAOSTAT, 2020). Global rubber prices increased substantially until the 2008 economic crisis, when they fell steeply, before rapidly recovering (Goh et al., 2016). Since then, global rubber prices have remained high, albeit fluctuating (Kenney-Lazar et al., 2018). Currently, countries in Southeast Asia account for over 78.5% of global rubber production, with rubber areas spanning over 9.5 Mha in the region (FAOSTAT, 2020). Indeed, during the last 50 years, land under rubber production has increased by 1500% in the region, mainly at the expense of forest (70%) and croplands (30%) (Hurni and Fox, 2018). During this period, there has also been a marked shift from smallholder-based rubber production to large plantations and outgrower schemes (Kenney-Lazar and Ishikawa, 2019). This is largely because rubber requires an average of seven years between planting and harvesting, and it is thus perceived as a long-term investment requiring access to secured land rights and sustained capital (Byerlee, 2014).

Depending on the country, a very diverse set of actors and processes have been linked to the establishment and operations of rubber plantations in the region (Fox and Castella, 2013). Multinational corporations such as Michelin have traditionally exerted a major role in rubber expansion (Byerlee, 2014). However, in recent decades, the Chinese market (i.e. the world largest consumer of rubber) and companies have become much more prominent (Fox and Castella, 2013), both through the direct engagement of Chinese firms and the Chinese government via its Opium Replacement Program (ORP) that offers alternative livelihood options to opium smallholders (Lu, 2017). Interestingly, many rubber LSLAs are domestic, as some national governments perceive rubber as an avenue to modernise and industrialise their agrarian systems (Dao, 2015).

However, many of the recent rubber LSLAs in Southeast Asia have been considered to be land grabs due to the informal processes employed for acquiring the land (Kenney-Lazar, 2012; Dao, 2015). Furthermore, many rubber-related LSLAs have also appropriated large quantities of water for irrigation, especially in the Mekong region (He and Martin, 2015), therefore also characterised as water grabs (Friend et al., 2019; Matthews, 2012).

Deforestation underpins the most severe environmental impacts of large-scale rubber production manifest in the region. For example, the significant deforestation association with rubber expansion throughout the region has reportedly threatened biodiversity and the provision of ecosystem services (Ahrends et al., 2015; Warren-Thomas et al., 2015). Furthermore, this deforestation has been linked to current and future carbon stock loss, despite the moderate potential of rubber plantations as carbon sinks (Blagodatsky et al., 2016; Warren-Thomas et al., 2018). Rubber plantations are also noted to affect hydrological cycles and water scarcity due to their similar water requirement with forest (but less than other crops), as well as contribute to water pollution through the high use of pesticides and fertiliser (Chiarelli et al., 2020; Giambelluca et al., 2016).

In terms of economic impacts, the large-scale production of rubber has contributed significantly to some national economies in the region. Currently, many investors and national governments frame rubber production, especially eco-friendly production, as a major opportunity for green growth (OECD, 2014; Kenney-Lazar et al., 2018). At the local level, similar to other large-scale industrial crop production systems, rubber plantations generate rural employment and income, directly improving the livelihoods of workers (Gordon, 2004; van Noordwijk et al., 2014; Vongkhamheng et al., 2016). However, on many occasions, the working conditions are precarious and dangerous (Gordon, 2004), plantation salaries are low (Portilla, 2017), and employment is geared towards migrants affecting labour geographies (Baird et al., 2019).

When it comes to social impacts, some large-scale mono-cultural rubber production systems have negative impacts on food security, for example, by reducing food availability through cropland conversion (e.g. paddy rice fields) (Li and Fox, 2012) and the loss of trees that are important food sources in the region (van Noordwijk et al., 2014). Furthermore, rubber LSLAs have triggered the loss of communal land and land dispossession for some local communities, having important ramifications for food security and local livelihoods (e.g. Kenney-Lazar, 2012; Kusakabe and Myae, 2019; Dao, 2015). Similar studies of other industrial crops have pointed to the gendered experiences with large-scale rubber production and associated land grabs, as well the lower quality jobs in plantations (Lamb et al., 2017; Park, 2019; Dao, 2018).

Oil palm in Latin America

Global palm oil production has skyrocketed across the tropics to meet the increasing demand from the food and biofuel industry (FAOSTAT, 2020) (Chapter 1). Traditionally, Southeast Asia, and especially Malaysia and Indonesia, has been by far the largest producing region. However, oil palm production has been expanding rapidly in other parts of the world such as Latin America, which now constitutes the fastest growing and the second largest producing region (7% of global production) (FAOSTAT, 2020) (Chapter 1). However, not all countries follow the same expansion model, with some countries mainly relying on smallholder schemes (e.g. Honduras, Mexico), others on large-scale production (e.g. Guatemala, Colombia, Brazil), and others on hybrid models (e.g. Ecuador, Peru, Costa Rica)(Castellanos-Navarrete et al., 2021). For example, 95% of oil palm production in Guatemala is concentrated in large estates, while in other countries the respective shares are lower: e.g. Brazil (73%), Colombia (72%), Costa Rica (67%), Ecuador (44%), Peru (44%), Mexico (40%), and Honduras (35%) (Castellanos-Navarrete et al., 2021). This prominence of large-scale commercial models in some countries seems to reflect "*pre-existing unequal agrarian structures, land tenure regimes posing few obstacles to land concentration, and, in some cases, oil palm policies biased towards large-scale producers*" (Castellanos-Navarrete et al., 2021: 296).

The drivers of oil palm expansion in Latin America are mostly linked to food security initiatives, energy security ventures, and a recent interest in new hubs of global capital accumulation through FDIs (CIFOR, 2015). Indeed the global biofuel boom (Box 3.1, Chapter 4) has driven oil palm expansion in countries such as Brazil, Colombia, and Peru, with the domestic biodiesel mandates (Colombia, Peru) and the preferential inclusion of some plantations for the social fuel seal (Brazil) facilitating significantly this expansion (CIFOR, 2015). The current efforts to transform part of Latin America into new hubs of global capital accumulation have further intensified FDIs for oil palm-based LSLAs in countries, such as Mexico, Colombia, and Guatemala (Furumo and Mitchell Aide, 2017; Pietilainen and Otero, 2019). Interestingly, many intra-regional actors are involved in LSLAs, especially trans-Latin American companies that are some of the major investors in the sector (Borras et al., 2012). These companies have strong lobbying power (CIFOR, 2015) and are known to establish strong alliances with central governments and inject international capital into the sector (Borras et al., 2012).

The bulk of the added oil palm output has been achieved through expansion in cultivated areas rather than improvement in yields (Carter et al., 2007; Furumo and Mitchell Aide, 2017). However, as huge areas were converted for oil palm production through LSLAs, there have been many instances of land grabbing in countries, such as Colombia (Marin-Burgos and Clancy, 2017; Potter 2020), Guatemala (Alonso-Fradejas, 2012), and Honduras (Leon Araya, 2019), among others. Although the recent expansion and the overall production of oil palm in Latin America are not as vast compared to Southeast Asia, oil palm production has significant socioeconomic and environmental impacts.

Similar to other industrial crops, deforestation underscores the main environmental impact of oil palm expansion in Latin America. Based on a large sample of oil palm plantations across Latin America (342,032 ha), a recent study found that oil palm plantations mainly replaced agricultural areas (76% of total land conversion of which cattle pastures accounted for 56%) and woody vegetation (e.g. forests, 21% of total land conversion) mainly in the Amazon and northern Guatemala (Furumo and Mitchell Aide, 2017). Through their outgrowers, core estates have significant deforestation spillages, which are usually much higher than other smallholders (Bennett et al., 2018). Thus, it has been argued that the current expansion patterns might have substantial negative impacts on biodiversity and carbon stocks (Savilaakso et al., 2014; Vijay et al., 2016), with large carbon debts estimated for palm oil-based biofuels in Latin America (Achten and Verchot, 2011). Such environmental trade-offs could be minimised by establishing new plantations on pasture land and improving pastureland productivity elsewhere (Garcia-Ulloa et al., 2012).

The economic impacts of large-scale oil palm production depend substantially on different actors. On the one hand, palm oil is a very lucrative agricultural commodity that creates large profits for agri-businesses (CIFOR, 2015). On the other hand, although large-scale oil palm projects generate rural employment and income, the actual job creation is rather low and is characterised by flexible labour regimes, employment uncertainty, low salaries, and risky working conditions (Castellanos-Navarrete et al., 2021; Brandao et al., 2019). Overall, it has been argued that these rural incomes cannot drive social change, as the neoliberal arrangements characterising current oil palm expansion dynamics in many parts of the region permeate many of the entrenched social inequalities (Castellanos-Navarrete et al., 2019).

Large-scale oil palm production has multiple context-specific social impacts throughout the region. For example, despite improvements in food access due to income from plantation employment, large-scale oil palm production can reportedly reduce food availability for the wider communities surrounding plantations (Hervas and Isakson, 2020). Furthermore, large-scale oil palm production can have major ramifications at the interface of food security and gender inequality, as it can decrease maize cultivation and productivity and reduce time for other household activities, including women's resting time (Mingorría et al., 2014). Furthermore, there have been social conflicts throughout the region due to the loss of land rights (Mingorría, 2017) and high rates of displacement

(Sabogal, 2013). Examples of social conflicts include local communities mobilisation against investors to safeguard their land rights and control deforestation (Potter, 2020).

Political ecology of LSLAs and land grabs related to industrial crops

Sections "Critical aspects of LSLAs and land-grabbing for industrial crops" through "Unpacking the interface of industrial crops, LSLAs and land-grabbing" clearly show that LSLAs and land grabs linked to industrial crops have been major agents of ecological, agrarian, socioeconomic, and institutional transformation. Arguably, the ecological transformation brought by LSLAs and land grabs in the form of extensive landscape conversion and intensified production underpin how agrarian and socioeconomic transformation manifest (Chapters 5, 9, and 11). For example, the modernisation of agricultural practices, the generation of formal employment opportunities in plantations, and the loss of traditional livelihoods are only some of the mechanisms through which the large-scale production of industrial crops catalyses changes in agrarian structure (Chapter 8). In some contexts, this agrarian change manifests through the emergence of new social classes such as landless plantation labourers, contract farmers, and local elites, among others (Chapters 7–8 and 11). Finally, ecological and agrarian transformation can catalyse socioeconomic transformation through, for example, changes in food security, gender equality, or social conflicts (Chapters 8 and 11).

Interestingly, institutional transformation can both facilitate industrial crop LSLAs and land grabs, as well as be their outcome (Chapter 1). On the one hand, the change or manipulation (often through illicit means) of policy frameworks was a pre-condition to obtain access to large tracts of land or to allow certain intensified production practices (Chapter 5). On the other hand, the actual large-scale production of industrial crops eventually transformed the institutions governing access to natural resources and essentially the distribution of the accruing benefits (Chapter 11).

Considering the above, it is fair to argue that industrial crop-based LSLAs and land grabs are not politically neutral (Chapters 1–2). In fact, the actors engaged in industrial crop LSLAs and land grabs are entangled in (and via) very diverse and context-specific political processes (see "Linking LSLAs to land-grabbing"), which underpin the actual impacts, and the mechanisms through which they emerge (see "Actors in LSLAs and land grabbing: Roles and vested interests" through "Impacts of LSLAs and land grabbing" and "Unpacking the interface of industrial crops, LSLAs and land-grabbing") (Table 3.2). Furthermore, the transformative power of LSLAs and land grabs does not affect social groups in a homogenous way but comes with clear winners and losers. This differentiation more often than not (re)produces inequalities and power relations among actors.

Consistent with the theme of this edited volume, here we argue that political ecology offers an ideal lens to unpack processes that are central to industrial crop-related LSLAs and land grabs (Peet et al., 2011; Robbins, 2012) (Chapters 1–2). Many studies have adopted such political ecology lenses to critically explore such processes in diverse global contexts (Ahmed et al., 2019a; Bennett et al., 2018; Marin-Burgos and Clancy, 2017; Elmhirst et al., 2017; Lamb et al., 2017; Kenney-Lazar et al., 2018) (Chapters 5, 8, 9, and 11). Below, we briefly synthesise some of the main information presented in "Critical aspects of LSLAs and land-grabbing for industrial crops" through "Unpacking the interface of industrial crops, LSLAs and land-grabbing" using the basic conceptual tools of political ecology, namely attention to (a) multi-scalar dynamics, (b) social differentiation, (c) marginalisation, and (d) discursive power (Chapters 2 and 12).

First, LSLAs and land grabs operate at multiple scales. International markets and consumers dictate the global demand for practically all major industrial crops, with policies and consumption patterns in distant places driving the production of industrial crops in national and local contexts (Box 3.1). Furthermore, some international actors such as investors, governments, or NGOs drive industrial crop LSLAs in targeted countries based on their individual vested interests, whether they revolve around profit generation, resource security, or development aspirations (see "Actors in LSLAs and land grabbing: Roles and vested interests"). National governments in a sense act as an "intermediary" mediating the processes unfolding at the international and the local level by developing amenable policy frameworks to both attract FDIs for LSLAs, as well as regulate them. Eventually, LSLAs are operationalised (and occasionally become land grabs) in local contexts through the actions of private companies and other facilitators (e.g. traditional authorities, elite actors). It is worth noting that the actors operating across all levels very often mobilise different narratives and discourses to justify their actions.

Second, many of the benefits and costs associated with industrial crop-based LSLAs and land grabs are socially differentiated, in that some groups gain or lose disproportionally (see "Actors in LSLAs and land grabbing: Roles and vested interests" and "Unpacking the interface of industrial crops, LSLAs and land-grabbing"). Winners can include (a) international investors, national governments, and local elites profiting from industrial crop production (Chapters 5–6, 9–11), and (b) selected members of the local communities benefiting from rural jobs and income (Chapters 8, 10–11). Losers are mainly different segments of local communities, usually due to losing access to natural resources or capturing relatively limited benefits compared to other groups (especially during land grabs) (Chapters 5, 8, and 11). Beyond local communities, other losers include smaller individual farmers that cannot compete with large producers (Chapter 9) or actors during boom-and-bust cycles (von Maltitz et al., 2014; Ahmed et al., 2017). It is worth mentioning that differentiation does not only manifest between groups but also within groups. For example, the distribution of costs and benefits can be differentiated within the same group (e.g. plantation workers, smallholders) across lines of gender, ethnicity, age, wealth, political power, or even location (see "Actors in LSLAs and land grabbing: Roles and vested interests" and "Unpacking the interface of industrial crops, LSLAs and land-grabbing") (Chapters 7–8 and 11).

Third, social marginalisation can be both a driver and an outcome of engaging with industrial crop LSLAs and land grabs. On the one hand, workers in industrial crop plantations often come from marginalised groups such as landless households or poor migrants (Chapters 8 and 11), as it is one of the few (if not the only) livelihood options available to them in the poor rural contexts that are usually the backdrops of industrial crop expansion in the global South. In some cases, marginalised groups only have access to unskilled, precarious, and low-salary plantation jobs (Chapters 7–8 and 11) (see "Actors in LSLAs and land grabbing: Roles and vested interests" and "Unpacking the interface of industrial crops, LSLAs and land-grabbing"), which can further reinforce social inequalities and marginalisation. On the other hand, industrial crop LSLAs (and especially land grabs) can marginalise some social groups, such as those with very weak land rights, little political power, or reliant on ever-scarcer ecosystem services from common lands in the periphery of large-scale industrial crop projects (see "Actors in LSLAs and land grabbing: Roles and vested interests").

Finally, different discourses and narratives have been mobilised across all scales to both justify and oppose industrial crop LSLAs. On the one hand, narratives linked to "economic growth", "rural development", "energy security", and "climate change mitigation" have been invariably mobilised by international, national, and local actors to justify industrial crop expansion in the past decades (see "Drivers and extent of LSLAs" and "Unpacking the interface of industrial crops, LSLAs and land-grabbing") (Chapter 1). Such narratives have been usually paired with constructs such as "marginal lands" and "comparative advantage" to further justify certain types of LSLAs and the development of amenable policy frameworks to attract and allocate land to them (see "Drivers and extent of LSLAs") (Chapters 4, 6, and 11). Conversely, counter-narratives such as "food vs. fuel" have been used to oppose LSLAs, particularly those related to bioenergy crops in food-insecure countries of the global South (Chapter 4).

Conclusion

This chapter synthesised the main drivers, actors, and impacts of industrial crop expansion through LSLAs in countries of the global South and the critical aspects leading to land grabbing. Overall, a complex set of context-specific factors and vested interests have led to the proliferations of LSLAs and land grabs in many countries. LSLAs and land grabs have become major agents of ecological, agrarian, socioeconomic, and institutional change in many parts of Africa, Asia, and Latin America. However, they are not politically neutral and are constantly (re)producing inequalities and power relations among stakeholders, with some winners and some losers. We argue that the current knowledge and appropriate solutions about industrial crop LSLAs and land grabbing should not be understood or addressed solely on theoretical grounds, but should become a subject of deep empirical inquiry. The field of Political Ecology with its attention to multi-scalar dynamics, social differentiation, marginalisation, and discursive power offers an ideal lens to guide such efforts.

Acknowledgements

We acknowledge the support of the Asahi Glass Foundation through a Research Grant for Young Scientists and a Continuation Grant and the Japan Science and Technology Agency (JST) for the Belmont Forum project FICESSA. AA acknowledges the support of a DAAD ClimapAfrica Postdoc Fellowship.

Notes

1 The International Land Coalition (ILC) indicates that a land grab occurs when an LSLA (a) violates human rights, (b) does not seek free and prior informed consent, (c) disregards socioeconomic and environmental impacts, and/or (d) is not based on democratic planning and participation (International Land Coalition, 2011). For the purpose of this chapter we define land grabbing as "*the control (whether through ownership, lease, concession, contracts, quotas, or general power) of larger than locally-typical amounts of land by any person or entity (public or private, foreign or domestic) via any means ('legal' or 'illegal') for purposes of speculation, extraction, resource control or commodification at the expense of peasant farmers, agroecology, land stewardship, food sovereignty and human rights*" (Eurovia, 2016: 2).

2 Similar to industrial crop-based LSLAs, tree plantations for eucalyptus, pine and other woody biomass species have expanded significantly in the past decades following, among others, the adoptions of the Bonn Challenge, the African Forest Landscape Restoration Initiative (AFR100), and the 20X20 initiative in Latin America (Messinger and Winterbottom, 2016; Schweizer et al., 2021).

3 According to the UN Committee on World Food Security, "*food security exists when all people, at all times, have physical, social and economic access to sufficient, safe and nutritious food to meet their dietary needs and food preferences for an active and healthy life. The four pillars of food security are availability, access, utilization and stability*" (FAO, 2009: 1).

References

Achten, W.M.J., Verchot, L.V., 2011. Implications of biodiesel-induced land-use changes for CO_2 emissions: case studies in tropical America, Africa, and Southeast Asia. Ecology and Society 16(4): 14. https://doi.org/10.5751/ES-04403-160414

Adams, E.A., Kuusaana, E.D., Ahmed, A., Campion, B.B., 2019. Land dispossessions and water appropriations: Political ecology of land and water grabs in Ghana. Land Use Policy 87, 104068.

Ahmed, A., Abubakari, Z., Gasparatos, A., 2019a. Labelling large-scale land acquisitions as land grabbing: Procedural and distributional considerations from two cases in Ghana. Geoforum, 105, 191–205.

Ahmed, A., Dompreh, E., Gasparatos, A., 2019b. Human wellbeing outcomes of involvement in industrial crop production in Ghana: Evidence from sugarcane, oil palm and jatropha sites. PLoS ONE, 14(4), e0215433.

Ahmed, A., Campion, B., Gasparatos, A., 2019c. Towards a classification of the drivers of jatropha collapse in Ghana elicited from the perceptions of multiple stakeholders. Sustainability Science, 14, 315–339.

Ahmed, A., Jarzebski, MP, Gasparatos, A., 2018a. Using the ecosystem service approach to determine whether jatropha projects were located in marginal lands in Ghana: Implications for site selection. Biomass and Bioenergy, 114, 112–124.

Ahmed, A., Kuusaana, E.D., Gasparatos, A., 2018b. The role of chiefs in large-scale land acquisitions for jatropha production in Ghana: Insights from agrarian political economy. Land Use Policy, 75, 570–582.

Ahmed, A., Campion, B.B., Gasparatos, A, 2017. Biofuel development in Ghana: Policies of expansion and drivers of failure in the jatropha sector. Renewable and Sustainable Energy Reviews, 70, 133–149.

Ahrends, A., Hollingsworth, P.M., Ziegler, A.D., Fox, J., M., Chen, H., Su, Y., Xu, J., 2015. Current trends of rubber plantation expansion may threaten biodiversity and livelihoods. Global Environmental Change, 34, 48–58. https://doi.org/10.1016/j.gloenvcha.2015.06.002

Alexander, P., Rounsevell, M.D.A., Dislich, C., Dodson, J.R., Engström, K., Moran, D., 2015. Drivers for global agricultural land use change: The nexus of diet, population, yield and bioenergy. Global Environmental Change, 35, 138–147.

Alonso-Fradejas, A., 2012. Land control-grabbing in Guatemala: The political economy of contemporary agrarian change. Canadian Journal of Development Studies, 33(4), 509–528.

Anseeuw, W., Boche, M., Breu, T., Giger, M., Lay, J., Messerli, P., Nolte, K., 2012. Transnational Land Deals for Agriculture in the Global South Analytical Report based on the Land Matrix Database. CDE/CIRAD/GIGA, Bern/Montpellier/Hamburg.

Arndt, C., Benfica, R., Thurlow, J., 2011. Gender implications of biofuels expansion in Africa: The case of Mozambique. World Development, 39, 1649–1662.

Baccarin, J.G., de Oliveira, J.A., Mardegan, G.E., 2020. The environmental, social and economic effects of recent technological changes in sugarcane on the State of São Paulo, Brazil. Journal of Agrarian Change, 20, 598–617.

Baird, I.G., Noseworthy, W., Tuyen, N.P., Ha, L.T., Fox, J., 2019. Land grabs and labour: Vietnamese workers on rubber plantations in southern Laos. Singapore Journal of Tropical Geography, 40, 50–70.

Behrman, J., Meinzen-Dick, R., Quisumbing, A., 2012. The gender implications of large-scale land deals. Journal of Peasant Studies, 39, 49–79.

Bennett, A., Ravikumar, A., Cronkleton, P., 2018. The effects of rural development policy on land rights distribution and land use scenarios: The case of oil palm in the Peruvian Amazon. Land Use Policy, 70, 84–93.

Blagodatsky, S., Xu, J., Cadisch, G., 2016. Carbon balance of rubber (Hevea brasiliensis) plantations: A review of uncertainties at plot, landscape and production level. Agriculture, Ecosystems & Environment, 221, 8–19.

Boamah, F., 2014. How and why chiefs formalise land use in recent times: The politics of land dispossession through biofuels investments in Ghana. Review of African Political Economy, 41, 406–423.

Boccanfuso, D., Coulibaly, M., Savard, L., Timilsina, G., 2018. Macroeconomic and distributional impacts of Jatropha based biodiesel in Mali. Economies, 6, 63.

Boerema, A., Peeters, A., Swolfs, S., Vandevenne, F., Jacobs, S., Staes, J. et al., 2016. Soybean trade: Balancing environmental and socio-economic impacts of an intercontinental market. PLoS ONE, 11(5), e0155222.

Borras, S., Franco, J., 2010. From threat to opportunity? Problems with the Idea of a "Code of Conduct" for land-grabbing. Yale Human Rights and Development Law Journal, 13, 507–523.

Borras, S.M., Franco, J.C., Gómez, S., Kay, C., Spoor, M., 2012. Land grabbing in Latin America and the Caribbean. The Journal of Peasant Studies, 39, 845–872.

Borras, S.M., Hall, R., Scoones, I., White, B., Wolford, W., 2011. Towards a better understanding of global land grabbing: An editorial introduction. The Journal of Peasant Studies, 38, 209–216.

Borras Jr., S.M., Mills, E.N., Seufert, P., Backes, S., Fyfe, D., Herre, R., Michéle, L., 2020. Transnational land investment web: Land grabs, TNCs, and the challenge of global governance. Globalizations, 17, 608–628.

Bosch, C., Zeller, M., 2019. Large-scale biofuel production and food security of smallholders: Evidence from Jatropha in Madagascar. Food Security, 11, 431–445.

Brandao, F., de Castro, F., Futemma, C., 2019. Between structural change and local agency in the palm oil sector: Interactions, heterogeneities and landscape transformations in the Brazilian Amazon. Journal of Rural Studies, 71, 156–168.

Byerlee, D., 2014. The fall and rise again of plantations in Tropical Asia: History repeated? Land, 3, 574–597.

Campion, B., Acheampong, E., 2014. The Chieftaincy Institution in Ghana: Causers and arbitrators of conflicts in industrial Jatropha investments. Sustainability 6, 6332–6350.

Carter, C., Finley, W., Fry, J., Jackson, D., Willis, L., 2007. Palm oil markets and future supply. European Journal of Lipid Science and Technology, 109, 307–314.

Castellanos-Navarrete, A., de Castro, F., Pacheco, P., 2021. The impact of oil palm on rural livelihoods and tropical forest landscapes in Latin America. Journal of Rural Studies. 81, 294–304.

Castellanos-Navarrete, A., Tobar-Tomás, W.V., López-Monzón, C.E., 2019. Development without change: Oil palm labour regimes, development narratives, and disputed moral economies in Mesoamerica. Journal of Rural Studies, 71, 169–180.

Chiarelli, D.D., Passera, C., Rulli, M.C., Rosa, L., Ciraolo, G., D'Odorico, P., 2020. Hydrological consequences of natural rubber plantations in Southeast Asia. Land Degradation & Development, 31, 2060–2073.

CIFOR, 2009. The impacts and opportunities of oil palm in Southeast Asia: What do we know and what do we need to know? Center for International Forestry Research (CIFOR), Bogor.

CIFOR, 2015. Managing oil palm landscapes: A seven-country survey of the modern palm oil industry in Southeast Asia, Latin America and West Africa. Center for International Forestry Research (CIFOR), Bogor.

Cotula, L., 2012. The international political economy of the global land rush: A critical appraisal of trends, scale, geography and drivers. The Journal of Peasant Studies, 39, 649–680.

Cotula, L., Vermeulen, S., Leonard, R. and Keeley, J., 2009, Land grab or development opportunity? Agricultural investment and international land deals in Africa. IIED/FAO/IFAD, London/Rome.

Dao, N., 2015. Rubber plantations in the Northwest: Rethinking the concept of land grabs in Vietnam. The Journal of Peasant Studies, 42, 347–369.

Dao, N., 2018. Rubber plantations and their implications on gender roles and relations in northern uplands Vietnam. Gender, Place & Culture 25, 1579–1600.

De Schutter, O., 2009. Large-scale land acquisitions and leases: A set of core principles and measures to address the human rights challenge. United Nations, New York.

De Schutter, O., 2011. How not to think of land-grabbing: Three critiques of large-scale investments in farmland. The Journal of Peasant Studies, 38, 249–279.

Degerickx, J., Almeida, J., Moonen, P.C., Vervoort, L., Muys, B. and Achten, W.M., 2016. Impact of land-use change to Jatropha bioenergy plantations on biomass and soil carbon stocks: A field study in Mali. GCB Bioenergy, 8(2), 443–455. https://doi.org/10.1111/gcbb.12288

Dell'Angelo, J., D'Odorico, P., Rulli, M.C., Marchand, P., 2017. The tragedy of the grabbed commons: Coercion and dispossession in the global land rush. World Development, 92, 1–12.

Dib, J.B., Krishna, V.V., Alamsyah, Z., Qaim, M., 2018. Land-use change and livelihoods of non-farm households: The role of income from employment in oil palm and rubber in rural Indonesia. Land Use Policy, 76, 828–838.

EcoRuralis, 2016. What is land grabbing? – A critical review of existing definitions. EcoRuralis, Cluj-Napoca.

Edelman, M., 2013. Messy hectares: Questions about the epistemology of land grabbing data. The Journal of Peasant Studies, 40, 485–501.

Edelman, M., Oya, C., Borras, S.M., 2013. Global land grabs: Historical processes, theoretical and methodological implications and current trajectories. Third World Quarterly, 34, 1517–1531.

Elmhirst, R., Siscawati, M., Basnett, B.S., Ekowati, D., 2017. Gender and generation in engagements with oil palm in East Kalimantan, Indonesia: Insights from feminist political ecology. The Journal of Peasant Studies, 44, 1135–1157.

EU, 2009. Directive 2009/28/EC of the European Parliament and of the Council of 23 April 2009 on the Promotion of the Use of Energy from Renewable Sources. Official Journal of the European Union, L140, 16–62.

EUROVIA, 2016. How do we define Land Grabbing? Towards a common understanding and definition of Land Grabbing around the world. European Coordination Via Campesina (EUROVIA), Brussels.

FAO, 2009. Reform of the UN Committee on Food Security: Final version. Food and Agricultural Organization (FAO), Rome.

FAOSTAT, 2020. Food and agriculture data. Food and Agriculture Organisation (FAO), Rome.

Fargione, J., Hill, J., Tilman, D., Polasky, S., Hawthorne, P., 2008. Land clearing and the biofuel carbon debt. Science 319, 1235–1238.

Filoso, S., do Carmo, J.B., Mardegan, S.F., Machado Lins, S.F., Gomes, T.F., Martinelli, L.A., 2015. Reassessing the environmental impacts of sugarcane ethanol production in Brazil to help meet sustainability goals. Renewable and Sustainable Energy Reviews, 52, 1847–1856.

Fonjong, L., 2017. Interrogating Large-Scale Land Acquisition and its Implications for Women's Land Rights in Cameroon, Ghana and Uganda. International Development Research Centre (IDRC), Ottawa.

Fox J., Castella J.C., 2013. Expansion of rubber (Hevea brasiliensis) in Mainland Southeast Asia: What are the prospects for smallholders? The Journal of Peasant Studies, 40, 155–170.

Friend, R.M., Thankappan, S., Doherty, B., Aung, N., Beringer, A.L., Kimseng, C., Cole, R., Inmuong, Y., Mortensen, S., Nyunt, W.W., Paavola, J., Promphakping, B., Salamanca, A., Soben, K., Win, Saw, Win, Soe, Yang, N., 2019. Agricultural and food systems in the Mekong region: Drivers of transformation and pathways of change. Emerald Open Research 1, 12.

Furumo, P.R., Mitchell Aide, T., 2017. Characterizing commercial oil palm expansion in Latin America: Land use change and trade. Environmental Research Letters, 12, 024008.

Garcia-Ulloa, J., Sloan, S., Pacheco, P., Ghazoul, J., Koh, L.P., 2012. Lowering environmental costs of oil palm expansion in Colombia. Conservation Letters, 5, 366–375.

Gasparatos, A., Lee, L., von Maltitz, G., Mathai, M., Puppim de Oliveira, J., Johnson, F.X., Willis, K., 2013. Catalysing biofuel sustainability: International and national policy interventions. Environmental Policy and Law, 43, 216–221.

Gasparatos, A., Romeu-Dalmau, C., von Maltitz, G., Johnson, F.X., Shackleton, C., Jarzebski, M.P., Jumbe, C., Ochieng, C., Mudombi, S., Nyambane, A., Willis, K.J., 2018. Mechanisms and indicators for assessing the impact of biofuel feedstock production on ecosystem services, Biomass and Bioenergy, 114, 157–173.

Gasparatos, A., von Maltitz, G., Johnson, F.X., Lee, L., Mathai, M., Puppim de Oliveira, J., Willis, K., 2015. Biofuels in Africa: Drivers, impacts and priority policy areas. Renewable and Sustainable Energy Reviews, 45, 879–901.

German, L., Schoneveld, G., Mwangi, E., 2013. Contemporary processes of large-scale land acquisition in Sub-Saharan Africa: Legal deficiency or elite capture of the rule of law? World Development, 48, 1–18.

Giambelluca, T.W., Mudd, R.G., Liu, W., Ziegler, A.D., Kobayashi, N., Kumagai, T., Miyazawa, Y., Lim, T.K., Huang, M., Fox, J., Yin, S., Mak, S.V., Kasemsap, P., 2016. Evapotranspiration of rubber (Hevea brasiliensis) cultivated at two plantation sites in Southeast Asia. Water Resources Research, 52, 660–679.

Gibbon, P., 2011. Experiences of Plantation and Large-Scale Farming in 20th Century Africa: DIIS Working Paper. The Danish Institute for International Studies (DIIS), Copenhagen.

Goh, H.H., Tan, K.L., Khor, C.Y., Ng, S.L., 2016. Volatility and market risk of rubber price in Malaysia: Pre- and post-global financial crisis. Journal of Quantitative Economics, 14, 323–344.

Gonda, N., 2019. Land grabbing and the making of an authoritarian populist regime in Hungary. The Journal of Peasant Studies, 46, 606–625.

Gordon, A., 2004. Dynamics of labour transformation: Natural rubber in Southeast Asia. J Journal of Contemporary Asia, 34, 523–546.

GRAIN, 2016. The global farmland grab in 2016: How big, how bad? GRAIN, Barcelona.

GRAIN, 2018. Failed farmland deals: A growing legacy of disaster and pain. GRAIN, Barcelona.

Green, J.M.H., Croft, S.A., Durán, A.P., Balmford, A.P., Burgess, N.D., Fick, S., Gardner, T.A., Godar, J., Suavet, C., Virah-Sawmy, M., Young, L.E., West, C.D., 2019. Linking global drivers of agricultural trade to on-the-ground impacts on biodiversity. Proceedings of the National Academy of Sciences, 116, 23202–23208.

Greenpeace, 2016. Palm oil's new frontier: How industrial expansion threatens Africa's rainforests. Greenpeace, Amsterdam.

Gyapong, A.Y., 2019. Land deals, wage labour, and everyday politics. Land, 8, 94.

Gyapong, A.Y., 2020. How and why large scale agricultural land investments do not create long-term employment benefits: A critique of the "state" of labour regulations in Ghana. Land Use Policy, 95, 104651.

Hall, D., 2011a. Land grabs, land control, and Southeast Asian crop booms. The Journal of Peasant Studies, 38, 837–857.

Hall, R., 2011b. Land grabbing in Southern Africa: The many faces of the investor rush. Review of African Political Economy, 38, 193–214.

Hall, R., Scoones I., Tsikata, D., 2017. Plantations, outgrowers and commercial farming in Africa: Agricultural commercialisation and implications for agrarian change. The Journal of Peasant Studies, 44, 515–537.

Hansen, M., Conteh, M., Shakya, M., Loewenstein, W., 2016. Determining minimum compensation for lost farmland: A theory-based impact evaluation of a land grab in

Sierra Leone. Institute of Development Research and Development Policy (IEE), Bochum.

He, P., Martin, K., 2015. Effects of rubber cultivation on biodiversity in the Mekong Region. CAB Reviews, 10, 1–7.

Hervas, A., Isakson, S.R., 2020. Commercial agriculture for food security? The case of oil palm development in northern Guatemala. Food Security, 12, 517–535.

Hess, T.M., Sumberg, J., Biggs, T., Georgescu, M., Haro-Monteagudo, D., Jewitt, G., … Knox, J.W. (2016). A sweet deal? Sugarcane, water and agricultural transformation in Sub-Saharan Africa. Global Environmental Change, 39, 181–194. https://doi.org/ 10.1016/J.GLOENVCHA.2016.05.003

Huggins, C.D., 2014. "Control Grabbing" and small-scale agricultural intensification: Emerging patterns of state-facilitated "agricultural investment" in Rwanda. The Journal of Peasant Studies, 41, 365–384.

Hunsberger, C. 2010. The politics of Jatropha-based biofuels in Kenya: Convergence and divergence among NGOs, donors, government officials and farmers. The Journal of Peasant Studies, 37, 939–962.

Hurni, K., Fox, J., 2018. The expansion of tree-based boom crops in mainland Southeast Asia: 2001 to 2014. Journal of Land Use Science, 13, 198–219.

International Land Coalition, 2011. Global assembly 2011. International Land Coalition, Rome.

Jarzebski, M.P., Ahmed, A., Boafo, Y.A., Balde, B.S., Chinangwa, L., Saito, O., … Gasparatos, A. 2020. Food security impacts of industrial crop production in sub-Saharan Africa: a systematic review of the impact mechanisms. Food Security, 12(1), 105–135. https://doi.org/10.1007/s12571-019-00988-x

Kenney-Lazar, M., 2012. Plantation rubber, land grabbing and social-property transformation in southern Laos. The Journal of Peasant Studies, 39, 1017–1037.

Kenney-Lazar, M., Ishikawa, N., 2019. Mega-plantations in Southeast Asia. Environment and Society, 10, 63–82.

Kenney-Lazar, M., Wong, G., Baral, H., Russell, A.J.M., 2018. Greening rubber? Political ecologies of plantation sustainability in Laos and Myanmar. Geoforum, 92, 96–105.

Kgathi, D.L., Mmopelwa, G., Chanda, R., Kashe, K., Murray-Hudson, M., 2017. A review of the sustainability of Jatropha cultivation projects for biodiesel production in southern Africa: Implications for energy policy in Botswana. Agriculture, Ecosystems & Environment, 246, 314–324.

Kirst, S., 2020. "Chiefs do not talk law, most of them talk power". Traditional authorities in conflicts over land grabbing in Ghana. Canadian Journal of African Studies, 54, 519–539.

Krieger, T., Meierrieks, D., 2016. Land grabbing and ethnic conflict. Homo Oeconomicus, 33, 243–260.

Kusakabe, K., Myae, A.C., 2019. Precarity and vulnerability: Rubber plantations in Northern Laos and Northern Shan State, Myanmar. Journal of Contemporary Asia, 49, 586–601.

Lamb, V., Schoenberger, L., Middleton, C., Un, B., 2017. Gendered eviction, protest and recovery: A feminist political ecology engagement with land grabbing in rural Cambodia. The Journal of Peasant Studies, 44, 1215–1234.

Lay, J., Nolte, K., 2018. Determinants of foreign land acquisitions in low- and middle-income countries. Journal of Economic Geography, 18, 59–86.

Leon Araya, A., 2019. The politics of dispossession in the Honduran palm oil industry: A case study of the Bajo Aguan. Journal of Rural Studies, 71, 134–143.

Li, T.M., 2011. Centering labor in the land grab debate. The Journal of Peasant Studies, 38, 281–298.

Li, T.M., 2015. Social impacts of oil palm in Indonesia: A gendered perspective from West Kalimantan. Center for International Forestry Research (CIFOR), Bogor.

Li, Z., Fox, J.M., 2012. Mapping rubber tree growth in mainland Southeast Asia using time-series MODIS 250 m NDVI and statistical data. Applied Geography, 32, 420–432.

Liu, C., Pang, J., Jepsen, M., Lü, X., Tang, J., 2017. Carbon stocks across a fifty year chronosequence of rubber plantations in tropical China. Forests, 8, 209.

Locke, A., Henley, G., 2013. Scoping report on biofuels projects in five developing countries. Overseas Development Institute, London.

Lu, J.N., 2017. Tapping into rubber: China's opium replacement program and rubber production in Laos. The Journal of Peasant Studies, 44, 726–747.

Maloa Nnoko-Mewanu, J., 2016. Who is not at the table: Land deal negotiations in Southwestern Cameroon. Iowa State University, Iowa.

Marin-Burgos, V., Clancy, J.S., 2017. Understanding the expansion of energy crops beyond the global biofuel boom: Evidence from oil palm expansion in Colombia. Energy, Sustainability and Society, 7, 21.

Marin-Burgos, V., Clancy, J.S., Lovett, J.C., 2015. Contesting legitimacy of voluntary sustainability certification schemes: Valuation languages and power asymmetries in the Roundtable on Sustainable Palm Oil in Colombia. Ecological Economics, 117, 303–313.

Matenga, C.R., Hichaambwa, M., 2017. Impacts of land and agricultural commercialisation on local livelihoods in Zambia: Evidence from three models. The Journal of Peasant Studies, 44, 574–593.

Matthews, N., 2012. Water grabbing in the Mekong Basin – an analysis of the winners and losers of Thailand's Hydropower Development in Lao PDR. Water Alternatives, 5, 392–411.

Matus, S.L.S., Cimpoies, D., Ronzon, T., 2013. Panorama of typologies of agricultural holdings. Food and Agriculture Organization (FAO), Rome.

McCarthy, J.F., 2010. Processes of inclusion and adverse incorporation: Oil palm and agrarian change in Sumatra, Indonesia. The Journal of Peasant Studies, 37, 821–850.

McCarthy, J.F., Vel, J.A.C., Afiff, S., 2012. Trajectories of land acquisition and enclosure: Development schemes, virtual land grabs, and green acquisitions in Indonesia's Outer Islands. The Journal of Peasant Studies, 39, 521–549.

Mechiche-Alami, A., Piccardi, C., Nicholas, K.A., Seaquist, J.W., 2019. Transnational land acquisitions beyond the food and financial crises. Environmental Research Letters, 14, 084021.

Messerli, P., Giger, M., Dwyer, M.D., Breu, T., Eckert, S., 2014. The geography of large-scale land acquisitions: Analysing socio-ecological patterns of target contexts in the global South. Applied Geography, 53, 449–459.

Messinger, J., Winterbottom, B., 2016. African forest landscape restoration initiative (AFR100): Restoring 100 million hectares of degraded and deforested land in Africa. Nature and Faune 30, 14–17.

Mingorría, S., 2017. Violence and visibility in oil palm and sugarcane conflicts: The case of Polochic Valley, Guatemala. The Journal of Peasant Studies, 45, 1314–1340.

Mingorría, S., Gamboa, G., Martín-López, B. et al., 2014. The oil palm boom: Socio-economic implications for Q'eqchi' households in the Polochic valley, Guatemala. Environment Development and Sustainability, 16, 841–871.

Mudombi, S., Von Maltitz, G.P., Gasparatos, A., Romeu-Dalmau, C., Johnson, F.X., Jumbe, C., … Willis, K.J. 2018. Multi-dimensional poverty effects around operational biofuel projects in Malawi, Mozambique and Swaziland. Biomass and Bioenergy, 114, 41–54. https://doi.org/10.1016/j.biombioe.2016.09.003

Murnaghan, K., 2017. A comprehensive evaluation of the EU's biofuel policy: From biofuels to agrofuels. Institute for International Political Economy (IPE), Berlin.

Nally, D., 2015. Governing precarious lives: Land grabs, geopolitics, and "food security". The Geographical Journal, 181, 340–349.

Nkansah-Dwamena, E., Bonnie Raschke, A., 2020. Justice and fairness for Mkangawalo people: The case of the Kilombero Large-scale Land Acquisition (Lasla) Project in Tanzania. Ethics, Policy & Environment, In Press. doi.org/10.1080/21550085.2020.1848187.

Nolte, K., Chamberlain, W., Giger, M., 2016. International Land Deals for Agriculture: Fresh insights from the Land Matrix: Analytical Report II. Centre for Development and Environment; Centre de coopération internationale en recherche agronomique pour le développement; German Institute of Global and Area Studies; University of Pretoria; Bern Open Publishing, Bern/Montpellier/Hamburg/Pretoria.

Notess, L., Veit, P., Andiko, I.M. Sulle, E., Larson, A.M., Gindroz, A.S., Quaedvlieg, J., Williams, A., 2020. Community land formalization and company land acquisition procedures: A review of 33 procedures in 15 countries. Land Use Policy. In Press. doi.org/10.1016/j.landusepol.2020.104461.

Obidzinski, K., Andriani, R., Komarudin, H., Andrianto, A., 2012. Environmental and social impacts of oil palm plantations and their implications for biofuel production in Indonesia. Ecology and Society 17, 25.

OECD, 2014. Towards green growth in Southeast Asia. Organisation for Economic Cooperation and Development (OECD), Paris.

Oliveira, G., Hecht, S., 2016 (Eds.). Soy production in South America: Globalization and new agroindustrial landscapes. The Journal of Peasant Studies, 43, 251–610.

Oxfam, International Land Coalition, Rights and Resources Initiative, 2016. Common ground: Securing land rights and safeguarding the earth. Oxfam, Oxford.

Oxfam, 2016. Burning land, burning the climate: The biofuel industry's capture of EU bioenergy policy. Oxfam, Oxford.

Park, C.M.Y., 2019. "Our lands are our lives": Gendered experiences of resistance to land grabbing in Rural Cambodia. Feminist Economics, 25, 21–44.

Peet, R., Robbins, P., Watts, M., 2011. Global political ecology. Routledge, London.

Pietilainen, E.P., Otero, G., 2019. Power and dispossession in the neoliberal food regime: Oil palm expansion in Guatemala. The Journal of Peasant Studies, 46, 1142–1166.

Portilla, G.S., 2017. Land concessions and rural youth in Southern Laos. The Journal of Peasant Studies, 44, 1255–1274.

Potter, L., 2020. Colombia's oil palm development in times of war and "peace": Myths, enablers and the disparate realities of land control. Journal of Rural Studies, 78, 491–502.

Pramudya, E.P., Hospes, O., Termeer, C.J.A.M., 2017. Governing the palm-oil sector through Finance: The changing roles of the Indonesian State. Bulletin of Indonesian Economic Studies, 53(1), 57–82.

Ragsdale, K., Read-Wahidi, M.R., Wei, T., Martey, E., Goldsmith, P., 2018. Using the WEAI+ to explore gender equity and agricultural empowerment: Baseline evidence among men and women smallholder farmers in Ghana's Northern Region. Journal of Rural Studies, 64, 123–134.

REN21, 2020. Renewables 2020: Global Status Report. REN21 Secretariat, Paris.

Robbins, P., 2012. Political ecology: A critical introduction. J. Wiley & Sons, Sussex.

Rogers, T.D., 2015. Agricultural transformations in sugarcane and labor in Brazil. Oxford Research Encyclopedia of Latin American. Oxford University Press, Oxford.

Romeu-Dalmau, C. Gasparatos, A., von Maltitz, G., Graham, A., Almagro-Garcia, J., Wilebore, B., Willis, K.J., 2018. Impacts of land use change due to biofuel crops on climate regulation services: Five case studies in Malawi, Mozambique and Swaziland. Biomass and Bioenergy, 114, 30–40.

Romijn, H.; Heijnen, S.; Rom Colthoff, J.; De Jong, B.; Van Eijck, J., 2014. Economic and social sustainability performance of Jatropha projects: Results from field surveys in Mozambique, Tanzania and Mali. Sustainability, 6, 6203–6235.

Rutten, R., Bakker, L., Alano, M.L., Salerno, T., Savitri, L.A., Shohibuddin, M., 2017. Smallholder bargaining power in large-scale land deals: A relational perspective. The Journal of Peasant Studies, 44, 891–917.

Sabogal, C.R., 2013. Spatial analysis of the relationship between oil palm cultivation and forced displacement in Colombia. Cuadernos Economia. 32, 683–718.

Sari, F.I.P., Mahardika, R.G., Roanisca, O., 2019. Water quality testing due to oil palm plantation activities in Bangka Regency. IOP Conference Series: Earth and Environmental Science, 012019.

Savilaakso, S., Garcia, C., Garcia-Ulloa, J., Ghazoul, J., Groom, M., Guariguata, M.R., Laumonier, Y., Nasi, R., Petrokofsky, G., Snaddon, J., Zrust, M., 2014. Systematic review of effects on biodiversity from oil palm production. Environmental Evidence, 3, 1–20.

Schlimmer, S., 2019. Negotiating land policies to territorialise state power. Revue Internationale des Études du Développement. 238, 33–59.

Schoneveld, G., German, L., 2014. Translating legal rights into tenure security: Lessons from the new commercial pressures on land in Ghana. The Journal of Development Studies, 50, 187–203.

Schoneveld, G.C., 2014. The geographic and sectoral patterns of large-scale farmland investments in sub-Saharan Africa. Food Policy, 48, 34–50.

Schoneveld, G.C., German, L.A., Nutakor, E., 2011. Land-based investments for rural development? A grounded analysis of the local impacts of biofuel feedstock plantations in Ghana. Ecology and Society, 16, 10.

Schut, M., Slingerland, M., Locke, A., 2010. Biofuel developments in Mozambique. Update and analysis of policy, potential and reality. Energy Policy, 38, 5151–5165.

Schweizer, D., Meli, P., Brancalion, P.H.S., Guariguata, M.R., 2021. Implementing forest landscape restoration in Latin America: Stakeholder perceptions on legal frameworks. Land Use Policy, 104, 104244.

Semie, T.K., Silalertruksa, T., Gheewala, S.H., 2019. The impact of sugarcane production on biodiversity related to land use change in Ethiopia. Global Ecology and Conservation, 18, e00650.

Smalley, R., 2013. Farming and commercial farming areas in Africa: A comparative review. Institute for Poverty, Land and Agrarian Studies (PLAAS), Cape Town.

Sulle, E., Nelson, F., 2009. Biofuels, land access and rural livelihoods in Tanzania. International Institute for Environment and Development (IIED), London.

Tufa, F.A., Amsalu, A., Zoomers, E., 2018. Failed promises: Governance regimes and conflict transformation related to Jatropha cultivation in Ethiopia. Ecology and Society, 23(4), 26.

Van Der Wulp, A.C.E., 2013. The role of the state in facilitating land grabs in Ethiopia. Wageningen University, Wageningen.

van Noordwijk, M., Bizard, V., Wangpakapattanawong, P., Tata, H.L., Villamor, G.B., Leimona, B., 2014. Tree cover transitions and food security in Southeast Asia. Global Food Security, 3, 200–208.

Vermeulen, S., Cotula, L., 2010. Over the heads of local people: Consultation, consent, and recompense in large-scale land deals for biofuels projects in Africa. The Journal of Peasant Studies, 37, 899–916.

Vijay V., Pimm S.L., Jenkins C.N., Smith S.J., 2016. The impacts of oil palm on recent deforestation and biodiversity loss. PLoS ONE, 11(7), e0159668. https://doi.org/10.1371/journal.pone.0159668

von Maltitz, G., Gasparatos, A., Fabricius, C., 2014. The rise, decline and future resilience benefits of jatropha in southern Africa. Sustainability, 6, 3615–3643.

von Maltitz, G., Gasparatos, A., Fabricius, C, Morris, A., Willis, K., 2016. Jatropha cultivation in Malawi and Mozambique: Impact on ecosystem services, local human wellbeing and poverty alleviation. Ecology and Society, 21(3), 3.

von Maltitz, G.P., Henley, G., Ogg, M., Samboko, P.C., Gasparatos, A., Ahmed, A., Read, M., Engelbrecht, F., 2019. Institutional arrangements of outgrower sugarcane production in southern Africa. Development Southern Africa, 36, 175–197.

von Maltitz, G.P., Setzkorn, K.A., 2013. A typology of Southern African biofuel feedstock production projects. Biomass and Bioenergy, 59, 33–49.

Vongkhamheng, C., Zhou, J., Beckline, M., Phimmachanh, S., 2016. Socioeconomic and ecological impact analysis of rubber cultivation in Southeast Asia. Open Access Library Journal, 3, 1–11.

Warren-Thomas, E., Dolman, P.M., Edwards, D.P., 2015. Increasing demand for natural rubber necessitates a robust sustainability initiative to mitigate impacts on tropical biodiversity. Conservation Letters, 8, 230–241.

Warren-Thomas, E.M., Edwards, D.P., Bebber, D.P. et al., 2018. Protecting tropical forests from the rapid expansion of rubber using carbon payments. Nature Communications, 9, 911.

Watson H.K., Diaz-Chavez R.A., 2011. An assessment of the potential of drylands in eight sub-Saharan African countries to produce bioenergy feedstocks. Interface Focus, 1, 263–270.

White, B., Borras Jr., S.M., Hall, R., Scoones, I., & Wolford, W. 2012. The new enclosures: critical perspectives on corporate land deals. Journal of Peasant Studies, 39(3–4), 619–647. https://doi.org/10.1080/03066150.2012.691879

Wicke, B., Smeets, E., Watson, H., Faaij, A., 2011. The current bioenergy production potential of semi-arid and arid regions in sub-Saharan Africa. Biomass and Bioenergy, 35, 2773–2786.

Wolford, W., Borras, S.M., Hall, R., Scoones, I., White, B., 2013. Governing global land deals: The role of the state in the rush for land. Development and Change, 44, 189–210.

4 Marginal land for bioenergy crop production

Ambiguities, contradictions, and cultural significance in policy and farmer discourses

Orla Shortall and Richard Helliwell

Introduction

In the early 2000s, biofuels, and bioenergy more broadly, were positioned as a politically acceptable alternative to conventional transport fuels. Bioenergy crops[1] were perceived as promising opportunities for climate change mitigation, energy security, and rural development, appearing to be a "win–win" technological solution for policy makers. By 2007 at least 64 countries had articulated biofuel targets (Smith, 2010). Driven by regulation and incentives, the use of biofuels in the European Union (EU) (primarily "first generation" biofuels made edible grain and oil crops) increased more than 20-fold between 2000 and 2011 (IEEP, 2014) (see Chapters 1 and 3). However, over the course of this decade, the once-promising "win–win" discourse surrounding biofuels had become hotly contested. As land use for bioenergy crops increased significantly, so too did concern about the impacts of this expansion. This started becoming apparent around 2007–2008 when an interlinked set of controversies surrounding biofuels emerged including those related to "food versus fuel", land use change, and environmental and social justice, fundamentally shifting policy discourse in response (Gamborg et al., 2012).

During this period of policy recalibration "marginal land" was promoted as discourse to overcome the controversies raised by the use of "prime" agricultural land for industrial crops, including bioenergy crops (Cai et al., 2011; Nuffield Council on Bioethics, 2011; SEEMLA, 2018) (see Chapter 5). Marginal land was generally defined as land where the productive cultivation of food and fodder crops is limited by biophysical factors (Richards et al., 2014). It was claimed that the use of marginal land for bioenergy crop production would (a) reduce or eliminate competition between industrial crops and food production (RFA, 2008); (b) prevent the conversion of ecologically valuable natural lands for industrial crop production (Searchinger et al., 2008); and (c) overcome land use conflicts due to the different stakeholders' claims to land (Nuffield Council on Bioethics, 2011). Furthermore, to make marginal land comparably productive, this discourse was coupled in policy narratives with emerging novel bioenergy crops such as switchgrass, miscanthus, coppiced trees, and jatropha. These crops, in contrast to traditional food and oil-based bioenergy crops, were positioned as being better suited to marginal land and able to thrive under otherwise challenging agronomic conditions (Karp and Richter, 2011).

However, some scholars have raised concerns about the discourse and practice surrounding the use of marginal land for bioenergy and industrial crop production (Ahmed et al., 2018; Shortall, 2013). Political ecology is a suitable methodological approach for exploring the use of marginal land for industrial crop production, as it is concerned with the interaction of nature and society and social control over natural resources (Watts and Peet, 2004) (see Chapters 1–2). Essentially, the discourse of marginal land improvement and modernisation, through the cultivation of novel non-food bioenergy crops "justifies" the inclusion of marginal lands into the market economy through the investment of national and transnational capital for industrial crop production (Davis, 2011) (Chapter 5). By describing land as "marginal" this discourse strips this land of its wider societal, environmental, and cultural uses and meanings, as well as its importance to those who need access to it. In a sense, if this land is currently used for subsistence farming and other livelihood activities outside of formal markets, then it is only "marginal" from the perspective of its role in the marketised and capitalist system. Subsequently, "marginal land" has been repeatedly contested on numerous grounds both as a policy discourse and in practice.

This chapter highlights the assumptions and value judgements that underpin the policy discourse around marginal land, using a political ecology lens. We explore how the idea of using marginal land for bioenergy crop production initially emerged (see sub-sections "Biofuel crop controversies" through "Marginal land for biofuel crop production"), particularly for biofuels, and how it has been contested in practice (see "Contesting marginal land"). Even though this chapter focusses on bioenergy and biofuel crops as a particular class of industrial crops[1] (see Chapter 1), we believe that the main insights provided in this chapter are similar and for the other types of industrial crops discussed throughout this edited volume.

Biofuel crop controversies

Concern about the conflicts surrounding the production of biofuels and food crops date from the 1970s and 1980s (Rathmann et al., 2010). However, these earlier concerns over the competition between food and biofuel production were not viewed as overly problematic, as it was assumed that biofuel production and consumption would be confined largely within national borders (Raman and Mohr, 2014). In reality, biofuel markets rapidly expanded after the year 2000, becoming a transnational phenomenon within a few years (Mol, 2007). The concerns over "food versus fuel" took on a global character when biofuel critics started juxtaposing the rights of communities in developing countries to food with the desire of consumers in developed countries to drive their cars (Tomei and Helliwell, 2016).

The "food versus fuel" controversy came to the fore in 2007–2008 when the escalating food prices caused riots in many countries in the global South. Many scholars and media commentators blamed this spike in prices partly or wholly on the diversion of staple food crops such as maize and wheat from the food market to the biofuels market, or the production of biofuels crops instead of other staple food crops (McMichael, 2010) (Chapter 10). An influential report published by the World Bank attributed as much as 75% of the increase in food prices to biofuels production (Mitchell, 2008). In October 2007 Jean Ziegler, the UN special rapporteur on the right to food, stated that "it is a crime against humanity to divert arable land to the production of crops which are then burned for fuel" (BBC, 2007). At the same time, many NGOs started identifying biofuel expansion as an important campaigning issue, pressing national governments for the removal of biofuel targets (Biofuelwatch et al., 2007; The Gaia Foundation et al., 2008).

Simultaneously various concerns over the environmental impacts of biofuel crops were emerging (Mol, 2007). For example, there was evidence that large tracts of forest and other highly biodiverse land were being cleared to make way for biofuels production in the global South (German et al., 2010) (Chapter 1). The direct land use change (DLUC) caused by biofuel crop production was linked to higher greenhouse gas (GHG) emissions than fossil fuels, if biofuel crops were grown on previously uncultivated natural land due to the release of the carbon stored in soil and vegetation (Fargione et al., 2008; Creutzig et al., 2015; Qin et al., 2016). Indirect land use change (ILUC) was also linked to such carbon debts even if biofuel crops were grown on previously cultivated land, as land in other parts of the world could be cleared to compensate for food production loss in biofuel producing countries (Searchinger et al., 2008). Other studies highlighted how biofuel crop production following the same processes as conventional food crop production (e.g. monocultures, intensive farming) has the same environmental impacts, such as biodiversity loss, soil erosion and degradation, water overexploitation, eutrophication, species invasion, and nitrous oxide emissions from fertiliser use (Correa et al., 2017; Doornbosch and Steenblik, 2008).

Industrial crop production for energy was also linked to social issues. Some international organisations framed the production of bioenergy crops for export to energy markets (primarily in the global North) as a positive opportunity for the rural poor in the global South through employment generation, transport infrastructure development, and improved access to markets (FAO, 2008). However, whether or not smallholders and rural communities in the global South benefit from biofuel crop production (as well as the magnitude of this benefit) depends significantly on how cultivation is carried out (German et al., 2010; Mudombi et al., 2018; Ahmed et al., 2019). Biofuel crops were also linked to "land grabbing", which manifested mainly in some countries in Africa, South America, and Asia, where land users were displaced from their land if they did not have any pre-existing property rights, or due to power differentials between local users, and corporate and/or government bodies trying to access land (Borras et al., 2010; Borras and Franco, 2010; Cotula et al., 2008) (see Chapters 3 and 5). While there was some evidence that employment in large-scale biofuels production projects raised living standards in some contexts (Cotula et al., 2008), such employment could also disadvantage local communities through precarious and abusive working practices (Franco et al., 2010) (Chapter 8). The local use of domestic bioenergy feedstocks was also framed as an opportunity to improve national and local energy security in the global South, given that many global south countries depended on imported fossil fuels for the transport sector (Gasparatos et al., 2015). However, there was mixed evidence about the potential benefits from small-scale bioenergy projects in local communities (Gasparatos et al., 2015), as the production and consumption of local energy resources were often dependent on investment in infrastructure, knowledge, and technological development from national governments and NGOs (German et al., 2010; Gilbert, 2011).

Despite these high-profile controversies and the emerging evidence about the negative impacts of biofuels, the EU continued to support their production and use with the 2009 Renewable Energy Directive (RED). The RED established a target mandating member states to achieve a 10% renewable fuel mix in the transport sector by 2020. Biofuel use was expected to fulfil most of this target (Bowyer, 2019), but the surrounding rhetoric and policy shifted partly in response to these controversies (Boucher, 2012). Notably, the use of marginal land for biofuel production emerged as an important policy discourse to reconcile ongoing support for biofuels with the controversies surrounding their production.

Marginal land for biofuel crop production

Origins of the "marginal land" discourse

The contemporary discussion on using marginal land to produce bioenergy and biofuel crops is actually situated in older efforts seeking to mobilise marginal land in response to different environmental, political, and socio-economic problems. The establishment of the UK Forestry Commission to co-ordinate national timber production following the World War I targeted uplands and marginal agricultural lands for conversion to coniferous plantations (Forestry Commission, 2015). In contrast, land retirement programs in the US initially linked to the New Deal, and subsequently to nature conservation, removed marginal land from use to manage soil erosion and agricultural over-production (Sylvester et al., 2016). Narratives of marginal land have also historically been entangled in colonial regimes (Chapter 2). For example, Davis (2011) identifies how French and British colonial authorities imagined North Africa and the Middle East as ecologically marginal and degraded due to "misuse" by local communities. This imagined marginality was entangled with colonial strategies to both "improve" and "protect" these lands, often through the appropriation of land for large-scale irrigation projects and the control of local populations in the name of curbing supposedly destructive local practices (Davis, 2011) (Chapter 2). Additionally, the use of marginal land for growing bioenergy crops was not particularly novel, as it was proposed as a possible option to boost rural development and energy security in Ireland during the 1980s (McElroy and Dawson, 1986). In this respect, the most recent mobilisation of the marginal land narrative in response to biofuel controversies marks the most recent chapter in this longer history.

Recent use of marginal land in the biofuel policy discourse

As discussed above, around 2008, attention turned to the use of marginal land for bioenergy crop production due to its perceived lower environmental and social impacts. For example, in their influential paper about direct land use change, Fargione et al. (2008) suggest that the use of abandoned and degraded land for biofuel production could halt ecosystem degradation. A subsequent influential report commissioned by the UK Government undertaken by the Renewable Fuels Agency (now defunct) established the expectations that the use of marginal and idle land could overcome many of the controversies described above (RFA, 2008). With influential academic and policy literature rapidly embracing marginal land as a "solution" to the controversies surrounding biofuel feedstock production, defining and locating this so-called marginal land became an important policy priority.

In the Renewable Fuels Agency review, marginal land was defined as "*land unsuited for food production, e.g. with poor soils or harsh weather environments; and areas that have been degraded, e.g. through deforestation*" (RFA, 2008: 33). Most subsequent definitions of marginal land have similarly accorded prominence to the biophysical characteristics of the land, its associated economic characteristics, location, use, and size (Richards et al., 2014). Studies began estimating the amount of abandoned (Field et al., 2007), degraded (Tilman et al., 2006), marginal (Gopalakrishnan et al., 2011), and other similar categories of land available for biofuel production globally. The RFA (2008) report further stated that there was enough available land to meet food and fuel demand up to 2020, and policies should be changed to ensure that biofuel production takes place on suitable land.

However, in the immediate wake of the biofuels controversies (see "Biofuel crop controversies"), the EU policy on the sustainability of biofuel crop production coalesced around GHG emissions from direct and indirect land use change as the main sustainability criterion (Palmer, 2014). Scholars have argued that this happened to the exclusion of other issues and framings (Levidow, 2013; Palmer, 2014). A series of life cycle assessments (LCAs) of GHG emissions for different biofuel crops were amended to include a land use change "factor", which was intended to capture the likelihood of additional GHG emissions from land use change. The use of marginal land was therefore potentially perceived more favourably within this process, due to assumptions that its utilisation minimised land use change. These and more recent EU policies have aimed to embed a preference for bioenergy crop production on marginal land rather than land with high carbon stocks or biodiversity value. For example, in December 2018, the EU Renewable Energy Directive 2018/2001 was amended to mandate that biofuels produced on land designated for nature conservation, forest land, highly biodiverse grassland, or land with high carbon stocks (including wetlands and peatland) will not count towards renewable energy targets. In addition, the directive states that biofuel cultivation on "degraded" land that was not in agricultural use in January 2008 and is either severely degraded or contaminated will receive a carbon sequestration bonus.

Production of novel bioenergy crops on marginal land

It is worth noting that the discourse on marginal/idle land that emerged from 2008 to 2009 evolved in tandem with the promotion of novel industrial crops for bioenergy generation, that were seen as more suitable for land of lower quality (Stoof et al., 2015). Coupling these novel non-food industrial crops with marginal land was perceived to overcome the "food versus fuel" controversy (Graham-Rowe, 2011; Tait, 2011). A report by the UK think tank the Nuffield Council on Bioethics (2011: 47) defined second-generation biofuels in terms of sustainability as follows:

> The unifying principles of development of new approaches to biofuels centre on abundant feedstocks that: can be produced without harming the environment or local populations, are in minimal competition with food production, need minimal resources, such as water and land, can be processed efficiently to yield high-quality liquid biofuels and are deliverable in sufficient quantities.

Many non-food biofuel feedstocks were largely perceived as offering higher GHG emission savings compared to first-generation feedstocks, due to their higher energy yield per unit area and carbon sequestration and storage potential (Georgescu et al., 2011). These feedstocks were also seen as less environmentally damaging due to lower input requirements (e.g. fertilisers, pesticides) (Karp and Richter, 2011) and as offering higher biodiversity (Verheyen et al., 2014) and soil remediation benefits (Fernando et al., 2018). Consequently, several EU countries developed policy targets for the production of non-food bioenergy crops, e.g. 350,000 ha in the UK by 2020 (DTI et al., 2007) and 100,000 ha in Denmark by 2020 (Ministeriet for Fødervarer Landbrug og Fiskeri, 2008). At the same time, in many developing countries in Asia (Ariza-montobbio et al., 2010) and Africa (Muok and Källbäc, 2008), jatropha was hailed as a pro-poor agricultural option due to its perceived drought tolerance and ability to grow on poor quality soil.

Eventually, the controversies surrounding food bioenergy crops (but also the expected benefits of non-food bioenergy crops) influenced the revision of the EU Renewable Energy Directive. The revised directive put a cap on the amount of renewable energy derived from food-based biofuels and stipulated by 2030 a 3.5% contribution from non-food "advanced" biofuels and crops with a low indirect land use change risk (including those which can be grown on abandoned or severely degraded land) (European Commission, 2019).

Contesting marginal land

Despite the expected benefits of growing industrial crops on marginal land, certain concerns have been raised about marginal land both as a discourse construct and for its practical implementations. At the discursive level, the term "marginal" land and associated terms (including idle, fallow, unused, underused, abandoned, degraded, barren, wasteland, available, free, and damaged land) are all very ambiguous, and thus subject to a wide range of interpretations and claims. Shortall (2013) conducted an analysis of the meaning of the term marginal land in UK policy documents and found three definitions: (a) land unsuitable for food production, (b) land of ambiguous lower quality, and (c) "economically marginal" land where agricultural production is likely to take place under different market conditions. It has been argued that this discursive ambiguity can be exploited to justify further the controversial promotion of biofuels production (Nalepa et al., 2016; Borras et al., 2011). However, as discussed throughout this edited volume, various assumptions about the political ecology of global land dynamics have been questioned in relation to the use of marginal land for industrial crop production, including for biofuels.

Below, we outline some of the most prominent such concerns, mainly related to the categorisation of marginal land (see "Contesting categorisations of 'marginal land'"), and its use for biofuel crop production (see "Contesting production of bioenergy crops on 'marginal land'"). The first category includes contestations such as that (a) categorising land as marginal is a political exercise that downplays other uses of land and (b) categorising land as marginal ignores many of the different economic, environmental, social, and cultural functions of this land. The second category contains contestations such as that (a) producing bioenergy crops on a commercial scale is not always feasible on marginal land and (b) using marginal land for biofuel production may not overcome the land use change problems posed by first-generation biofuels.

Contesting categorisations of "marginal land"

First, many scholars have questioned whether "marginal" land actually exists, in terms of (a) the accuracy of the methods used to estimate its extent and (b) the disregard of the current land uses. The National Non-Food Crops Centre published a report arguing that the "*basic premise recommended by the Gallagher Review, that biofuel crop production should be segmented to appropriate idle or marginal land, is unlikely to stand up as a viable option when put to close scrutiny*" (Booth et al., 2009: 113). Studies estimating the amounts of "available" and "marginal" land (both globally and in individual countries) consider land through a macroscopic lens and base their estimates on satellite imagery or maps grouping land by biophysical characteristics or assumed present use (Nalepa and Bauer, 2012; Young, 1999). Nalepa and Bauer (2012) argue that these methods are often flawed, as they do not provide enough fine-grain information about the actual use of the land. When estimating the extent of "marginal land" in this manner, figures become decontextualised and depoliticised so that land "availability" becomes a technical exercise of correctly categorising and aggregating land, stripped of attachment, meanings, and alternative market and non-market values (Robbins, 2001). Degraded land, rather than being a-historical concept, is arguably the result of social marginalisation, which can in turn be reinforced by land degradation (Blaikie and Brookfield, 1987) (Chapter 2). Political ecology theory maintains that power is not only exerted through control over resources, but also through control over the production of knowledge (Martinez-Alier, 2009). Thus, the scientific exercise of categorising land as "marginal" becomes a political exercise of deciding what resources matter for whom.

Second, the disregard of other uses of land when categorising it as marginal has given rise to tensions in many countries of the global South where the concept of marginal land has been operationalised to allow industrial crop development, including for bioenergy production. For example, the Ethiopian government created a land bank to facilitate foreign direct investment (FDI) in food and industrial crop production. This land was designated as unused and marginal and was estimated using geospatial technologies such as Google Maps, but large swathes of this land were in fact used by pastoralists and local communities practicing shifting cultivation (Nalepa et al., 2016). In Mozambique, a study estimated between 10 and 19 Mha of available land for agricultural development, including industrial crops, which were also shown to be used for livestock farming and subsistence agriculture (Borras et al., 2011). The Indian government classified 13.4 Mha of land as "wasteland" and created a legal framework to lease this land to companies and government enterprises, many for jatropha production, removing the ability of pastoralist communities to use this land (Ariza-montobbio et al., 2010). In Cambodia, the government created Economic Land Concessions to allow land cultivation by private investors (including for jatropha), designating the land as unused and degraded, and eventually dispossessing smallholders from communal management (Neef et al., 2013). German et al. (2013) explore how land was allocated for biofuels development in Zambia, Mozambique, Tanzania, and Ghana through discourses of land underutilisation. Ahmed et al. (2018) show how environmental impact assessments did not take into account the full range of ecosystem services that supposedly marginal land provides to local communities and its implications for the siting of jatropha projects in Ghana. "Marginal lands" may be particularly important to women in rural communities if they do not have property or inheritance rights (Borras and Franco, 2010; Ndi, 2019). All of the above experiences highlight how rather than being a latent or "idle" resource primed for exploitation, marginal land is already embedded in existing systems of agricultural and community use which become disrupted and displaced through these policy mobilisations.

These experiences have led some scholars to argue that the marginal land discourse is fundamentally a post-colonial tool to integrate areas and communities in the global market economy, for the benefit of multinational companies and national governments (Nalepa and Bauer, 2012). Similar discourses over the development and exploitation of "available" resources were used to justify colonialism (Davis, 2011). Such concerns are also evident within the discourse of marginal land for land improvement and development. In particular, it has been argued that the production of biofuel feedstock on marginal land could rehabilitate previously degraded soil (RFA, 2008) and introduce infrastructure and resources to remote or marginalised areas (Borras et al., 2011). One of the rationales for developing this land is to generate employment for local communities. However, this happens through their displacement and the commodification of their land as an economic resource, rather than a place they are connected to (and which has value for them) (Borras et al., 2011). Even though it was claimed that biofuels production in Mozambique could improve national energy security and reduce the use of environmentally damaging charcoal, most biofuel was exported, with what fuel remained domestically mainly used as transport fuel by the middle class rather than for power generation in poor rural and urban settings lacking access to electricity (Borras et al., 2011). Therefore, communities in the global South must often accept poor working conditions for the "efficient" production of raw materials, while the more lucrative process of value-addition happens in the global North (De Schutter, 2011). Thus this discourse of improving marginal land is therefore potentially entangled in practices comparable to land enclosures in the UK (up to the 19th century) that justified the dispossession of land from the rural poor in the name of efficiency and modernisation (Paul, 2009).

This disjuncture between the high-level policy categorisation of marginal or "idle" land and how the land is used and understood in practice is not limited to countries of the global South. Mapping exercises commissioned by the UK government have delineated marginal land based on an arable-centric definition of land quality and, as a result, identify large tracks of temporary and permanent grassland as marginal land within their land assessments (Haughton et al., 2009; Lovett et al., 2009; Turley et al., 2010). Unsurprisingly, livestock farmers using this land for commercial sheep, beef, and dairy farming did not share this assessment (Helliwell, 2018). Similarly, by highlighting how marginal land embeds value assumptions about how land should be used in ways that favour arable cultivation as an indicator of "productive" land, it either erases or marginalises the cultural and economic significance of livestock rearing, which remains the major agricultural activity on the grasslands and uplands (Qi et al., 2018).

In the global North, such discourses around marginal land sometimes influence farmers and land-owners towards (and the adoption of) non-food bioenergy crops. For example, Helliwell (2018) highlighted how a UK policy and academic definition of marginal land built around "universal" bio-physical characteristics only partially overlap with the practical and cultural understandings of farmers over land quality. Notably, the symbolic and cultural significance of land (and its management) influences on-farm decision-making and management practices in ways that render non-food bioenergy crops unattractive *because* they have been explicitly associated with marginal land in UK policy discourses (Helliwell, 2018). Through the eyes of farmers growing such crops, there was a tacit admission that this was done on land of poor quality, but at the same time, the farmers' pride in their land meant it was never marginal *enough* to be considered for non-food bioenergy crop production (Helliwell, 2018). The importance of these aesthetic and symbolic characteristics of land has been noted more widely, both in relation to non-food bioenergy crop production (Skevas et al., 2014) and more broadly in relation to agri-environmental schemes and the phenomena of roadside farming[2] When situated within the context of farmers' values, connection to land, and wider community norms and values, the marginal land policy narrative has caused resistance to consider the crops (Helliwell, 2018). For example, despite the nearly 20 years of policy support in the UK (including grants to fund the establishment of industrial non-food bioenergy crops), the cultivation area has decreased since 2009 from an already-low level (DEFRA, 2017).

Contesting production of bioenergy crops on "marginal land"

First, the assumption that commercially viable industrial crop production is feasible on marginal land has also been questioned. While it has been claimed that jatropha and many "second generation" biofuel crops can grow with lower resource input than food crops, this has often not been the case in practice (Nalepa and Bauer, 2012). Indeed, the earlier policy discourse on marginal land claimed that its use posed a threat to the future viability of non-food bioenergy crops because cultivation on marginal land could likely lower yields and economic returns, diminishing interest from farmers (Royal Commission on Environmental Pollution, 2004). Arguably, these earlier concerns manifested in the case of jatropha, which, although can survive on dry land with low fertility, does not produce high oil seed yields, requiring in some cases scarce water resources (Ariza-montobbio et al., 2010; Borras et al., 2011; von Maltitz et al., 2014). In many instances, marginal land has simply proven not to be readily available for industrial crop production but conversely has been found to be practically difficult and economically risky to manage. Therefore, even though the discourse of marginal land is often used, in practice, the land that is actually used for industrial crop production can be of higher quality and with access to scarce water resources (Ariza-montobbio et al., 2010; Borras et al., 2011; Nalepa et al., 2016) (Chapter 5).

Second, evidence also suggests that many "marginal lands" have environmental significance (Ahmed et al., 2018). In this sense, rather than preventing direct land use change, the production of bioenergy crops (and industrial crops more broadly) on marginal land can have negative environmental outcomes. For example, Ahmed et al. (2018) found that communal land converted for jatropha production in Ghana provided regulating ecosystem services such as flood regulation, water retention during rainfall, and maintenance of soil fertility, which were degraded to some extent after the initiation of jatropha cultivation. Similarly, in Mozambique, the cultivation of industrial crops on marginal land has shifted into national forestry areas having important environmental ramifications (Borras et al., 2011). In 1987, the EU introduced a set-aside policy that 10% of farmland needed to be left idle to combat over-production and the negative environmental impacts of conventional agriculture (Garzon, 2006). However, an exception was introduced for biofuel crop production in 1993, followed by a bioenergy crop aid of 45 EUR/ha, until set-aside was abolished in 2008 following the Common Agricultural Policy Health Check (Fonseca et al., 2010). In the UK, in addition to set aside, farmers routinely enrolled their marginal land in agri-environment schemes (Helliwell, 2018; Marr and Howley, 2018). These schemes provide funding to farmers and land managers in return for utilising the land in ways that support biodiversity, enhance landscape conservation, and minimise water, air, and soil degradation. Cultivating this land with perennial bioenergy crops raised the prospect of disrupting these existing revenue streams and established environmental management practices. In Denmark, the use of riparian buffer zones and low-lying nutrient-rich marginal land was encouraged for perennial bioenergy crop production, but policies were changed because of disputes over environmental priorities, the landscape impact of tall willow crops, and the different preferences for biodiversity associated with open landscapes (Shortall et al., 2019).

Finally, it is worth noting the different responses to the controversies surrounding the discourse and use of "marginal land" outlined above. Some scholars have argued for revising the concept, using better and more precise definitions, and undertaking more research to identify better the types of industrial crops that can be grown on "marginal lands" (Monti and Alexopoulou, 2017; Richards et al., 2014). Others have pointed to the potential of advanced plant-breeding techniques (e.g. genetic modification) to improve the yields of perennial bioenergy crops on marginal land, by increasing their resistance to pests and diseases and decreasing their nutrient and water requirements (Karp and Richter, 2011; Karp and Shield, 2008). However, these claims have been contested on the grounds of being technologically oriented promises raising expectations about two (as yet) largely unproven technologies (Levidow and Papaioannou, 2013).

Conclusions

Industrial crops for bioenergy generation have been widely promoted through multiple policies aiming to reduce GHG emissions by reducing the use of fossil fuels, promoting energy security in developed and developing countries, and incentivising rural development through additional markets for such crops. However, this has raised major controversies since the mid-2000s due to the environmental consequences of the underlying production methods, including the conversion of ecologically important land. The "food versus fuel" controversy further suggested that bioenergy crop production may have had impacts on food production and food prices, by diverting food crops onto the biofuel market. Bioenergy crops were also criticised in relation to many social issues such as land grabbing and exploitative labour relations. One proposed response was to produce such crops on marginal land, rather than prime agricultural land. It was perceived that using "marginal land" of lower quality could overcome much of this controversy, by not interfering with existing value chains concentrated on prime land. The concept of marginal land was promoted in tandem with many different non-food industrial crops, which were framed as suitable for production on such land.

However, a series of concerns have been raised about the discourse and practices surrounding the use of marginal land for industrial crop production. At the discursive level "marginal land", and related terms, are ambiguous. It has been claimed that this ambiguity can be exploited to justify the controversial production of industrial crops for biofuels. Furthermore, the accuracy of the methods employed to characterise marginal land, and assess its extent, have been disputed due to their reliance on high-level data, which may ignore the actual uses of the land on the ground. Because land has cultural, economic, and environmental value, it has been questioned if (and under what circumstances) land can indeed be considered "marginal". Furthermore, the possible use of "marginal land" outside of marketed supply chains might downplay its actual use for industrial crop production, while at the same time, the commercial viability of industrial crop production on such poor quality land has been questioned.

From a political ecology perspective, the assumptions that the social and environmental impacts of industrial crop production can be overcome if grown on non-prime land have been criticised as naïve. They have also been characterised as blind to the continuation of the same power dynamics and structures, which have caused the current controversies related to biofuel production on "prime land". Worse, the marginal land discourse can legitimise these power dynamics and structures by marginalising the claims and interests of those who use this land. From this point of view, "marginal land" can be seen as a fundamentally flawed concept because the very idea of land having negligible value is problematic given its undisputed economic, ecological, and cultural values, which vary substantially between contexts. If industrial crop production is to be promoted on non-prime agricultural land, there must be a real effort to take into account the power structures, existing land uses, and local priorities for this land. Reconciling these viewpoints, if at all possible, would be necessary for ensuring the sustainable production of industrial crops on non-prime agricultural land.

Notes

1 For the purpose of this chapter, we refer to bioenergy crops as those industrial crops used for energy generation. Bioenergy can be generated through different processes including the direct combustion of the entirety or parts of a bioenergy crop or the development of energy carriers such as liquid or gaseous fuel. The latter includes liquid biofuels such as bioethanol and biodiesel used in the transport sector. Liquid biofuels are broadly divided into: (a) first-generation biofuels produced through relatively conventional chemical processes such as the fermentation or transesterification of sugar, starch, or oil crops into bioethanol or biodiesel; and (b) second-generation biofuels derived through more advanced processes based on the decomposition of the lignin, hemicellulose, and cellulose in biomass, and the subsequent fermentation of the derived sugars into bioethanol. First-generation biofuels are usually derived from conventional food and industrial crops such as maize, sugarcane, soybeans, oil palm, and rapeseed, among many others, raising many of the concerns related to "food vs. fuel" discussed throughout the chapter. Second-generation biofuels are derived from trees (e.g. willow, poplar, alder), grasses (e.g. miscanthus, switchgrass), and agricultural waste (e.g. stalks) and have been partly promoted for their perceived ability to be grown on marginal land and/or reduce "food vs. fuel" competition, as discussed in this chapter (Muok and Källbäc, 2008; Valentine et al., 2012).

2 Roadside farming denotes the phenomenon of farmers investing more time and money in visible fields to demonstrate their skill as farmers (Burton, 2004, 2012; Burton et al., 2008).

References

Ahmed, A., Dompreh, E., and Gasparatos, A. (2019). Human wellbeing outcomes of involvement in industrial crop production: Evidence from sugarcane, oil palm and jatropha sites in Ghana. *PLoS ONE*, 14(4): e0215433.

Ahmed, A., Jarzebski, M.P., and Gasparatos, A. (2018). Using the ecosystem service approach to determine whether jatropha projects were located in marginal lands in Ghana: Implications for site selection. *Biomass and Bioenergy*, 114: 112–124. https://doi.org/10.1016/j.biombioe.2017.07.020

Ariza-montobbio, P., Lele, S., Kallis, G., et al. (2010) The political ecology of Jatropha plantations for biodiesel in Tamil Nadu, India. *Journal of Peasant Studies* 37(November): 875–897. DOI: 10.1080/03066150.2010.512462.

BBC News. (2007). Biofuels "crime against humanity." *BBC News.* http://news.bbc.co.uk/1/hi/7065061.stm (accessed 16 April 2021)

Biofuelwatch, Carbon Trade Watch, Corporate Europe Observatory, et al. (2007) *Agrofuels towards a reality check in nine key areas.* Edinburgh: Biofuelswatch.

Blaikie, P. and Brookfield, H. (1987) *Land degradation and society.* London: Methuen.

Booth, E., Walker, R., Bell, J., Mccracken, D., Curry, J. and Biddle, A. (2009) *An assessment of the potential impact on UK agriculture and the environment of meeting renewable feedstock demands.* Edinburgh: National Non-Food Crops Centre.

Borras, S.M., Fig, D. and Monsalve Suarez, S. (2011) The politics of agrofuels and mega-land and water deals: Insights from the ProCana case, Mozambique. *Review of African Political Economy* 38(128): 215–234.

Borras, S.M. and Franco, J.C. (2010) Contemporary discourses and contestations around pro-poor land policies and land governance. *Journal of Agrarian Change* 10(1): 1–32. DOI: 10.1111/j.1471-0366.2009.00243.x.

Borras, S.M., McMichael, P. and Scoones, I. (2010) The politics of biofuels, land and agrarian change: Editors' introduction. *The Journal of Peasant Studies* 37(4): 575–592. DOI: 10.1080/03066150.2010.512448.

Boucher, P. (2012) The role of controversy, regulation and engineering in UK biofuel development. *Energy Policy* 42: 148–154. DOI: 10.1016/j.enpol.2011.11.058.

Bowyer, C. (2019) *Anticipated Indirect Land Use Change Associated with Expanded Use of Biofuels and Bioliquids in the EU – an Analysis of the National Renewable Energy Action Plans*. Brussels: Institute for European Environmental Policy.

Burton, R.J.F. (2004) Seeing through the "good farmer's" eyes: Towards developing an understanding of the social symbolic value of "productivist" behaviour. *Sociologia Ruralis* 44: 195–215.

Burton, R.J.F. (2012) Understanding farmers' aesthetic preference for tidy agricultural landscapes: A Bourdieusian perspective. *Landscape Research* 37(1): 51–71.

Burton, R.J.F., Kuczera, C. and Schwarz, G. (2008) Exploring farmers' cultural resistance to voluntary agri-environmental schemes. *Sociologia Ruralis* 48: 16–37.

Cai, X.M., Zhang, X.A. and Wang, D.B. (2011) Land availability for biofuel production. *Environmental Science and Technology* 45: 334–339.

Correa, D.F., Beyer, H.L., Possingham, H.P., Thomas-Hall, S.R., and Schenk, P.M. (2017). Biodiversity impacts of bioenergy production: Microalgae vs. first generation biofuels. *Renewable and Sustainable Energy Reviews*, 74(February), 1131–1146. https://doi.org/10.1016/j.rser.2017.02.068

Cotula, L., Dyer, N. and Vermeulen, S. (2008) *Fuelling exclusion? The biofuels boom and poor people's access to land*. London: International Institute for Environment and Development and Food and Agriculture Organization.

Creutzig, F., Ravindranath, N.H., Berndes, G., Bolwig, S., Bright, R., Cherubini, F., Chum, H., Corbera, E., Delucchi, M., Faaij, A., Fargione, J., Haberl, H., Heath, G., Lucon, O., Plevin, R., Popp, A., Robledo-Abad, C., Rose, S., Smith, P., ... Masera, O. (2015). Bioenergy and climate change mitigation: An assessment. *GCB Bioenergy* 7: 916–944.

Davis, D. (2011) Imperialism, orientalism, and the environment in the Middle East: History, policy, power, and practice. In D. Davis & E.I. Burke (Eds.), *Environmental imaginaries of the Middle East and North Africa* (pp. 1–22). Athens: Ohio University Press.

De Schutter, O. (2011) The green rush: The global race for farmland and the rights of land users. *Harvard International Law Journal* 52(2): 504–556.

DEFRA (2017) *Crops Grown For Bioenergy in the UK: 2017*. London: DEFRA.

Doornbosch, R. and Steenblik, R. (2008) *Biofuels: Is the cure worse than the disease?* Paris: OECD Roundtable on Sustainable Development.

DTI, DfT and Defra (2007) *UK biomass strategy*. London: Department of Trade and Industry (DTI).

European Commission (2019) Sustainability criteria for biofuels specified. Available at: europa.eu/rapid/press-release_MEMO-19-1656_en.htm (accessed 21 October 2019).

FAO (2008) *The state of food and agriculture. Biofuels: Prospects, risks and opportunities*. Rome: FAO.

Fargione, J., Hill, J., Tilman, D., et al. (2008) Land clearing and the biofuel carbon debt. *Science (New York, N.Y.)* 319(5867): 1235–1238. DOI: 10.1126/science.1152747.

Fernando, A.L., Costa, J., Barbosa, B., et al. (2018) Biomass and bioenergy environmental impact assessment of perennial crops cultivation on marginal soils in the Mediterranean Region. *Biomass and Bioenergy* 111: 174–186. DOI: 10.1016/j.biombioe.2017.04.005.

Field, C.B., Campbell, J.E. and Lobell, D.B. (2007) Biomass energy: The scale of the potential resource. *Trends in Ecology and Evolution* 23(2): 65–72. DOI: 10.1016/j.tree.2007.12.001.

Fonseca, M.B., Burrell, A., Gay, H., Henseler, M., Kavallari, A., M'Barek, R., Domínguez, I.P., and Tonini, A. (2010). *Impacts of the EU biofuel target on agricultural markets and land use: a comparative modelling assessment.* Brussels: European Commission.

Forestry Commission (Producer). (2015) 100 years of state forestry in the UK. Forestry Commission, Bristol. Retrieved from www.fsc-uk.org/en-uk/about-fsc/what-is-fsc/case-studies/forest-management/100-years-of-state-forestry-in-the-uk (accessed 16 April 2021)

Franco, J., Levidow, L., Fig, D., et al. (2010) Assumptions in the European Union biofuels policy: Frictions with experiences in Germany, Brazil and Mozambique. *The Journal of Peasant Studies* 37(4): 661–698. DOI: 10.1080/03066150.2010.512454.

Gamborg, C., Millar, K., Shortall, O., and Sandøe, P. (2012). Bioenergy and Land Use: Framing the Ethical Debate. *Journal of Agricultural and Environmental Ethics* 25: 909–925. https://doi.org/10.1007/s10806-011-9351-1

Garzon, I. (2006) *Reforming the common agricultural policy: A history of paradigm change.* London: Palgrave MacMillan.

Gasparatos, A., Maltitz, G.P. von, Johnson, F.X., Lee, L., Mathaih, M., Oliveira, J.A.P. de, and Willis, K.J. (2015). Biofuels in sub-Sahara Africa: Drivers, impacts and priority policy areas. *Renewable and Sustainable Energy Reviews* 45: 879–901.

Georgescu, M., Lobell, D.B. and Field, C.B. (2011) Direct climate effects of perennial bioenergy crops in the United States. *Proceedings of the National Academy of Sciences* 108(11): 4307–4312.

German, L., Schoneveld, G. and Mwangi, E. (2013) Contemporary processes of large-scale land acquisition in Sub-Saharan Africa: Legal deficiency or elite capture of the rule of law? *World Development* 48: 1–18.

German, L., Schoneveld, G., Skutch, M., Andriani, R., Obidzinski, K., Pacheco, P., Komarudin, H., Andrianto, A., Lima, M., and Norwana, A.D. (2010). The local social and environmental impacts of biofuel feedstock expansion. A synthesis of case studies from Asia, Africa and Latin America. *CIFOR Infobriefs* 34: 1–12.

Gilbert, N. (2011) Local benefits: The seeds of an economy. *Nature* 474: S18–S19.

Gopalakrishnan, G., Negri, M.C. and Snyder, S.W. (2011) A novel framework to classify marginal land for sustainable biomass feedstock production. *Journal of Environmental Quality* 40: 1593–1600. DOI: 10.2134/jeq2010.0539.

Graham-Rowe, D. (2011) Beyond food versus fuel. *Nature* 474: S6–S8.

Haughton, A.J., Bond, A.J., Lovett, A.A., Dockerty, T., Sünnenberg, G., Clark, S.J., … Finch, J.W. (2009) A novel, integrated approach to assessing social, economic and environmental implications of changing rural land use: A case study of perennial biomass crops. *Journal of Applied Ecology* 46(2): 315–322.

Helliwell, R. (2018) Where did the marginal land go? Farmers perspectives on marginal land and its implications for adoption of dedicated energy crops. *Energy Policy* 117: 166–172. doi.org/10.1016/j.enpol.2018.03.011

IEEP (2014) *Re-examining EU biofuels policy: A 2030 perspective.* Brussels: Institute for European Environmental Policy.

Karp, A. and Richter, G.M. (2011) Meeting the challenge of food and energy security. *Journal of experimental botany* 62: 3263–3271. DOI: 10.1093/jxb/err099.

Karp, A. and Shield, I. (2008) Bioenergy from plants and the sustainable yield challenge. *New Phytologist* 179(1): 15–32. DOI: 10.1111/j.1469-8137.2008.02432.x.

Levidow, L. (2013) EU criteria for sustainable biofuels: Accounting for carbon, depoliticising plunder. *Geoforum* 44: 211–223. DOI: 10.1016/j.geoforum.2012.09.005.

Levidow, L., and Papaioannou, T. (2013). State imaginaries of the public good: Shaping UK innovation priorities for bioenergy. *Environmental Science and Policy*, 30: 36–49. https://doi.org/10.1016/j.envsci.2012.10.008

Lovett, A., Sünnenberg, G.M., Richter, G.M., Dailey, G., Richer, A.B. and Karp, A. (2009) Land use implications of increased biomass production identified by GIS-based suitability and yield mapping for Miscanthus in England. *Bioenergy Research* 2(1–2): 17–28. doi:10.1007/s12155-008-9030-x

Maltitz, G. von, Gasparatos, A., and Fabricius, C. (2014). The Rise, Fall and Potential Resilience Benefits of Jatropha in Southern Africa. *Sustainability* 6: 3615–3643.

Marr, E.J., and Howley, P. (2018). Woodlots, wetlands or wheat fields? Agri-environmental land allocation preferences of stakeholder organisations in England and Ontario. *Land Use Policy* 75: 673–681.

Martinez-Alier, J. (2009) Social metabolism, ecological distribution conflicts, and languages of valuation. *Capitalism Nature Socialism* 20(1): 58.

McElroy, G.H. and Dawson, M.W. (1986) Biomass from short-rotation coppice willow on marginal land. *Biomass* 10: 225–240.

McMichael, P. (2010) Agrofuels in the food regime. *Journal of Peasant Studies* 37(4): 609–629. DOI:10.1080/03066150.2010.512450.

Mitchell, D. (2008) *A note on rising food prices*. Washington D.C.: World Bank.

Mol, A.P.J. (2007) Boundless biofuels? Between environmental sustainability and vulnerability. *Sociologia Ruralis* 47(4): 297–315. doi:10.1111/j.1467-9523.2007.00446.x

Monti, A. and Alexopoulou, E. (2017) Non-food crops in marginal land: An illusion or a reality? *Biofuels, Bioproducts and Biorefining* 11: 937–938.

Muok, B. and Källbäc, L. (2008) *Feasibility study of jatropha curcas as a biofuel feedstock in Kenya*. Nairobi: African Centre for Technology Studies.

Ministeriet for Fødervarer Landbrug og Fiskeri. (2008). *Landbrug og Klima*. Copenhagen: Ministeriet for Fødervarer Landbrug og Fiskeri.

Mudombi, S., Maltitza, G.P. Von, Gasparatos, A., Romeu-Dalmau, C., Johnson, F.X., Jumbe, C., Ochieng, C., Luhanga, D., Lopesh, P., Balde, B.S., and Willis, K.J. (2018). Multi-dimensional poverty effects around operational biofuel projects in Malawi, Mozambique and Swaziland. *Biomass and Bioenergy* 114: 41–54.

Nalepa, R. and Bauer, D.M. (2012) Marginal lands: The role of remote sensing in constructing landscapes for agrofuel development. *Journal of Peasant Studies* 39(2): 403–422. DOI: 10.1080/03066150.2012.665890.

Nalepa, R.A., Short Gianotti, A. and Bauer, D.M. (2016) Marginal land and the global land rush: A spatial exploration of contested lands and state-directed development in Contemporary Ethiopia. *Geoforum* 82: 237–251.

Ndi, F.A. (2017). Land grabbing: A gendered understanding of perceptions and reactions from affected communities in Nguti Subdivision of South West Cameroon. *Development Policy Review* 37: 348–366.

Neef, A., Touch, S. and Chiengthong, J. (2013) The politics and ethics of land concessions in rural cambodia. *Journal of Agricultural and Environmental Ethics* 26: 1085–1103.

Nuffield Council on Bioethics (2011) *Biofuels: Ethical issues.* London: Nuffield Council on Bioethics.

Palmer, J.R. (2014) Biofuels and the politics of land-use change: Tracing the interactions of discourse and place in European policy making. *Environment and Planning A* 46: 337–352. DOI: 10.1068/a4684.

Paul, H. (2009) No idle threat to the marginalised: The focus on "marginal and idle " land for biofuels (agrofuels). *The Ecologist* (July 2008): 1–4.

Qi, A., Holland, R.A., Taylor, G., and Richter, G.M. (2018). Grassland futures in Great Britain – Productivity assessment and scenarios for land use change opportunities. *Science of the Total Environment*, 634: 1108–1118.

Qin, Z., Dunn, J.B., Kwon, H., Mueller, S., and Wander, M.M. (2016). Soil carbon sequestration and land use change associated with biofuel production: empirical evidence. *GCB Bioenergy*, 8: 66–80.

Raman, S. & Mohr, A. (2014) Biofuels and the role of space in sustainable innovation journeys. *Journal of Cleaner Production* 65: 224–233. doi.org/10.1016/j.jclepro.2013.07.057

Rathmann, R., Szklo, A. and Schaeffer, R. (2010) Land use competition for production of food and liquid biofuels: An analysis of the arguments in the current debate. *Renewable Energy* 35: 14–22. DOI: 10.1016/j.renene.2009.02.025.

RFA (2008) *The Gallagher Review of the indirect effects of biofuels production.* Sussex: Renewable Fuels Agency.

Richards, B.K., Stoof, C.R., Cary, I.J., et al. (2014) Reporting on marginal lands for bioenergy feedstock production: A modest proposal. *Bioenergy Research* 7: 1060–1062. DOI: 10.1007/s12155-014-9408-x.

Robbins, P. (2001) Fixed categories in a portable landscape: The causes and consequences of land-cover categorization. *Environment and Planning A* 33: 161–179.

Royal Commission on Environmental Pollution (2004) *Biomass as a renewable energy source.* London: Royal Commission on Environmental Pollution.

Searchinger, T., Heimlich, R., Houghton, R., et al. (2008) Use of U.S. croplands for biofuels increases greenhouse gases through emissions from land-use change. *Science* 319(5867): 1238–1240.

SEEMLA (2018) SEEMLA: Sustainable exploitation of biomass for bioenergy from marginal lands. Available at: seemla.eu/en/home/ (accessed 24 April 2018).

Shortall, O.K. (2013) "Marginal land" for energy crops: Exploring definitions and embedded assumptions. *Energy Policy* 62: 19–27. DOI: 10.1016/j.enpol.2013.07.048.

Shortall, O.K., Tegner, H., Sandøe, P., et al. (2019) Biomass and bioenergy room at the margins for energy-crops? A qualitative analysis of stakeholder views on the use of marginal land for biomass production in Denmark. *Biomass and Bioenergy* 123(January): 51–58. DOI: 10.1016/j.biombioe.2019.01.042.

Skevas, T., Swinton, S.M. and Hayden, N.J. (2014) What type of landowner would supply marginal land for energy crops? *Biomass and Bioenergy* 67: 252–259. doi.org/10.1016/j.biombioe.2014.05.011

Smith, J. (2010) *Biofuels and the globalisation of risk.* London: Zed Books.

Stoof, C.R., Richards, B.K., Woodbury, P.B., et al. (2015) Untapped potential: Opportunities and challenges for sustainable bioenergy production from marginal lands in the Northeast USA. *Bioenergy Research* 8(2): 482–501. DOI: 10.1007/s12155-014-9515-8.

Sylvester, K.M., Gutmann, M.P. and Brown, D.G. (2016) At the margins: Agriculture, subsidies and the shifting fate of North America's Native Grassland. *Population and Environment* 37(3): 362–390. doi:10.1007/s11111-015-0242-7

Tait, J. (2011) The ethics of biofuels. *Global Change Biology Bioenergy* 3: 271–275. DOI: 10.1111/j.1757-1707.2011.01107.x.

The Gaia Foundation, Biofuelwatch, the African Biodiversity Network, Salva La Selva, Watch Indonesia and EcoNexus. (2008) *Agrofuels and the Myth of Marginal Lands*. London: The Gaia Foundation.

Tilman, D., Hill, J. and Lehman, C. (2006) Carbon-negative biofuels from low-input high-diversity grassland biomass. *Science* 768(December): 1598–1601.

Tomei, J. and Helliwell, R. (2016). Food versus fuel? Going beyond biofuels. *Land Use Policy* 56: 320–326. doi.org/10.1016/j.landusepol.2015.11.015

Turley, D., Taylor, M., Laybourn, R., Hughes, J., Kilpatrick, J., Proctor, C., ... Edgington, P. (2010) *Assessment of the availability of "marginal" and "idle" land for bioenergy crop production in England and Wales: Project NF0444. Funded by the UK Department for Energy and Climate Change*. London: Department of Energy and Climate Change (DECC).

Valentine, J., Clifton-Brown, J., Hastings, A., et al. (2012) Food vs. fuel: The use of land for lignocellulosic "next generation" energy crops that minimize competition with primary food production. *GCB Bioenergy* 4(1): 1–19. DOI: 10.1111/j.1757-1707.2011.01111.x.

Verheyen, K., Buggenhout, M., Vangansbeke, P., et al. (2014) Potential of short rotation coppice plantations to reinforce functional biodiversity in agricultural landscapes. *Biomass and Bioenergy* 67: 435–442. DOI: 10.1016/j.biombioe.2014.05.021.

Watts, M.J. and Peet, R. (2004) Liberating political ecology. In R. Peet and M.J. Watts (Eds.), Liberation ecologies. London: Routledge.

Young, A. (1999) Is there really spare land? A critique of estimates of available cultivable land in developing countries. *Environment, Development and Sustainability* 1: 3–18.

Part II
Ecological transformation

5 Transforming nature, crafting irrelevance

The commodification of marginal land for sugarcane and cocoa agroindustry in Peru

Patricia Urteaga-Crovetto and Frida Segura-Urrunaga

Introduction

Land commodification and land grabbing for large-scale agro-industrial projects have become some of the most challenging current threats to sustainability in many parts of the world. Land commodification for large-scale agro-industrial projects, including industrial crop production, has increased exponentially in the past few decades (Chapters 2–4). In particular, land conversion and grabbing for the production of food and industrial crops has affected millions of hectares around the world (Borras et al., 2020, 2011a, 2011b, 2012a; Li, 2014; HLPE, 2011; Cotula et al., 2009) (Chapter 3). This has in turn increased the vulnerability of land systems, natural resources, and local communities vis-à-vis the market offensive (Tsing, 2002; Sawyer and Gomez, 2012; Manirabona and Vega Cardenas, 2019) (Chapter 1). At the same time, the transformation of nature through these processes has imposed symbolic changes in the local milieu (Li, 2014), usually overriding the local understanding and meanings of land, and essentially nature (Chapter 4). Although some differences between the various conceptions of land and nature may coexist, there are also gradual converges as discussed below (Diez, 2017; Burneo, 2011, 2012; Huamán 2014; Ahmed et al., 2019; Tsing, 2002).

Land grabbing has been linked to many different driving forces related to colonization, food/energy security concerns, high-return agricultural production demands, mineral extraction, land reform policies, and market speculation (Cotula, 2014; Borras et al., 2012a, 2012b; Hollander, 2008, 2010) (Chapter 3). The literature on land grabbing has mainly focused on land appropriation through accumulation (Borras et al., 2012b; White et al., 2012; Hall, 2013). From a Marxist perspective, the attention has been on "value grabbing", as a means of unveiling the political character of distributional relations, and particularly "*the appropriation of value through rent*", which differs from processes of mere capital accumulation (Andreucci et al., 2017: 28). Concerns have also been raised over "benefit grabbing", which in some cases explains better the inequitable distribution of benefits among local actors in the context of large-scale land acquisition (Ahmed et al., 2019; Hall et al., 2015; Huamán, 2019).

Scholars have also pointed to the importance of looking at the sophisticated multiscale mechanisms and technicalities composing the land market (Nader, 1972; Li, 2006, 2014; Tsing, 2000) and legal thinking within markets (Riles, 2011). However, according to Wolford et al. (2013: 192) the focus should not be only "*on land or on specific acts of grabbing; rather we need to analyse a host of processes [...] to understand the means by which large-scale dispossession, appropriation and extraction come to be seen as not just possible but even necessary*".

Whereas land might be perceived as a "*provisional assemblage of heterogenous elements including material substances, technologies, discourses and practices*" that transforms diachronically (Li, 2014: 590), land grabbing may be conceptualized as the capitalist control of extensive land areas and natural resources usually for extractive purposes (Borras et al., 2012a, 2012b). Thus, the unending transform- ation of land into a resource or a commodity that is measurable, limited, subject to speculation, and with constantly incremental values (Li, 2014) configures a pattern of the "economy of appearances" (Tsing, 2000). Indeed, not always made with the aim of promoting agricultural or livestock production, land investments are often traded on stock markets[1] (Tsing, 2000; Li, 2014; Cotula, 2014;Visser, 2017). Li (2014) notes the virtual and a-spatial spectacularity of the promotion of transnational investments associated with land grabbing projects.As discussed below, this virtual economy energizes land transactions by describing land investments as essential and urgent, land resources as scarce, companies as sustainable enterprises, and exorbitant profit margins as highly desirable (Li, 2014; Bakker, 2010). This increasingly virtual character of land grabbing has been receiving growing attention as it is key to understand not only the trans- formation of nature but also its apparently ineluctable commodification.

As mentioned above, large land investments are often promoted through crisis narratives, such as the looming challenges posed by food and energy insecurity, and are hidden under the guise of sustainability.[2] They are also supported by specific environmental legal frameworks, large infrastructure, special economic zones, financial instruments, as well as the rules, procedures, and incentives provided by the international community (White et al., 2012; Li, 2014). For example, the World Bank's project investment publications (Li, 2014; White et al., 2012; Cotula, 2011) indicate that Sub-Saharan Africa, Latin America, and the Caribbean supposedly have the highest land availability for large investments for maize, soybeans, wheat, sugarcane, and oil palm (Li, 2014) (Chapter 9). To promote land commodification, international financial institutions (IFI) such as the World Bank estimate land availability in areas located six hours from the nearest market, with the amount of available land essentially tripling compared to other areas (Li, 2014).

Legal instruments are another key aspect of "forming" land value. International land transactions are usually legal transactions. Even though land appropriation is often done outside legal borders, large-scale land acquisitions generally require a mantle of legality (Cotula, 2011; Burneo, 2013; Huamán, 2014; Ahmed et al., 2019). The World Bank assumes that national laws can be potent instruments for protecting vulnerable communities,[3] but in practice, formal requirements such as land registration do not prevent land conflicts as the underlying social imbalances usually remain unresolved[4] (Riles, 2011; Burneo; 2013, 2020; Urteaga-Crovetto, 2016; OXFAM, 2014). To decode land value crafting, it is important to understand (a) the "legal" arguments for the appropriation of land, (b) the modalities of land purchase contracts, and (c) the nature of land transfers (Cotula, 2011). According to Cotula (2011), trade agreements for large-scale land acquisitions are very general documents that, on the one hand, grant long-term rights over extensive land areas and preferential water rights to investors, and on the other hand, offer low or marginal profits to the state. At the same time, the investment promises and/or job generation potential are very vague, with little concern for social and environmental issues. Such trade agreements also push for centralized land control that eventually excludes local decision-making processes. Furthermore, these negotiations are usually undertaken under time pressure, between parties with unequal power, and often lack transparency. It is obvious that the description of the land and its value is essential in these processes, and once land is transferred, its value rises in tandem with the negotiation of the land as a commodity. This includes many parallel processes that grant and grab water rights, and "improve" the quality of the land.

Many of the processes and discourses outlined above are observed in Latin America (Chapter 3). Indeed, land grabbing is very prevalent in the region and has highly interlinked local, national, and transnational dimensions (Eguren, 2011; Zegarra et al., 2007; Dammert Bello, 2017; Borras et al., 2011b). Brazil, Argentina, Colombia, and Peru have the highest farmland availability in the region, with huge amounts of land accumulated for the production of fruits, vegetables, sugarcane, palm oil, soy, cocoa, minerals, and oil (Zegarra et al., 2007; Borras et al., 2011a, 2011b; Eguren, 2011) (Chapter 9).

In Peru, land and water commodification has had grave consequences for ecosystems and local communities, particularly because their relationship with the land was deeply altered, with some communities even being dispossessed (van der Ploeg, 2006; Chirif, 2015; Dammert Bello, 2017; Segura, 2019) (Chapters 3–4). For example, since the mid-1990s, several companies have acquired thousands of hectares of land on the Peruvian coast for biofuel production (Burneo, 2011; Urteaga, 2017; Del Castillo, 2013; Eguren, 2011), transforming land tenure and water access to facilitate their large-scale production (see similar examples in Eswatini, Chapter 11). Similarly, in the Amazon, serious conflicts followed the acquisition of large tracts of land for industrial crop production (cocoa and oil palm), legal and illegal logging, and the extraction of minerals, gas, and hydrocarbons (Urteaga et al., 2019). The aggressive accumulation of land has been quite extensive, stripping away the land from thousands of local farmers and indigenous communities[5] (Herman and Mayrhofer, 2016; Urteaga-Crovetto, 2016; Dammert Bello, 2017). Similar situations have occurred in protected areas and reserves for isolated indigenous communities (OXFAM, 2014).

One of the major research issues here is how so many thousands of hectares of land, with initially no commercial value,[6] ended up becoming an object of dispute between local communities and corporate investors, and eventually transformed into a high-valued commodity in the global market.

The aim of this chapter is to tackle this question by illustrating the transformation of land into a commodity in two areas of industrial crop production in Peru. Thus, our interest in this chapter does not lie so much in the appropriation of land value (for which there is ample literature as discussed above), but on the "magical process" whereby land value is formed. By grasping the processes mediating the creation of value in the two study sites, this approach may shed light on the "magic" entangled in the transformation of nature.[7] This process is important not so much because it masks social reality, as Marxists contend, but because through it, nature and the social are co-constituted. This is achieved by imagining a "virtual reality" that simultaneously naturalizes it and has concrete effects on the environment and local communities. According to Visser (2017: 187), this is essentially a process of "asset making"; that is, "*an 'advanced' stage of commodification – or, in the case of a natural resource like land, we might more precisely speak of 'resource making'*".

The "Methodology" section outlines the conceptual framework of this chapter, the basic characteristics of the study sites, and the process of data collection. The "Frameworks regulating agro-industrial expansion in Peru" section outlines the main institutional transformation that enabled large-scale land acquisition in marginal areas in Peru. The sub-sections "Biofuels in Piura: Commodifying land through water rights" and "Cocoa in Loreto: Commodifying land at the expense of the Amazon forest" outline the process through which "irrelevant" land was transformed into a valuable market commodity in two areas experiencing land grabbing for sugarcane and cocoa production. Finally the "Industrial crops as agents of ecological transformation in Peru" sub-section discusses the mechanisms used to assign agricultural and market value to the land, which essentially acted as a major agent of ecological transformation in the study areas.

Methodology

Conceptual framework: Transforming nature and crafting of irrelevance

Since the 1980s, farmland in Peru has been considered a scarce good (ONERN, 1985), and for this reason, most of the commercialized land is not agricultural (Burneo, 2020; Herman and Mayrhofer, 2016; Eguren, 2011). To trade this land in the global market and make it appealing to investors, two processes are necessary: (a) to transform it into agricultural land, and (b) to upgrade its value (Andreucci et al., 2017; Visser, 2017; Li, 2014; Wolford et al., 2013). Although institutional and normative frameworks are necessary to proceed with land transformation, in many cases (including the study sites in this chapter, "Study sites"), informal and semi-formal mechanisms are also observed (see "Frameworks regulating agro-industrial expansion in Peru" for such mechanisms in the study sites).

First, irrelevance is crafted when the land is designated as marginal, uncultivated, fallow, barren, unproductive, wasteland, primary forest, or deserted, among many other epithets (Wolford et al., 2013) (Chapter 4). As global demand makes agricultural land a scarce good, "irrelevant" land is transformed into agricultural land to be offered as a commodity in the international market. This results in previously undervalued land acquiring productive potential and a market value through various processes (Wolford et al., 2013; Visser, 2017). Although different in nature, the transformation of nature and the crafting of value are inseparable parts of the process of land grabbing (Chapter 3).

Second, land value is upgraded through the process of value crafting. Visser (2017) describes a typology of asset making (see also Li, 2014) that may be useful in understanding how these elements unfold on the ground (see the "Biofuels in Piura: Commodifying land through water rights" through "Cocoa in Loreto: Commodifying land at the expense of the Amazon forest" subsections): (a) potential for profit, (b) scarcity of the object, (c) the liquidity of the object, (d) standardization of the object, and (e) the legitimacy of the object.

Regarding type (a), some of the features that can make land profitable include the physical conditions of land itself, close location to infrastructure, climatic conditions, and water provision, with the latter being essential in the imagery that attempts to create land value.[8] For this reason, scholars have warned that value making implies land grabbing is usually accompanied by water grabbing (Borras et al., 2011a, 2011b; Franco et al., 2013; Mehta et al., 2012) (Chapters 3 and 11). Technology can be used to increase profit potential, for example, by describing the land as productive and often establishing these standards of productivity by comparing the land to other regions. The consolidation of land into large plots is key for guaranteeing economies of scale and decreasing purchase costs (Visser, 2017). Regarding types (b) and (d), discourses of scarcity and standardization include the description of the land in terms of commodities that are easily traded in the global market. Type (c) points in the same direction, as it requires processes that facilitate the transferability of land. These include easy and inexpensive land transactions and any other processes that make the land more appealing to the market (e.g. clarity of information, free and secure land markets, increasing demand), that is, fungibility, easy selling and transferring requirements, complete information of the asset, guarantees to trade in a stock market, property rights, and large land extent. Finally, regarding type (e), legitimacy refers to the moral quality of the processes regarding the land and the business to acquire it. That is, it evaluates whether land transactions or acquisitions are violating human rights, democratic values, international investment law and standards, among others (Visser, 2017).

Study sites

The chapter focuses on two case studies to explain how the processes of land value formation intersect in the context of land acquisitions. The studies are located in geographical areas of Peru with different characteristics (Figure 5.1).

Figure 5.1 Location of study sites.

The first study site is located in the Chira Valley (Piura department), a semi-desert area on the northern coast. Maple Etanol Peru S.R.L. operated in this area until 2015 (Urteaga-Crovetto, 2016). Most of the land that Maple bought from the regional government included primary forest containing carob trees and many different species of wild flora (20), birds (21), mammals (2), and reptiles (2) (CIPCA and RAA, 2010: 7). In this semi-arid dry woodland, the tropical and humid climate exhibits irregular rainfall during the summer months, and temperatures usually ranging between 24 and 30°C. These conditions are particularly suitable for sugarcane production.

The second study site is located in Tamshiyacu (Loreto province), an area of tropical rainforest in the northeastern Amazon basin, where Cacao del Peru Norte S.A.C. (CDPN) operated until 2017. This area was established by mestizos for the migratory management of small agroforestry plots, as well as for fishing, hunting, and the extraction of forest products (Hiraoka, 1986a, 1986b, 1985a, 1985b, 1985c; Padoch et al., 1988). The climate is humid and warm, with maximum average annual temperatures ranging between 20.3 and 32.5°C and an average annual rainfall of 2,556.2 mm. These climatic conditions are very promising for non-irrigated cocoa cultivation (Quintero and Díaz, 2004). The natural vegetation consists of heterogeneous forest, distributed across different strata, among which the *"varillales"* stands out. This type of forest grows on white sand and has high conservation value, as it shelters a community of specialized plants that are adapted to poor and water-saturated soils (Pitman et al., 2015). Part of the land acquired for cocoa production included this type of forests, where it is illegal to cultivate crops (Segura, 2019). It is worth mentioning that according to the Research Institute of the Peruvian Amazon (Escobedo et al., 1994; Vriesendorp et al., 2015), the soils of Tamshiyacu are suitable for forestry activities, but have an extremely low capacity to support crops such as cocoa, which requires soil rich in organic matter, with good drainage and a pH between 4.5 and 6.5 (Romero, 2016).

Data collection and analysis

This chapter synthesizes information from various data sources, including expert interviews, site observation, and archival work. Fieldwork was conducted in 2012 in the Chira Valley, and in 2018 in Tamshiyacu. Table 5.1 summarizes the different data collection instruments and related information.

Results and discussion

Frameworks regulating agro-industrial expansion in Peru

Before understanding how the processes of land transformation and value crafting (see "Conceptual framework: Transforming nature and crafting of irrelevance") unfolded in the two study sites, it is first important to appreciate the main aspects of the underlying institutional framework in Peru.

During the 1980s and particularly the 1990s, a series of normative, institutional, and policy frameworks facilitated agro-industry promotion in Peru (Urteaga, 2017; Del Castillo, 2013; Eguren, 2011), with agribusinesses expanding in the coastal and Amazon areas. It was then that the main legal and institutional frameworks facilitating the acquisition of *"tierras eriazas"* for agricultural exports were designed. In July 1991, the Legislative Decree 653 (or Law on the

Table 5.1 Data collection methods

Method	Data	Comments
Expert interviews	• Semi-structured interviews • 17 interviews in Tamshiyacu and Iquitos (cocoa study) • 21 interviews in Piura and Lima (sugarcane study)	• Interviews with local community and authorities, smallholders, company workers, and representatives to elicit their understanding and visions of land • Interviews with national and regional state authorities to learn about policies promoting agribusiness and land protection • Interviews with local authorities to elicit their vision on how to upgrade value to the land • Interviews with experts in agronomy, hydrology, soil, and forest management to elicit their views on the impact of land transformation
Site observation	• Sullana: November 2012 and April 2013 • Tamshiyacu: February 2018	• Visual appreciation of the converted land, new plantations, and facilities in relation to local communities and ecosystems
Document analysis	• State archives • Legal documents • National and international project reports	• State actions in relation to the public land sales, water rights allocation, and soil protection • Laws related to the promotion of agribusiness, water rights, and soil protection to analyze how legal frameworks contribute to value crafting • Financial information, including differences between the price of land purchase and the subsequent valuation of the land

Promotion of Investments in the Agricultural Sector) was issued. Its Article 2(d) promotes investment for agricultural, forestry, or agro-industrial production. Its Article 24 defines "*eriazos*" as the land that is not arable due to the complete or partial lack of water, such as hills and meadows with natural pastures, dry forests, protected areas, and areas of archaeological and cultural heritage. Much of the land that was not productive or deemed underutilized was declared state land, marginal, uncultivated, fallow, barren, or deserted (see Chapter 4). In the same vein, Law N° 26505 (or Land Law) was adopted in 1995 and the Law on "*Eriazas*" Lands and Irrigation Projects (D.L. 994) in 2008, both of which completed the legal constellation for agro-industry.

Upon request from companies, land for agro-industrial development was usually sold through direct sale or public auctions, after which property rights were allocated.[9] However, it was not deemed necessary to make an assessment of land tenure rights, which later generated a range of legal conflicts between external (new) and local (old) land holders (Urteaga, 2017). The accumulation and control over resources such as land and water, which are organized as *enclosures*[10] for production (Corrigan and Sayer, 1985; Harvey, 2005; van der Ploeg, 2006), excluded other forms of use and management (see Chapter 4). Some dry forest lands are legally considered *"eriazas"* and have been sold as such by regional governments. For example, due to its considerable demand for water, such land in Piura was transformed into agricultural land for planting sugarcane for ethanol (Urteaga, 2013) at the expense of informal local users of the land (Urteaga-Crovetto, 2016; Urteaga, 2017) (see "Study sites" and "Biofuels in Piura: Commodifying land through water rights"). According to the provisions of the above policies, in order to establish a plantation on land with forest cover, the regional government should grant a permit for changing the land use once the productive capacity of the land is demonstrated.

Water is another essential element for transforming "irrelevant" land. Legislation requires technical studies to demonstrate water availability, but there are multiple ways to circumvent these requirements. For example, as it will be discussed in later in the chapter that in Chira the formulae used to calculate water availability in the local hydraulic system led to serious opposition from local water users, including from a group of state officials opposing the regional and central government support of biofuels projects (Urteaga-Crovetto, 2016). The new environment for agro-industrial projects demanded not only water infrastructure but also the construction of new roads and bridges to facilitate transportation. The regional government assumed the costs of the construction (Urteaga, 2017).

Arguably, a series of informal and semi-formal mechanisms unfold in the process of transforming marginal land into agricultural land. Informal techniques include *de-facto* deforestation, unauthorized effluent discharge, land appropriation, intentional fires, illegal groundwater extraction, and wildlife predation. Some of the more commonly observed semi-formal mechanisms include institutional techniques such as lobbying, mediation for the flexibilization of national and regional legal requirements, local and national media support, discourses of development and agricultural modernization, questionable agreements with local authorities, and simplification of ecological and biological diversity, among others (see Chapters 2–4).

In fact, the process of transforming "irrelevant" land into agricultural land entails fragmenting and covering land features such as biodiversity, which could indicate the ecological relevance of land, jeopardizing market value creation. These elements covertly seek to facilitate the flow of capital. In this sense, the *"essentialist ontological conception of nature [with its] will to subsume diversity in universality, to subject heterogeneous being to the measure of a universal equivalent"* (Leff, 2015: 48) is concomitant to recode the complexity of nature in terms of exchangeable market values. This is extremely visible in both study sites as outline in sections "Biofuels in Piura: Commodifying land through water rights" and "Cocoa in Loreto: Commodifying land at the expense of the Amazon forest".

Biofuels in Piura: Commodifying land through water rights

Since the mid-1990s, the growing global interest for land for agriculture and sustainable energy production in the northern coast of Peru has been boosted by a legal framework that allows the purchase of both agricultural and unproductive land in the region (Urteaga, 2017) (see "Frameworks regulating agro-industrial expansion in Peru"). The *Decreto Legislativo* 994 (2003) declared that the development of irrigation projects on unproductive land with agricultural capacity, through private investment projects, was in the public's interest and allowed the trading of this land in the global market.

Maple, a transnational company registered in Ireland since 1994, with subsidiaries in Peru (IFC, 2007a), actively sought large tracts of land for sugarcane ethanol production. In 2007, the World Bank's International Finance Corporation (IFC) granted Maple a USD 30 million loan to pursue its projects (IFC, 2007b), including hydrocarbon extraction and sugarcane ethanol production in the Amazon and on the coast of Piura, respectively. In particular, "*the drilling and well work-over programs and related activities aimed to extend production of its existing hydrocarbon fields; exploration and related activities in Maple's hydrocarbon concessions; and the development of a greenfield ethanol project*" (IFC, 2007b: online). The IFC's actual project description described the land as follows: "*the proposed ethanol project is expected to be developed on 10,672 hectares of unused and uncultivated semi-desert land in northwestern Peru. The ethanol project will also involve the construction of storage and shipping facilities in the port of Paita to accommodate the sale of ethanol from the project*" (IFC, 2007b: online).

In September 2005, the Special Project Chira Piura (Proyecto Especial Chira Piura, PECHP) granted a reserve of "unproductive land" located on the left side of the Chira river to the Piura Regional Government (CIPCA and RAA, 2010).[11] Unproductive or marginal land was defined in the legal framework[12] as land not exploited due to the lack or excess of water (see Chapter 4). In 2006, the regional government of Piura granted Maple more than 10,000 ha of this unproductive and semi-desert land, which basically consisted of dry woods or "*bosque seco*" (CIPCA and RAA, 2010). Part of this area was known as "*despoblado*", which means locally "uninhabited region". This name, however, contrasts with the multiple uses this land had for local communities, with the area usually occupied and used by herders, smallholders, and peasant communities, among others (Chapter 4).

Despite the lack of physical and legal land use planning, Maple acquired this land including two local cemeteries and part of the territory of a peasant community (Urteaga-Crovetto, 2016). Maple acquired an additional 3,500 ha, while the Romero Group bought 3,800 ha from private owners (Maple Etanol Peru S.R.L., 2014). In the first phase of the project, Maple planted 7,500 ha, while for the second phase, a further 2,300 ha were scheduled (Maple Representative, Personal Communication, 15/04/13). After the land transaction, smallholder farmers, whose land in Vichayal and La Huaca was included in the area sold to Maple, were evicted (El Tiempo, 2006).

Another problem arose regarding the transformation of the semi-deserted area into agricultural land, as this land was in a juridical limbo. Registration had not been deemed necessary because the "*eriaza*" land had no absolute exchange value (Burneo, 2011). This had also major implications on water use. The Chira Valley was known to be a water-abundant region, with several water infrastructural facilities built since 1950 (Revesz and Oliden, 2011). The Chira-Piura Special Project (PECHP) owned 150,000 ha of land and the 885 Mm³ Poechos dam to irrigate them. The hydraulic infrastructure has been essentially the

> *symbol of progress and the means to transform previous socio-spatial configurations –* "*deserted lands*" *or* "*despoblado*" *– into* "*productive*" *private land and water regimes. As tokens of modernity, hydraulic works transformed Piura into an emporium of agricultural exports. In 2004, the coastal areas of Piura consisted of approximately 103,474 ha of agricultural land, of which 98.5% were irrigated.*
>
> (Urtega-Crovetto, 2016: 4)

In 2009, approximately 80% of the land in Piura was cultivated. However, local water users were concerned about water availability. All of the available water resources in the irrigation system were already allocated to the planned crop production in the Chira and Piura valleys, which left little water for future agricultural activities in the broader area (Urteaga-Crovetto, 2016). With a single hectare of sugarcane projected to consume 17,000 to 20,000 m³ of water per year with year-round irrigation, water availability ineluctably emerged as a contested issue (Urteaga-Crovetto, 2016). Corporate users and the regional government insisted that water was abundant and that cutting-edge technology, such as drip irrigation systems and private reservoirs, would save water. However, local irrigation users demonstrated otherwise (Urteaga-Crovetto, 2016).

Without the entire volume of water deemed necessary for the agricultural and industrial ethanol complex, marginal land could not be transformed into agricultural land, which was essential for land value crafting (see "Conceptual framework: Transforming nature and crafting of irrelevance"). Dionisio Romero Paoletti, the manager of a bioethanol company in the Chira Valley, would say: "*we are transforming uncultivated desert lands into productive lands by means of technical irrigation systems*" (La República, 2008: online). As a result, they engaged in a discursive battle to demonstrate that the Chira Valley had enough water and efficient technical systems to irrigate thousands of hectares of sugarcane (Urteaga-Crovetto, 2016). Maple considered water as practically a gift, with one of its representatives pointing to water pricing as an incentive to start sugarcane production in Piura: "*Water is for free, it does not cost anything. It is easy to produce high water demand crops*" (Deforge-Lagier, 2009: 43). Despite strong local opposition, Maple finally was given access to 186 Mm³ of water for its agro-industrial complex (Urteaga-Crovetto, 2016).

In the meantime, Maple purchased approximately 10,684.15 ha and 4,637 m² for ethanol production. The total cost reached USD 640,588 (Gallo, 2013), which at the time of the transaction was valued at USD 59.95 per hectare of "unproductive land". The company mentioned that it would also pay the regional government USD 500,000 per year over a 20-year period as an investment upon the start of ethanol production (CIPCA and RAA, 2010). However, there was a widespread consensus in Piura that the regional government had knowingly undervalued the marginal land, particularly because officials knew that irrigation could transform it into expensive agricultural land (Deforge-Lagier, 2009; Urteaga-Crovetto, 2016). Local officials from the PECHP pointed out that the cost per hectare responded to the quality of the marginal land at the moment of the transaction (Deforge-Lagier, 2009). However, both the Comptroller General Office and the National Congress had concerns regarding the low price of the land, especially when it could be easily transformed into agricultural land. Even Marisol Espinoza, a congress representative at the time, expressed doubts:

> *During more than 30 years of the PECHP, many conditions have changed. In 1969, by Law Decree 17463, the government of Velasco Alvarado declared this project as of national interest. By then those lands were marginal but now there are homesteads, districts, cemeteries, cultivation and grazing activities on them [...]. Despite that, the Piura governmental officials disregarded local inhabitants. They just decided the sale of those lands as marginal lands "free of occupiers", as clauses 5 and 11 of the contract with Maple Ethanol S.R.L. stipulate.*
>
> (Espinoza, 2008: 12)

Moreover, an administrative report from the Comptroller General's Office admitted that seven years before (1999), the PECHP had calculated the value of the same marginal land at USD 416.00 per hectare, which made it even more difficult to justify why Maple had purchased the land at approximately USD 60.00 per hectare, that is, one-seventh of the commercial value (Deforge-Lagier, 2009). For Deforge-Lagier (2009: 43), the explanation lies in the resolution of the government to promote such projects: "*This low price for land can be considered as implicit subsidies [...] Therefore, regional government through PECHP, owner of most of the lands operated in the project, incites private company to invest thanks to implicit subsidies. Besides the trade of marginal land, water license had been requested transforming those arid lands in agricultural lands at very low price*".

This whimsical characterization of the land purchased by Maple as marginal is an example of the crafting of irrelevance. Not only were the land and resources considered to be irrelevant in terms of exchange value, but also the local communities and users were symbolically "erased" from the characterization of the land forcing them to leave the "*bosque seco*". Water was another contentious issue, as it was essential for converting marginal into agricultural land (see Chapter 11). The regional and national water authorities granted Maple water rights for sugarcane production for ethanol, which came to the detriment of local irrigation users, reinforcing inequalities in the Chira Valley.

Cocoa in Loreto: Commodifying land at the expense of the Amazon forest

During the 1990s and 2000s, the widespread land commodification for large-scale agriculture also revived interest in land within the Amazon forest, especially for industrial monocrops such as cocoa and oil palm. Even though the Amazon forest produces huge volumes of water (about 20 million tons per day), its soils have limited capacity to support large-scale agriculture[13] (Bernex, 2015; Rodríguez, 1995). In early 1980s, 80% of the Loreto region was declared suitable for forestry, 10% for nature conservation, and only 1.7% for perennial crops and 1.5% for temporary crops[14] (ONERN, 1985; Consejo Nacional del Ambiente, 2005; Rodríguez et al., 1994). Critiques on agro-industrial development in the tropical forest center precisely on the fragility of Amazon soils.

Despite the scientific evidence on the limited capacity of the Amazonian soil to support commercial agriculture (ONERN, 1985), national regulations downgraded the requirements for environmental studies related to the productive capacity of the soil. For example, in 2008, the Legislative Decree 1089 and its regulations established that the determination of soil capacity to support an intensive use could be demonstrated using as a reference only the maps of the property to be titled. Local fieldwork was deemed unnecessary. By April 2012, the new flexible legal framework worked as an incentive for at least 12 applications to purchase land, which was presented to the Regional Government of Loreto. Four companies of the United Cacao Limited SEZC (United Cacao) group requested 106,000 ha for agro-industrial oil palm projects (Sociedad Peruana de Ecodesarrollo, 2013, 2012). The Ministry of Agriculture rejected their request to change the use of areas whose highest use capacity was forestry, to agricultural use[15] (Salazar and Rivadeneyra, 2016). Faced with the inability to acquire land from the State, United Cacao launched from mid-2012 another strategy to acquire land for its subsidiary, CDPN. The company would acquire titled land for agriculture in the town of Tamshiyacu, in Loreto. Of particular interest was the land of a former cattle association known as *"Los Bufaleros"*, which was formed by 60 individuals providing 49 ha of land each. During fieldwork (February 2018), former members of *"Los Bufaleros"* pointed out that the government of Loreto did not carry out any soil studies to evaluate agricultural capacity before granting them ownership of their land.

Through their contacts in the regional government of Loreto, the CDPN representatives were aware of the existence of the *"Los Bufaleros"* land, which was registered in the official registries (Salazar and Rivadeneyra, 2016). The company representatives intimidated the *"Bufaleros"* members to purchase their land at a low cost by telling them that the ownership of their land would revert to the regional government of Loreto because they had not given it a "productive use", that is, they had not cleared it for agricultural purposes[16] (Salazar and Rivadeynera, 2016; Dammert Bello, 2017). Faced with this threat, some members saw no better alternative than to sell their land to the company, as they thought they were at risk of losing everything (Salazar and Rivadeynera, 2016; Dammert Bello, 2017. By February 2013, the CDPN had acquired at least 50 plots of land from "*Los Bufaleros*" members for USD 1,900 per plot of 49 ha. In sum, the CDPN paid approximately USD 96,700 for 2,450 ha of land with primary forest, valuing it at USD 39.46 dollars per hectare. This is a gross undervaluation, considering that the value of the ecosystem goods and services lost due to the deforestation of these 2,150 ha of rainforest (see below) amounts to USD 60,713,512 (Kometer and Pautrat, 2014). In addition, in 2014, the National Forest Service indicated that the value of the timber alone amounted to USD 2,234,380 (Corte Superior de Justicia de Loreto, 2019).

Between October 2012 and August 2013, the CDPN deforested nearly 1,000 ha of forest on the land previously owned by "*Los Bufaleros*" (NASA imagery from November 22, 2013 show extensive deforestation), without land use change authorization or environmental impact studies, and by September 2015, deforestation reached 2,276 ha (Finer et al., 2015). Since November 2013, national and international media, as well as environmental groups, denounced CDPN for illegally deforesting and changing land use from forest to agriculture without due authorization. According to "*Los Bufaleros*" members, until October 2012, their plots were covered with primary forest, as they were located in an area classified for highest forest use or protection, where deforestation and agricultural activities are forbidden (Finer and Novoa, 2016; Finer et al., 2016). However, the state issued contradictory decisions over the legality of land use change in the area acquired by CDPN,[17] thus contributing to consolidating the illegal planting (Segura, 2019). A series of contradictions among regional and national decision-makers about the responsibilities of the company, together with the support of part of the press and local authorities, eventually contributed to the commodification of forest land[18] (Segura, 2019).

Amidst the complaints about deforestation and the illegal mechanisms for changing the use of the forest soil, on December 2, 2014, United Cacao Limited SEZC (the parent company) was admitted to trading on the Alternative Investment Market (AIM) of the London Stock Exchange (under the symbol CHOC). In its admission document to the AIM, United Cacao declared to be respectful of environmental and investment legal standards and stated its commitment to creating jobs for local communities. Also, the CDPN declared that the cost of its 3,523 ha property in Tamshiyacu amounted to USD 1,651,507[19] (United Cacao Limited SEZC, 2014), despite the fact that the 2,450 ha of the eventually deforested land that CDPN bought from "*Los Bufaleros*" cost USD 96,000. On the same day of its admission to the AIM, in the initial public offering, United Cacao raised USD 10,000,000 (McFarland, 2018), with the promise of expanding cocoa cultivation and its land bank in Tamshiyacu. Although United Cacao had declared to the AIM investors that it possessed all necessary administrative permits for its plantation in Tamshiyacu, in May 2016, environmental organizations filed a complaint with the AIM against United Cacao for violating investment law. Because of this complaint, in February 2017, United Cacao was permanently excluded from the AIM, resulting in losses amounting to USD 42,000,000 (Thoumi, 2017).

In 2017, the CDPN was acquired by investors from France and Singapore, and in 2018, changed its name to Tamshi S.A.C. The new company has attempted to change the former company's bad image, by proposing to restore the environmental impact caused by the deforestation of 2,276 ha and compensating for the losses (Agencia Agraria de Noticias, 2019). The underlying goal has been to obtain the environmental permits to allow cocoa export and trade in international markets.

Overall, the cocoa case aptly represents the crafting of irrelevance through the simplification and erasing of the complexity of the tropical forest ecosystem to build up the commercial value of the land at the expense of not only biodiversity but also local users. By changing the land use of the tropical primary forests in Tamshiyacu, the company was not only able to appropriate the land, but also to ease the crafting of the commercial value of the land in the global markets.

Industrial crops as agents of ecological transformation in Peru

Sections "Frameworks regulating agro-industrial expansion in Peru" and "Cocoa in Loreto: Commodifying land at the expense of the Amazon forest" clearly indicate that industrial crop production in Peru is a major agent of ecological transformation. The ecological transformation observed in the two study sites was mediated by two main processes: (a) the process of making land "irrelevant" in order to acquire it at a low cost and transform it into agricultural land, and (b) the process of value crafting to improve the market value of land. As described below, for both processes, a diverse set of mechanisms was put into play.

The introduction and "Frameworks regulating agro-industrial expansion in Peru" outlined a set of formal, informal, and semi-formal mechanisms usually mobilized to upgrade the value of undervalued land, including the provision of water rights, the change of land or soil use, and the allocation of property rights for agricultural use. Table 5.2 identifies the mechanisms mobilized to convert "irrelevant" land into agricultural land in the two study sites.

In both cases, the respective companies followed similar procedures. Initially, they employed to their benefit formal mechanisms such as national and local policies related to agribusiness, including a series of normative and institutional frameworks designed to prompt agro-industrial development across Peru (see "Frameworks regulating agro-industrial expansion in Peru"), e.g. flexible legal regimes to hire workers. Afterwards, investors searched for uncultivated and "unused" land in the coastal desert and "unproductive" forest land in the Amazon. They were particularly interested in the allocation of property land rights through direct sales. In the Chira case, this proceeded primarily through a request to the regional agricultural office and secondarily through private contracts to increase the extent of the land. In Tamshiyacu, it was impossible to change the forestry use of the requested area (due to a regulation of the Forestry and Wildlife Law of 2011), which impeded the direct sale or public auction.

Table 5.2 Mechanisms mobilized to convert "irrelevant" land into agricultural land in the study sites

Category	Mechanism	Type	Study site	
			Sugarcane	Cocoa
Formal mechanisms	Agribusiness promotion policies (e.g. flexible legal regime for hiring workers, tax incentives)	Political/legal	x	x
	Declaration of state land as marginal land (or similar)	Legal	x	x
	Allocation of property land rights for state land through direct sale or public auction	Legal	x	
	Allocation of property land rights through private contracts		x	
	Technical studies on water availability in water-stressed areas	Technical	x	
	Declaration of water reserves in favor of the regional government, and transfer of water rights to investors	Legal	x	
	Allocation of water rights (licenses)★	Legal	x	
	Construction of infrastructure (e.g. wells, reservoirs, bridges)	Physical	x	
	Studies on the productive capacity of soil	Technical	x	x
	Studies on environmental impact	Technical	x	x
	Shares traded on the stock market	Legal	x	x
Informal mechanisms	De-facto deforestation	Physical		x
	Soil use change without authorization from the subnational agricultural office	Physical		x
	Commencement of agricultural activities without a prior environmental impact assessment	Physical		x
	Land appropriation through intimidation	Physical		x
	Intimidation during the negotiations with local tenants to acquire the land	Physical	x	x
	Wildlife predation	Physical	x	x
Semi-formal mechanisms	Inaccurate land planning in charge of regional governments	Physical & legal	x	
	Official sale of land (including prohibited areas)	Legal	x	
	Simplification of biodiversity	Legal	x	x
	Flexibilization of national and regional legal requirements (i.e. land planning, change of soil use)	Legal	x	x
	Lobbying activities	Social	x	x
	Local and national media support	Social	x	x
	Discourses of agricultural development and modernization through technology replicated by authorities	Social	x	x
	Agreements with local authorities	Social		x

Note: ★ For an analysis of the close links between land grabbing and water grabbing see Borras et al., 2011b, 2011a; Franco et al., 2013; Kay and Franco, 2012; Mehta et al., 2012; Perrone and Hornberger, 2014; Woodhouse, 2012.

This prompted the company to sign private contracts with each owner of the "*Los Bufaleros*" Association.

The transformation of marginal land to agricultural land was further facilitated through the provision of water rights to the company (Chira case) and the removal of primary forest (Chira and Tamshiyacu cases). In the Chira site, the Piura regional government acquired a water reserve from the National Water Authority, which was later transferred to the company by means of water rights. As part of Maple's agro-industrial complex, two water reservoirs were built (see "Biofuels in Piura: Commodifying land through water rights"). Both companies argued that they had complied with environmental rules. In the Tamshiyacu site, water rights were not needed due to heavy rains (see "Cocoa in Loreto: Commodifying land at the expense of the Amazon forest"). In the Chira site, clearing permits and soil use change were eventually secured, whereas in Tamshiyacu non-legal mechanisms were used to justify the lack of such permits, after which the transformation of "irrelevant" land into agricultural land was concluded. Once the above was achieved, both companies traded in the stock market.

Up to that point, the mechanisms mobilized by the two companies to obtain agricultural land were defensible, as they followed national law, albeit manipulating some of its aspects. However, empirical evidence suggests that some informal and semi-formal mechanisms were also mobilized, which were beyond the legal and normative frameworks. Informal mechanisms in Tamshiyacu included de-facto deforestation, soil use change without due authorization, agricultural activities without a prior environmental impact assessment, land appropriation, and intimidation. Furthermore, in both areas, damages to wildlife and negotiations with local holders to convince them to sell their land occurred. In Tamshiyacu, these negotiations often occurred through intimidation.

Semi-formal mechanisms included the inaccurate physical and legal land planning processes in charge of the regional governments for the areas that were sold. In fact, the official land sale in Chira included areas that were private and public property, which caused unending legal claims. Furthermore, biodiversity was simplified, national and regional legal requirements were loosened to ease the land transaction, lobbying activities were undertaken, local and national media support was obtained, and discourses on development, modernization, and technology were deployed locally and nationally. In addition, in Tamshiyacu, there were some questionable agreements with local authorities.

The process of transforming agricultural land is followed by the process of crafting the market value of land. Table 5.3 shows the elements mobilized for this process of value crafting, following the typology proposed by Visser (2017): (a) potential for profit, (b) scarcity, (c) liquidity, (d) standardization, and (e) legitimacy.

With respect to potential for profit, in both cases, the description of the land was accompanied by information on the infrastructure that could facilitate exports, the low land value, the use of cutting-edge technology for agriculture and irrigation, and discourses on economies of scale and improvement of technical methods for efficient agricultural use.

Table 5.3 Elements related to value crafting in the study sites

Category	Element	Study site	
		Piura	Loreto
Profit potential	Proximity of infrastructure for export	x	
	Water availability	x	
	Secure water access	x	
	Low value of the land	x	x
	Low value of water	x	
	Technical systems for land productivity (e.g. irrigation)	x	x
	Discourses on water availability	x	
	Discourses on economy of scale and job creation	x	x
	Discourses on efficiency and improvement of technical methods for agricultural use	x	x
Scarcity	Physical conditions of land	x	
	Productivity of land	x	
Liquidity	Modality of land contracts	x	
	Land registration	x	x
	Flexible national and/or regional legal requirements	x	x
	Large tracts of land	x	x
	Property rights	x	x
	Transfers of property rights from the regional government and/or local holders to investors	x	x
	Administrative permits for plantation	x	x
	Respect for investment law	x	x
Standardization	Processes of physical and legal planning	x	
	Standardization of land by covering or reducing biodiversity and ecological complexity	x	x
	Change of land use	x	x
Legitimacy	Environmental and social licenses	x	x
	Intimidation		x
	Illegal deforestation		x
	Illegal land use change		x
	Criticism of corporate behavior by NGOs and local users	x	x
	Criticism of corporate behavior by the media	x	x

Regarding scarcity, in the Chira case, the physical characteristics of the land and their contribution to high productivity were both remarked. This included a series of discourses touching on water availability and access, low water and land prices, and availability of (and secure access to) water (Urteaga-Crovetto, 2016). Some of these discourses were also observed in Tamshiyacu, but arguably to a lower extent.

The liquidity of the agricultural land in both study sites was guaranteed through two types of land contracts, namely private and private-public. Careful steps were taken by both companies to legally register the land, although not completely successful. As mentioned above, in both areas, some of the requirements for land acquisition were loosened to guarantee the transaction. In addition, both companies sought to acquire large tracts of land from the regional government and from local landowners. Although not without any obstacles, in general, the property rights were secured for future land transactions and transfers, while administrative permits for the plantation were procured. Even though both companies assured their compliance with environmental and investment law, United Cacao was eventually accused of failure to comply, which led to its expulsion from the AIM of the London Stock Exchange.

Regarding the standardization of the land, as already mentioned, the legal and physical status of the land was not accurate. This led both companies to misinformed decisions on the legal status of the land, which later would generate them legal complaints. In order to guarantee standardization, the description of the land basically "erased" biodiversity and ecological complexity, as it was particularly evident through the requests and eventual success in changing soil use.

Finally, when it comes to the legitimacy of the land, both companies requested environmental and social licenses, which in the Chira case were finally granted. However, in Tamshiyacu, corporate intimidation against local landowners compelled them to sell land, which arguably is an illegitimate means to land acquisition. Similarly, both the unauthorized deforestation and land use change that characterized land acquisition in Tamshiyacu are arguably illegal. Both cases also show the opposition of NGOs and local land and water users toward the new land investment, with the local media articulating negative opinions toward the project.

Conclusions

The global land rush implies new legal and institutional frameworks financed by IFIs, upon which national and regional governments welcome agribusinesses in local contexts that were previously not integrated into markets. States increasingly promote neoliberal agendas by supporting such agro-industrial projects, implicitly compromising its alleged neutrality, with major ramifications for local communities and the environment. Peru has been experiencing this transition. The government has put in place a series of institutional frameworks that promote agro-industrial development through industrial crops such as sugarcane and cocoa. However, many of these projects exhibit the characteristics of land grabbing, as well as nature transformation and land value crafting.

Throughout this process, the virtual economy in the global markets has played a critical role in creating irrelevance. A mercantile vision of land has been implicit in this process that requires the erasing of the environmental and social value of land, by remarking its irrelevance. In focusing exclusively on the potential of land for agricultural use, it was possible to obscure the complexity of dry and tropical forest ecosystems, underscoring the commercial value of the agricultural land. The commodification of this land and the assignment of low price was an essential first step for designating land as irrelevant. Once the land transaction proceeded, the land acquired a significant new value that was eventually bolstered in the market. Water rights facilitated value addition to a dry forest described as marginal land in the Chira Valley. Meanwhile, by replacing the ecological complexity of the Tamshiyacu tropical forest with a universal commercial value, it was possible to trade this land in stock markets. Multiple legal and illegal mechanisms were put into play to achieve this, usually with the complicity of the state.

Crafting the irrelevance of the land also meant asserting the insignificance of local communities and their knowledge of the local environment. In both Chira and Tamshiyacu, this manufacturing of the triviality of local users led to land and water dispossession, and their eventual eviction from the land they previously occupied. Underlying these processes of land grabbing were opposing visions of the land that eventually led to the ecological transformation of the areas converted for sugarcane and cocoa cultivation.

Notes

1 That being the case, it is difficult to imagine the way in which the self-regulation proposed by the World Bank and other international actors could prevent the negative impacts of land accumulation (White et al., 2012).

2 This is particularly evident in "green" businesses such as biofuels. Bakker (2010: 726–727) emphasizes the capacity of neoliberalism to transform nature for profit under the cloak of sustainability: "*The neoliberalization of socio-nature must thus be understood as, simultaneously, a disciplinary mode of regulation, and an emergent regime of accumulation that redefines and co-constitutes socio-natures*".

3 For example, a recent World Bank study on the effects of large-scale land acquisitions for agriculture in developing countries found that "*laws and policies surrounding land acquisition were often ambiguous, incomplete or contradictory. The implementation of the legal and policy frameworks appeared to be inconsistent, often hurting those who were already vulnerable – especially those using land that was not demarcated or registered or those whose resource rights were unregistered*" (World Bank, 2014: 15–16).

4 Such as example is the serious 2009 conflict in Bagua in the Northern Amazon basin, due to the attempt of the national government to reduce legal requirements to sell indigenous communal lands. As a result, more than 30 indigenous people and policemen died (Comisión Multipartidaria encargada de Estudiar y Recomendar la Solución a la Problemática de los Pueblos Indígenas, 2008).

5 For example, OXFAM (2014) notes that between 2004 and 2008, oil concessions covered three-quarters of the Amazon basin, mainly affecting indigenous territories, water sources, and wetlands. In 2011, mining concessions in the Marañón River basin overlapped with indigenous territories to a relatively larger degree compared to other basins. According to OXFAM (2014: 20), "*in the rainforest of the Peruvian east and northeast, there is a substantial overlap between the lands ceded to usufruct to oil and gas companies, and the lands reserved for indigenous communities. The overlap occurs predominantly between oil exploration concessions and community lands, although drilling itself overlaps with native lands located throughout the region*".

6   Here, it is important to distinguish between price, value, and asset. While markets determine the actual prices, values include other aspects such as social and labor relations (Devine, 1993). Uncultivated land and untapped natural resources could *"have no value but a positive market price"* (Devine, 1993: 139). The so-called "free goods" or assets such as wasteland or uncultivated land are pseudo-commodities, which have both use and exchange values (Andreucci et al., 2017).

7   Scholars have drawn on the idea of magic to explain social phenomena specifically linked to colonial processes in Latin America. Examples include shamanist imaginary as consubstantial to colonial economy in Ecuadorian Amazonia (Taussig, 1987) and Venezuelan dependency on the petroleum industry as part of nation state formation process (Coronil, 1997).

8   Visser (2017: 196) outlined examples in Russia and Ukraine where water issues affected resource value formation: *"investors found out that the general weather conditions and microclimate, and in particular the availability of water, are much more critical for productivity than expected"*.

9   For more information on the connections between property rights and land grabbing in the Amazon, refer to (Foweraker, 2002; Schmink, 1982; Schmink and Wood, 1987).

10 Corrigan and Sayer (1985: 97) *define enclosures* as processes that took place in the 18th century to appropriate small portions of land and create landholdings of private property. The authors use this concept to explain transformation processes in England since the 16th century because of capitalism (Corrigan and Sayer, 1985). Harvey (2005) indicates how the appropriation and enclosure of common goods occurred as a condition of capitalist production, see also (Heynen and Robbins, 2005).

11 Transnational companies acquired land not only from the government, but also directly from small- and medium-scale farmers. For example, between 1994 and 2008, many smallholders in the Chira Valley sold approximately 36,000 ha of land, of which 13,600 ha were bought by five companies (Burneo, 2011).

12 See: Art. 24° of Decreto Legislativo 653; Art. 3° of Decreto Legislativo 994; Art. 5° of Decreto Legislativo 1064.

13 This is due to the fact that in the Amazon, the soil organic matter is shallow, which restricts nutrient supply for crop production. In addition, one of the main problems of these soils is the release of aluminum during erosion, which can be toxic to plants (Consejo Nacional del Ambiente, 2005).

14 A more recent and detailed analysis for the Alto Amazonas province indicates that only 2.1% of the total area is suitable for temporary crops and forest production with low to medium agro-ecological quality (Gobierno Regional de Loreto, Municipalidad Provincial Alto Amazonas e Instituto de Investigaciones de la Amazonía Peruana, 2015).

15 Some of this state land included at least 26,000 ha of the Permanent Production Forest of Loreto suitable for forestry. The Forestry and Wildlife Law of 2011 (Law 29763) forbids the change of the forest land use to agriculture (Sociedad Peruana de Ecodesarrollo, 2012).

16 As noted by former members of *"Los Bufaleros"* and by Finer and Novoa (2016), when the CDPN acquired this land in early 2013, they maintained primary forests.

17 For a detailed analysis of the controversy regarding the land use change due to CDPN activities, see Segura (2019).

18 It is worth noting that in 2019, two former CDPN functionaries were criminally convicted for the illegal trafficking of forest species and obstructing authority. They

were also sentenced to a civil reparation of USD 4,464,310 (Corte Superior de Justicia de Loreto, 2019).

19 As declared by United Cacao (2014), the valuation of CDPN land includes "adaptation costs", but details about these costs have not been specified. One possibility is that they include cost associated with the removal of forest cover, wood disposal, and the payment of workers during the deforestation stage.

References

Agencia Agraria de Noticias (2019) *Tamshi y Gerfor Loreto firman convenio para recuperar 1.704 hectáreas de bosques amazónicos.* agraria.pe/noticias/tamshi-y-gerfor-loreto-firman-convenio-para-recuperar-1-704--20435 (accessed: 11 June 2019).

Ahmed, A., Abubakari, Z. and Gasparatos, A. (2019) Labelling large-scale land acquisitions as land grabs: Procedural and distributional considerations from two cases in Ghana, *Geoforum.* DOI: 10.1016/j. geoforum.2019.05.022

Andreucci, D., García-Lamarca, M., Wedeking, J. and Swyngedouw, E. (2017) "Value grabbing": A political ecology of rent. *Capitalism Nature Socialism* 28(3): 28–47. DOI: 10.1080/10455752.2016.1278027

Bakker, K. (2010) The limits of "neoliberal natures": Debating green neoliberalism. *Progress in Human Geography* 34(6): 715–735.

Bernex, N. (2015) El Amazonas "capital de las sílabas de agua". Roca, F.; Álvarez, J.; Bernex, N.; Campos Baca, L.; Dourojeanni, M.; García, J.; Kauffmann, F.; Nieto, A.; Olivera, Q.; Recharte, J.; Sabogal, A.; Torres, F y Villa, M. *LA AMAZONÍA: sílabas del agua, el hombre y la naturaleza.* Lima: Banco de Crédito del Perú, 109–137.

Borras, S. M., Franco, J. C., Kay, C. and Spoor, M. (2011a) "*Land Grabbing in Latin America and the Caribbean Viewed from Broader International Perspectives*". Paper presented at Dinámicas en el mercado de la tierra en América Latina y el Caribe, November 14–15. FAO: Santiago de Chile.

Borras, S. M. Jr., Fig, D. and Suárez, S. M. (2011b) The politics of agrofuels and mega-land and water deals: Insights from the ProCana case, Mozambique. *Review of African Political Economy* 38: 215–234.

Borras, S. M., Franco, J. C., Gomez, S., Kay, C. and Spoor, M. (2012a) Land grabbing in Latin America and the Caribbean. *The Journal of Peasant Studies* 39(3/4): 845–872.

Borras, S. M., Gomez, S., Kay, C. and Wilkinson, J. (2012b) Land grabbing and global capitalist accumulation: Key features in Latin America. *Canadian Journal of Development Studies* 33(4): 2–16.

Borras, S. M. Jr., Mills, E. N., Seufert, P., Backes, S., Fyfe, D., Herre R. and Michéle, L. (2020) Transnational land investment web: land grabs, TNCs, and the challenge of global governance, *Globalizations*, 17:4, 608–628, DOI: 10.1080/ 14747731.2019.1669384Burneo, M. L. (2012) Elementos para volver a pensar lo comunal: Nuevas formas de acceso a la tierra y presión sobre el recurso en las comunidades campesinas de Colán y Catacaos. *Anthropologica*, XXXI(31): 15–41.

Burneo, M. L. (2013) Espacio regional, recursos naturales y estudios sobre Piura. Revista Argumentos. https://argumentos-historico.iep.org.pe/articulos/espacio-regional-recursos-naturales-y-estudios-sobre-piura/ (accessed: 8 April 2019).

Burneo, M. L. (2020) Técnicas territoriales para la apropiación del bosque seco peruano: El caso de los comuneros de Catacaos frente al avance de la agroindustria en un contexto de emergencia humanitaria. *Territorios* (42-Especial): 1–29. DOI: 10.12804/revistas. urosario.edu.co/territorios/a.7736

Burneo, Z. (2011) *El proceso de concentración de la tierra en el Perú*. Roma: Coalición Internacional para el Acceso a la Tierra, CIRAD, CEPES.

Centro de Investigacion y Promocion del Campesinado and Red de Accion en Agricultura Alternativa – CIPCA and RAA (2010) Situacion de Biocombustibles en el Departamento de Piura. Piura: CIPCA and RAAA.

Chirif, A. (2015) "*Territorios indígenas en la coyuntura actual*". Paper presented at Políticas de desarrollo, territorio y consulta previa, organized by Fórum Solidaridad Perú, Tarapoto, July 1–3. www.servindi.org/actualidad/134946 (accessed: 13 June 2019).

Comisión Multipartidaria encargada de Estudiar y Recomendar la Solución a la Problemática de los Pueblos Indígenas (2008) *Informe sobre los Decretos Legislativos vinculados a los Pueblos Indígenas, promulgados por el Poder Ejecutivo en mérito a la Ley 29157*. Lima: Congreso de la República del Perú.

Consejo Nacional del Ambiente (2005) *Indicadores ambientales Loreto*. Serie Indicadores Ambientales N° 7. Lima: Consejo Nacional del Ambiente.

Coronil, F. (1997) *The magical state. Nature, money, and modernity in Venezuela*. Chicago: The University of Chicago Press.

Corrigan, P. and Sayer D. (1985) *The great arch. English state formation as cultural revolution*. Oxford: Basil Blackwell.

Corte Superior de Justicia de Loreto (2019) Sentencia del 25 de julio recaída en el Exp. 00740-2014-41-1903-JR-PE-04, sobre el delito tráfico ilegal de productos forestales maderables y obstrucción del procedimiento, seguido en contra de Antonio Ruben Espinoza, Cubas Ramirez Giovanny y Vega Delgado Ernesto. www.pj.gob.pe/wps/wcm/connect/1d9f9d004aefb0ada20ea69507b119bf/SENTENCIA+INTEGRAL+TAMSHYYACU.pdf?MOD=AJPERES&CACHEID=1d9f9d004aefb0ada20ea69507b119bf (accessed: 15 January 2020).

Cotula, L. (2011) *Land deals in Africa: What is in the contracts?*. London: International Institute for Environment and Development (IIED).

Cotula, L. (2014) *Addressing the human rights impacts of "land grabbing"*. Brussels: Directorate-General for External Policies. Policy Department, European Union.

Cotula, L., Vermeulen, S., Leonard, R. and Keeley, J. (2009) *Landgrab or development opportunity? Agricultural investment and international land deals in Africa*. London, Rome: FAO, IIED and IFAD.

Dammert Bello, J. L. (2017) *Acaparamiento de tierras en la Amazonía peruana: el caso de Tamshiyacu*. Lima: WCS, OXFAM.

Deforge-Lagier, S. (2009) "*Impacts of Agrofuel Production on Land-use and Water in Semi-arid Areas: The Case of Piura-Chira, Peru*". Unpublished MSc thesis, UNESCO-IHE Institute for Water Education, Delft.

Del Castillo, L. (2013) «La comunidad y la irrigación de Olmos. Una relación nada justa», *Aguas robadas: despojo hídrico y movilización social*, Aline, A. and Rutgerd, B. (eds.), Quito: Justicia Hídrica, Abya Yala, Instituto de Estudios Peruanos, 83–102.

Devine, J. (1993) The law of value and Marxian political ecology. Jesse V., Ross D. and Ron F. (eds.) *Green on red: Evolving ecological socialist* (Socialist Studies/Études Socialistes, vol. 9), Winnepeg/Halifax, Canada: Society for Socialist Studies/Fernwood Publishing, 133–154.

Diez, A. (2017) Propiedad y territorio como (diferentes) bienes comunes. El caso de las tierras de comunidades en la costa norte peruana. *Utopía, Revista de Desarrollo Económico Territorial* 11: 17–39.

Eguren, F. (2011) *Acaparamiento de tierras. Reflexiones a partir de estudios de casos*. Lima: Food and Agricultural Organization (FAO).

Escobedo, R., Bendayán, L., Rojas, C., Rodríguez, F. and Marquina, L. (1994) *Estudio detallado de suelos de la zona «Fernando Lores», Tamshiyacu (Región Loreto)*. Documento Técnico 5. Iquitos: Instituto de Investigaciones de la Amazonía Peruana. repositorio. iiap.org.pe/handle/IIAP/256

Espinoza, M. (2008) *Informe de Gestión Parlamentaria. Legislatura 2007–2008.*. Lima: Congreso de la República. In: www4.congreso.gob.pe/congresista/2006/curibe/informe/INFORME-GESTION-2008.pdf (accessed: 2 May 2019).

Finer, M. and Novoa, S. (2016) *United Cacao Continúa la Deforestación de Bosque Primario en Tamshiyacu (Loreto, Perú)*. MAAP: 27. maaproject.org/2016/cacao-peru-norte/ (accessed: 20 February 2020).

Finer, M., Novoa, S. and Cruz, C. (2016) *Proyecto United Cacao se ubica en tierra clasificada Forestal*. MAAP: 38. maaproject.org/2016/forestal/ (accessed: 4 June 2019).

Finer, M., Olexy T. and Novoa, S. (2015) *Se reanuda el desbosque para cacao en Tamshiyacu (Loreto, Perú)*. MAAP 13. maaproject.org/2015/09/tamshiyacu3/ (accessed: 1 June 2019).

Foweraker, J. (2002) [1981]. *The Struggle for land: A political economy of the pioneer frontier in Brazil from 1930 to the present day*. Cambridge, London, New York, New Rochelle, Melbourne, Sydney: Cambridge University Press. (Cambridge Latin American Studies).

Franco, J., Mehta, L. and Veldwisch, G. J. (2013) The global politics of water grabbing. *Third World Quarterly* 34(9):1651–1675.

Gallo, L. (2013) "Los biocombustibles y el Agua, Piura-Peru". Research report. Lima: WOTRO.

Gobierno Regional de Loreto, Municipalidad Provincial Alto Amazonas e Instituto de Investigaciones de la Amazonía Peruana (2015) *Estudio de zonificación ecológica económica de la provincia de Alto Amazonas, departamento Loreto*. www.dar.org.pe/archivos/docs/Libro_ZEE.pdf (accessed: 10 August 2019).

Hall, D. (2013) Primitive accumulation, accumulation by dispossession and the global land grab. *Third World Quarterly* 34(9): 1582–1604.

Hall, R., Edelman, M., Borras Jr, S. M., Scoones, I., White, B. and Wolford, W. (2015) Resistance, acquiescence or incorporation? An introduction to land grabbing and political reactions "from below". *The Journal of Peasant Studies* 42(3–4): 467–488.

Harvey, D. (2005) *A brief history of neoliberalism*. Oxford: Oxford University Press.

Herman, M-O. and Mayrhofer, J. (2016) *Burning land, burning the climate. The biofuel industry's capture of EU bioenergy policy*. Oxford: OXFAM.

Heynen, N. and Robbins, P. (2005) The neoliberalization of nature: Governance, privatization, enclosure and valuation, capitalism nature. *Socialism* 16(1): 5–8. DOI: 10.1080/1045575052000335339

Hiraoka, M. (1985a) Floodplain farming in the Peruvian Amazon. *Geographical Review of Japan* Ser. b, 58(1): 1–23.

Hiraoka, M. (1985b) Changing floodplain livelihood patterns in the Peruvian Amazon, Tsukuba Studies. *Human Geography* 9: 243–275.

Hiraoka, M. (1985c) Mestizo subsistence in Riparian Amazonia. *National Geographic Research* 1(2): 236–246.

Hiraoka, M. (1986a) Zonation of mestizo riverine farming systems in Northeastern Peru. *National Geographic Research* 2(3): 354–71.

Hiraoka, M. (1986b) Cash cropping, wage labor, and urbanward migrations: changing floodplain subsistence in the Peruvian Amazon. Parker, E. (ed.) *The amazon caboclo: Historical and contemporary perspectives*. Studies in Third World Societies, N° 32.

Williamsburg Virginia: Department of Anthropology, College of William and Mary, 199–242.

HLPE (2011) *Tenencia de la tierra e inversiones internacionales en agricultura. Un informe del Grupo de Expertos de Alto Nivel sobre Seguridad Alimentaria y Nutrición*. Rome: UN Food and Agriculture Organization.

Hollander, G. (2008) "*Toward a Political Ecology of the Emerging Global Ethanol Assemblage*". Paper presented to the Berkeley Workshop on Environmental Politics, University of California, Berkeley. April 4, 2008

Hollander, G. (2010) Power is sweet: Sugarcane in the global ethanol assemblage. *The Journal of Peasant Studies* 37(4): 699–721.

Huamán, A. (2014) Tenencia y valor de la tierra en la comunidad campesina de Colán: Nuevas formas de apropiación, dimensiones del valor y tensiones comunales a partir de la entrada de la agroindustria de etanol. *Anthropía* 12: 69–80. revistas.pucp. edu.pe/index.php/anthropia/article/view/11283 (accessed: 07 February 2020).

Huamán, A. (2019) El paradigma modernizador de la agroindustria de caña para etanol: la agricultura por contrato y la ilusión del progreso. *Debate Agrario* 49: 129–165.

IFC (2007a) *Maple: Summary of proposed investment*, Washington, DC: International Finance Corporation (IFC). disclosures.ifc.org/#/projectDetail/ESRS/26110 (accessed: 12 September 2019).

IFC (2007b) *Maple: Environmental documents*. Washington, DC: International Finance Corporation (IFC). disclosures.ifc.org/#/projectDetail/SPI/26110 (accessed: 12 September 2019).

Kay, S. and Franco, J. (2012) *The global water grab: A primer*. Amsterdam: Transnational Institute.

Kometer, R. and Pautrat, L. (2014) *Valoración de los bienes y servicios ambientales perdidos por la deforestación en áreas de Tamshiyacu y Nueva Requena*. Lima: Sociedad Peruana de Ecodesarrollo. www.biofuelobservatory.org/Documentos/Informes-de-la-SPDE/ Valorizacion-de-danos-por-deforestacion-Loreto-y-Ucayali-2014.pdf (accessed: 20 December 2019).

La República (2008) *Alistan proyecto de etanol*. 5 June. larepublica.pe/economia/221042-alistan-proyecto-de-etanol/ (accessed: 11 June 2019).

Leff, E. (2015) Political ecology: A Latin American perspective. *Desenvolvimento e meio ambiente* 35: 29–64. DOI: 10.5380/dma.v35i0.44381

Li, T. M. (2006) "*Neo-Liberal Strategies of Government through Community: The Social Development Program of the World Bank in Indonesia*". International Law and Justice Working Papers. New York: Institute for International Law and Justice, New York University School of Law.

Li, T. M. (2014) What is land? Assembling a resource for global investment. *Transactions of the Institute of British Geographers* 39(4): 589–602.

Manirabona, A. and Vega Cardenas, Y. (2019) *Extractive industries and human rights in an era of global justice: New ways of resolving and preventing conflicts*. Canada: LexisNexis.

McFarland, B. J. (2018). Government domestic budgetary expenditures. In McFarland, B. J. (ed.) *Conservation of tropical rainforests. A review of financial and strategic solutions*. Windham, NH: Palgrave Macmillan, 241–293.

Mehta, L., van Veldwisch, G. and Franco, J. (2012) Water grabbing? Focus on the (re) appropriation of finite water resources. *Water Alternatives* 5(2):193–207.

Nader, L. (1972) Up the anthropologists. Perspectives gained from studying up. Hymes, D. (ed.) *Reinventing anthropology*. New York: Pantheon Books, 285–311.

ONERN (1985) *Mapa de Clasificación de Tierras del Perú*. Lima: Oficina Nacional de Evaluación de Recursos Naturales (ONERN) and Ministerio de Agricultura y Pesquería.

OXFAM (2014) *Geografías de conflicto. Superposiciones de mapas de usos de la tierra para industrias extractivas y agricultura, en Ghana y en el Perú*. Massachusetts, Washington: Oxfam.

Padoch, C., Chota, J., De Jong, W. and Unruh, J. (1988) Market-oriented agroforestry at Tamshiyacu. In Denevan, W. and Padoch, C. (eds.) *Swidden-fallow agroforestry in the Peruvian Amazon*. New York: The New York Botanical Garden, 90–96.

Perrone, D. and Hornberger, G. M. (2014) Water, food, and energy security: Scrambling for resources or solutions? *WIREs Water* 2014(1): 49–68. DOI: 10.1002/wat2.1004

Pitman, N., Vriesendorp, C., Rivera, L., Wachter, T., Alvira, A., del Campo, A., Gagliardi-Urrutia, G., Rivera, D., Trevejo, L., Rivera, D. and Heilpern, S. (2015) *Perú: Tapiche-Blanco. Rapid biological and social inventories*. Report N° 27.

Maple Etanol Peru S .R. L. (2014) *Proinversión: "Inversión Privada para el Desarrollo" Maple Etanol S.R.L.* www.proinversion.gob.pe/RepositorioAPS/0/0/EVE/FORO_INVERSIONPRIVADA/14_GFerreyros.pdf (accessed: 14 June 2019).

Quintero, M. and Díaz, K. (2004) El mercado mundial de cacao. *Agroalimentaria* 18: 47–59. erevistas.saber.ula.ve/index.php/agroalimentaria/article/view/1312 (accessed: 16 July 2019).

Revesz, B. and Oliden, J. (2011) Piura: Transformación del territorio regional. *Ecuador Debate* 84: 151–176.

Riles, A. (2011) *Collateral knowledge. Legal reasoning in the global financial markets*. Chicago: The University of Chicago Press.

Rodríguez, F. (1995) *El recurso del suelo en la Amazonía peruana, diagnóstico para su investigación*. Segunda Aproximación. Documento Técnico 14. Iquitos: Instituto de Investigaciones de la Amazonía Peruana.

Rodríguez, F., Bendayán, L. and Rojas, C. (1994) *Estudios de inventario y evaluación de suelos en la región de Loreto*. Documento Técnico 14. Iquitos: Instituto de Investigaciones de la Amazonía Peruana.

Romero, C. (2016) *Estudio del Cacao en el Perú y el Mundo. Situación Actual y Perspectivas en el Mercado Nacional e Internacional al 2015*. Lima: Ministerio de Agricultura y Riego.

Salazar, M. and Rivadeneyra, D. (2016) *Amazonía arrasada. El grupo Melka y la deforestación por palma aceitera y cacao en el Perú*. Lima: Convoca y Oxfam. www.convoca.pe/especiales/AMAZONIA/ (accessed: 17 June 2019).

Sawyers, S. and Gomez, E. T. (eds.) (2012) *The politics of resource extraction. Indigenous peoples, multinational corporations and the state*. London: Palgrave, Macmillan, UNRISD.

Schmink, M. (1982) Land conflicts in Amazonia. *American Ethnologist* 9(2): 341–357.

Schmink, M. and Wood, C. (1987) The political ecology of Amazonia. Little, P. Horowitz, M. and Nyerges, A. (eds.) *Land at risk in the third world: Local level perspectives*. Boulder: Westview.

Segura, F. (2019) *El Estado y la agroindustria en la Amazonía: análisis del caso Tamshiyacu, Loreto*. Tesis para optar por el grado de Magíster en Desarrollo Ambiental. (Master thesis on Environmental Development). Lima: Pontificia Universidad Católica del Perú.

Sociedad Peruana de Ecodesarrollo (2012) Adjudicación de terrenos para palma aceitera amenazan bosques primarios en la Amazonía peruana. Lima: Sociedad Peruana de Ecodesarrollo. www.biofuelobservatory.org/Documentos/Cartas/Informes/Adjudicacion-de-terrenos-para-palma-aceitera.pdf (accessed: 13 August 2019).

Sociedad Peruana de Ecodesarrollo (2013) Carta a la Dirección General Forestal y de Fauna Silvestre. Carta N° 122-2013/SPDE. Lima: Sociedad Peruana de Ecodesarrollo.

Taussig, M. (1987) *Shamanism, colonialism, and the wild man: A study in terror and healing.* Chicago: University of Chicago Press.

Thoumi, G. (2017) Engage the Chain Case Study #3: United Cacao. Aggressive Expansion Leads to Regulatory Violations and Insolvency. *Valuewalk.* www. valuewalk.com/2017/11/united-cacao-insolvency/ (accessed: 20 June 2019).

El Tiempo (newspaper) (2006) *Desalojan a Agricultores para proyecto de Etanol* 14/12/06.

Tsing, A. (2000) Inside the economy of appearances. *Public Culture* 12(1): 115–144. DOI: 10.1215/08992363-12-1-115

Tsing, A. (2002) Land as Law: Negotiating the meaning of property in Indonesia. In Richards, F. (ed.) *Land, property and the environment.* Oakland, CA: Institute for Contemporary Studies. 94–137.

United Cacao Limited SEZC (2014) *Admission to AIM by Strand Hanson Limited, VSA Capital Limited and Kallpa Securities Sociedad Agente de Bolsa S.A.* London: United Cacao Limited SEZC.

Urteaga, P., Segura, F. and Sanchez, M. (2019) *Derecho Humano al Agua, Pueblos Indígenas y Petróleo.* Lima: CICAJ, Departamento Académico de Derecho, Pontificia Universidad Católica del Perú.

Urteaga-Crovetto, P. (2016) Between water abundance and scarcity: Discourses, biofuels, and power in Piura, Peru. *Antipode* 48: 1059–1079. DOI: 10.1111/anti.12234.

Urteaga, P. (2017) Biocombustibles y agua. La transformación del espacio en Piura, Perú. *Anthropologica.* 35(38): 7–39. DOI: 10.18800/anthropologica.201701.001

van der Ploeg, J. D. (2006) *El futuro robado. Tierra, agua y lucha campesina.* Lima: IEP, Walir.

Visser, O. (2017) Running out of farmland? Investment discourses, unstable land values and the sluggishness of asset making. *Agricultural Human Values* 34: 185–198. DOI: 10.1007/s10460-015-9679-7

Vriesendorp, C., Pitman, N., Alvira, D. Stallard, R., Crouch, T. and O'Shea, B. (2015) Resumen de los sitios del inventario biológico y social. In Pitman, N., Vriesendorp, C., Rivera Chávez, L., Wachter, T., Alvira Reyes, D., del Campo, A., Gagliardi-Urrutia, G., Rivera González, D., Trevejo, L., Rivera González, D. and Heilpern, S. (eds.) *Perú: Tapiche-Blanco. Rapid biological and social inventories*, Report 27. Chicago: The Field Museum.

White, B., Borras, S. M. Jr., Hall, R., Scoones, I. and Wendy, W. (2012) The new enclosures: Critical perspectives on corporate land deals. *The Journal of Peasant Studies* 39(3–4): 619–647.

Wolford, W., Borras, S. M. Jr., Hall, R., Scoones, I. and White, B. (2013) Governing global land deals: The role of the state in the rush for land. *Development and Change* 44(2): 189–201.

Woodhouse, P. (2012) New investment, old challenges: Land deals and the water constraint in African agriculture. *The Journal of Peasant Studies* 39(3/4): 777–794.

World Bank (2014) *Enabling the Business of Agriculture, 2015 Progress Report.* Washington: The World Bank Group.

Zegarra, E., Oré, M. T. and Glave, M. (2007) El Proyecto Olmos en un territorio árido de la costa peruana. En: BENGOA, J. (ed.) *Movimientos sociales y desarrollo territorial rural en América Latina* . Santiago: RIMISP.

6 Cashews in conflict

The political ecology of cashew pomiculture in Guinea-Bissau

Brandon D. Lundy

Introduction

Global demand for nuts (and their byproducts) continues to push production in far-flung places. By 2024, the forecasted value of the nuts and seeds market worldwide is set to exceed USD 1.3 trillion. Cashews (*Anacardium occidentale* L.) form a billion-dollar global industry on their own as a specialty-nut, fruit juice, and alcohol (from the cashew apple), and source for natural phenol for chemical production (from the cashew nut shell liquid) (see Box 6.1). In 2016, cashew nuts (with shell) had a global gross production value of USD 2.5 billion (constant 2004–2006) (FAOSTAT 2019); cashew apples had a gross production value of USD 326 million (constant 2004–2006) (FAOSTAT 2019); and cashew nut shell liquid (CNSL) had a 2018 market size of USD 250.2 million (GlobeNewswire 2019).

The cashews value chains are truly globalized. For example, the packaged nuts sold anywhere in the US (the largest consuming country of cashews in the world)[1] originate from multiple producing regions such as India, Vietnam, Brazil, and many countries throughout Africa. In fact, most of the global cashew nuts supply comes from Africa (e.g., Cote d'Ivoire, Tanzania, Guinea-Bissau, Benin, Mozambique) and Asia (e.g., Vietnam, India, Philippines, Indonesia) (Mordor Intelligence 2021). And yet, almost all cashew processing, and hence, value addition, takes place in Asia, with India and Vietnam being the largest importers of raw cashew nuts (with shell), which are then processed and resold on the international market (FAOSTAT 2019), with Vietnam being the biggest processor and exporter of cashew nuts in the world for the 13th consecutive year in 2018 (Nguyen 2018). This means that there are long value chains with raw cashew nuts bought up in ports around the world, shipped to India and Vietnam where they are processed and packaged, and then re-shipped to the US and Europe (Mordor Intelligence 2021; Tessmann 2018).

While cashews spread rapidly around the globe, their demand as an international commodity was slow to materialize. It was only in 1905 that the US made its first imports, with commercial shipments from India reportedly beginning in 1923 and the first load of 45 metric tons of cashew kernels arriving infested with weevils (Ohler 1979: 16). It was not until the 1940s and World

Box 6.1 Cashews' components and end uses

Cashews are native to the Americas, specifically northern Brazil. They were first recorded by a French naturalist monk in 1558 and spread throughout the tropics during the European Age of Exploration. Due to its extensive root system (including a deep taproot), the cashew tree can tolerate a wide range of moisture levels and soil types. There are many varieties of cashew trees, but only a few have been selected for production, and are thus named. However, there is a high level of variability in harvests across cashew-producing areas, including in Africa, due to indiscriminate farming practices that often utilize inferior germplasm. Due to the strenuous production methods and the handling of some products (see below) mostly by poor farmers, some media have labeled cashews as "blood cashews" (Wilson 2015, see also Schwartz et al. 1945).

The cashew fruit consists of the cashew seed and the cashew apple, which has a light reddish to yellow color, and comprises a custard-like pulp filled interior with sweet and astringent juice. The cashew kernel found inside the seed is roasted for consumption, while the outer cashew nut shell liquid (CNSL) is either burned off or collected for industrial processing. Raw cashew nuts and processed cashew kernels are the main commercial products of the cashew tree. However, the yields of the cashew apple are estimated to be 8–10 times higher than the weight of the raw nuts, which is suggestive of its future commercial potential (Lundy 2009; Strom 2014; Drift 2002). The hardwood is used as timber, firewood, and to make charcoal (Orwa et al. 2009). Essentially, the cashew tree is an industrial crop with many valued products, as explained below.

Cashew nuts are the fifth most expensive nut in the world once processed, with their global average wholesale price reaching USD 1.42 per kg in 2019 (FAOSTAT 2019). They are found around the world as a snack or ground additive into sauces, curries, and sweets. Cashew nuts can be also used to manufacture cashew milk (a dairy milk alternative) and cashew oil for cooking and salad dressing. Cashews are increasingly recognized as a "superfood" rich with beneficial oils, plant-based protein, iron, calcium, phosphorus, magnesium, and vitamin C content, which is about five times more than an orange (Lundy 2009; Morton 1987). The cashew apple is also used in sweets, juiced, mixed into alcoholic beverages, naturally fermented as "wine", and distilled into alcohol.

The cashew kernel is surrounded by a double shell that contains the CNSL, whose composition varies depending on the processing. Cardanol, a phenolic lipid obtained from anacardic acid (the main component of CNSL), is used in resins, coatings, frictional materials, and surfactants used as pigment dispersants for water-based inks and epoxy coatings (Tullo 2008). Anacardic acid is an allergenic and skin irritant that is used in different parts of the world to treat skin disorders such as corns, eczema, psoriasis, warts, and leprosy (Lim 2012).

Most recently, both the cashew nut shell and CNSL have been identified as *"agro-wastes rich in valuable and functional renewable products ... [including as] sources for use in the production of biorenewable chemicals, materials and energy"* (Mgaya et al. 2019; see also Lundy 2009: 235).

War II that industrialization began to encourage cashew production once US scientists working in Brazil discovered that CNSL is a natural source of phenol, an alcohol used to synthesize a petroleum substitute for ammunition, rubber, and a wide variety of other wartime necessities (Maia et al. 1971; see also Meaney-Leckie 1991: 318) (Box 6.1). Between 1975 and 1990, global production doubled from 500,000 metric tons to more than 1 million metric tons of raw cashew nuts. By 2017, more than 4 million metric tons of cashew nuts were produced globally, with more than 36 countries involved in cashew production (FAOSTAT 2019).

The drivers of this cashew expansion are multifaceted and complex, and often tied directly to each country's political ecology. These factors can be as diverse as the geography, climate, history, livelihoods, cultural identity, development initiatives, government policy, viable agricultural alternatives, increased global demand, and more (Boafo 2019; Monteiro et al. 2017; Tola and Mazengia 2019).

Guinea-Bissau can offer a telling example of how these factors intertwine to influence cashew production. The country has become a top ten global producer of raw cashew nuts for more than a decade, relying on two major varieties, (a) the red *caju-di-terra*, known for its small nuts and sweet juice especially prized as fermented cashew wine (*vinhu-di-caju*), and (b) the *caju-di-Maçambique* that has larger nuts, is yellowish in color, and has more fibrous and less sweet fruit (Catarino et al. 2015: 463).

The governmental and international development rhetoric in Guinea-Bissau buoys the economic theory of *comparative advantage*, that under free trade, an agent will produce more and consume less of a good if they can produce that good at a lower relative opportunity cost (i.e., lower relative marginal cost prior to trade) (Ricardo 1817)[2] (Chapter 2). This development policy approach can lead to over-specialization and national-level monocultures, often improving production efficiency, while simultaneously increasing producers' risks such as exposure to diseases and pests or global market fluctuations. For example, the pressures exerted by a monocultural model of production pose explicit ecological concerns, as they can directly reduce overall biodiversity and ecosystem resilience (Vasconcelos et al. 2015) (Chapter 3). Between 1992 and 2015, Guinea-Bissau increased the extent of woody crop production by 38% in terms of land cover, while it reduced the production of herbaceous crops by 9% and the extent of tree-covered areas by 2.7% (FAOSTAT 2019). This suggests that many smallholders across the country are not only converting rice fields and other crop land to cashew orchards, but they are also converting forest land. It is worth noting that even though the cashew tree is adaptable to most of the soil and climatic conditions of Guinea-Bissau (Catarino et al. 2015), it is susceptible to climatic risks (e.g., droughts and floods due to climate change), over 60 known insect pests (Araújo 2013; Orwa et al. 2009), and diseases such as anthracnose and gummosis (Araújo 2013; Freire et al. 2002), which apparently spread in Guinea-Bissau. Unfortunately, localized research in this area remains sparse, making cashew cultivation both ecologically and economically vulnerable.

To illustrate, in 2006 Guinea-Bissau increased the export tax by approximately USD 0.20 per kg, partially as a political move to garner votes in the 2006 national elections. However, this frustrated international buyers, resulting in huge decreases in cashew exports, thousands of metric tons of raw cashew nuts rotting in Bissau warehouses, and the stepping down of Prime Minister Gomes. This had major knock-on effects for local livelihoods, with farmers trying to barter as many as five kilograms of raw cashew nuts for a single kilogram of rice (Lundy 2012). This and the preceding suggest both the complexity and the latent fragility of cashew production in Guinea-Bissau for both the political economy and political ecology.

This chapter employs a comprehensive political ecology framework in order to better understand the historical context, environmental factors, political milieu, and livelihood and development implications of cashew production in Guinea-Bissau. The "Methodology" section outlines the methodology used in this chapter and the "Results and discussion" section discusses the political ecology of cashews in Guinea-Bissau, focusing on observed patterns at the national (see " Political Ecology of smallholder-based cashew pomiculture in Guinea-Bissau") and the local levels (see " Perceptions of cashews"), using an extended historical exposition and primary data analysis, respectively. This focus demonstrates how cashew production in Guinea-Bissau is an agent of ecological, agrarian, and socioeconomic transformation shaping the country's livelihoods, environments, and global engagements (see "Cashews as an agent of ecological transformation in Guinea-Bissau").

Methodology

Research approach

Political ecology is defined as the academic field that *"attempts to explain the interactions between biophysical and social processes and how such interactions affect people's livelihoods … at varying geographic scales over time"* (Oppong and Kalipeni 2005: 5). Political ecology typically considers historical depth, context, and scale through social (Moseley 2017; Paulson and Gezon 2005) and structural (Atkinson 1991) lenses that factor conflict, value, hybridity, and justice (Blaikie 2012; Escobar 2006; Martinez-Alier 2003, 2014).

Therefore, the political-ecological approach to cashews in Guinea-Bissau used in this chapter considers the distribution of their positive and negative effects, the structuring of access to resources, and the right to be included in policy decisions. Thus, this chapter employs an integrated political ecology theory designed to capture the dialectic interactions of various systems (e.g., environmental, political, economic, cultural) and their effects on people's livelihoods.

Study site

Over the last several decades the cashew boom in Guinea-Bissau, West Africa has induced "*social and environmental*" effects "*yet to be analyzed and understood*" (Catarino et al. 2015: 459), as traditional slash-and-burn subsistence agriculture has been replaced by this cash crop. Guinea-Bissau is one of the world's largest producers of raw cashew nuts (Figure 6.1), with the sector being tremendously important for the national economy and rural development. However, the effects of the cashew boom have yet to be adequately assessed from a political ecology perspective despite the fact that the: "*Strong dependence on a single cash crop also renders the country vulnerable to market fluctuations, entailing risks to local producers and the national economy. … which may impact the living standards and food security of Bissau-Guineans*" (Catarino et al. 2015: 459).

Guinea-Bissau has a primarily rural, agricultural population. It is characterized by a tropical climate with wet and dry seasons and contains three main vegetation zones: (a) coastal mangroves, palm groves, and forest; (b) transitional grass and woodland savanna; and (c) interior woodlands and shrubby and herbaceous steppes. The Institute for Biodiversity and Protected Areas (IBAP) has invested considerable efforts to protect important natural and cultural heritage through a system of protected areas that now cover almost 26% of the national territory. These protection efforts have been necessary according to IBAP (2016: 53) because:

> *Every year, parts of the forest are burnt to free land for agriculture. Crops are then planted – rice, millet, sorghum or maze [sic.] – and take advantage of the plentiful rain to grow. The soil offers only meagre harvests and cereals are rarely sown more than 2 years in a row on the same patch. The land is then left fallow, to be used again 5 to 7 years later, or planted with fruit trees (cashew, banana or citrus). Slash-and-burn agriculture is one of the principal causes of deforestation. The wildlife, deprived of its natural habitat, invades the orchards, coming into conflict with the farmers.*

At the same time, human–environmental relationships are shaping socio-economic futures. According to the Climate Change Adaptation program of the United Nations Development Programme (UNDP 2019), Guinea-Bissau is already feeling the impacts of diminishing rainfall and the gradual rise in temperatures, with observed declines in the capacity of some aquifers and more

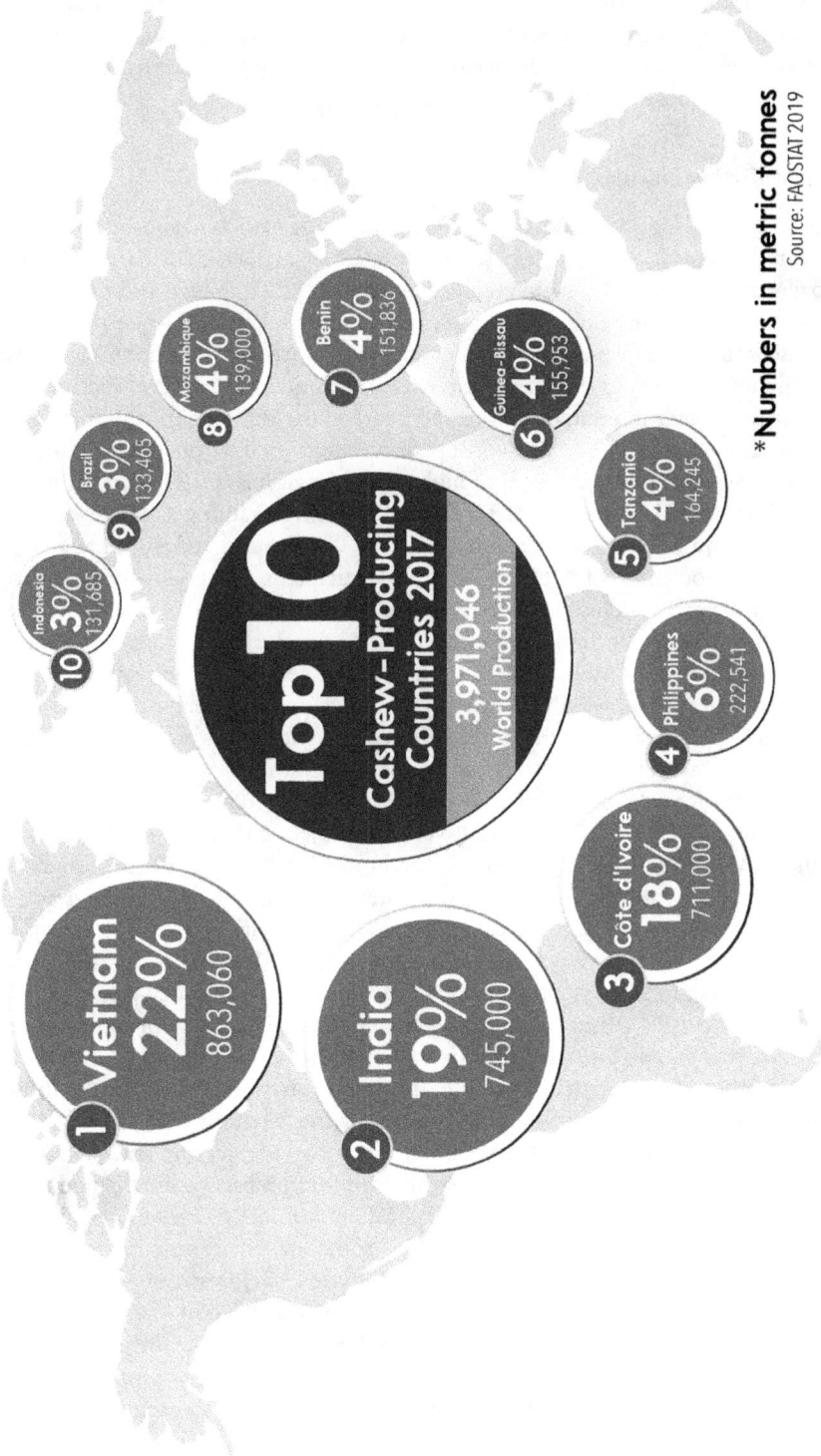

Top 10
Cashew–Producing
Countries 2017

3,971,046
World Production

*Numbers in metric tonnes
Source: FAOSTAT 2019

1 Vietnam 22% 863,060

2 India 19% 745,000

3 Côte d'Ivoire 18% 711,000

4 Philippines 6% 222,541

5 Tanzania 4% 164,245

6 Guinea-Bissau 4% 155,953

7 Benin 4% 151,836

8 Mozambique 4% 139,000

9 Brazil 3% 133,465

10 Indonesia 3% 131,685

Figure 6.1 The ten leading producers of cashew nuts in 2017.
Source: FAOSTAT 2019.

frequent inundation by saltwater. All the above indicate the multiple complex factors shaping cashew pomiculture in Guinea-Bissau, strongly suggesting that it is an agent of ecological, agrarian, and socioeconomic transformation that needs to be understood through a comprehensive lens.

Data collection and analysis

The primary data presented in the "Perceptions of cashews" sub-section was collected during a research trip to Guinea-Bissau in July 2018 to investigate broader perceptions of climate change. We surveyed 272 college and university students and alumni using a 36-question, 5-point Likert-scale questionnaire. We also collected 287 pieces of student and alumni artwork for content analysis in which we prompted university students and alumni to draw two pictures of their natural environment, one from the past and the other about the future.

We also organized 12 focus group discussions (FGDs) on environmental needs assessments with hundreds of high school English language teachers from throughout Guinea-Bissau at the Tchico Te Teacher Training College during the "2nd Annual Conference and National Convention, English Language Teachers' Association of Guinea-Bissau". Participants were self-divided into four rooms, each containing three diversified groups. The moderators provided definitions of key environmental concepts (i.e., sustainability, biodiversity, land management, energy use) aloud in English and asked the participants to draw what it meant for Guinea-Bissau. Each group then presented their drawings and continued to discuss about the different environmental issues affecting Guinea-Bissau more broadly with prompts from the moderators.

Furthermore, we conducted six expert semi-structured interviews with (a) an agronomist, (b) biology teacher and wetlands project coordinator, (c) Director of the IBAP, (d) the head student of the Environmental Sciences degree program at the University of Lusófona, (e) a group interview with the first class of graduating students from the University of Lusófona Environmental Sciences major, and (f) a group interview with the members of Voz di Paz (Voice of Peace), an NGO working to identify priorities for peace in Guinea-Bissau. We conducted these interviews in a mix of English, Portuguese, and Kriol, with interpreters provided by Tchico Te Teacher Training College when necessary.

The survey data is analyzed through descriptive statistics, with only the most relevant responses to cashew pomiculture and the environment presented in this chapter. Additionally, each respondent drew two images, the natural environment in the past and the future. This allowed for comparisons of change over time and analysis of the perceptions of local environments. Eleven descriptive codes (i.e., natural environment, built environment, environmental management, environmental harmony, scarcity, pollution, deforestation/desertification, commercialization, biodiversity, tradition, people) were marked present or absent for each drawing and were coded individually by two different coders (Cohen's kappa of 0.8 or higher, corresponding to a strong or almost perfect level of agreement). Then, specifying attributes were added to each respondent including age, sex, profession, place of birth, and religion. This data was then analyzed based on frequencies and McNemar's test used on paired nominal data with matched pairs, when comparing drawings about the past and future, to test the null hypothesis of marginal homogeneity (i.e., that two marginal probabilities for each outcome are the same). Rejecting the null means that the marginal proportions are significantly different from each other. We also tested for correlation between select coded variables in the drawings such as "environmental change" represented by loss of biodiversity, scarcity, pollution, and deforestation.

The FGDs and expert interviews are analyzed through a qualitative approach based on a Key-Word-in-Context (KWIC) analysis using the keyword "cashew" and then identifying thematic patterns within the broader contexts. The resulting themes are used as the subheadings in "Perceptions of cashews" to organize the findings.

Apart from the primary data outlined above, this chapter draws from insights generated through a total of 13 months of ethnographic research on livelihoods in Guinea-Bissau over the past 13 years. These include many hours of participant observation, FGDs, unstructured and semi-structured interviews, and surveys. These insights include firsthand experiences observing and participating in the many facets of cashew cultivation, harvesting, processing, and shipping, and further inform the contextual background and analysis in this chapter.

Results and discussion

Political ecology of smallholder-based cashew pomiculture in Guinea-Bissau

The political ecology of cashews in Guinea-Bissau encompasses small-scale smallholder and larger *ponteiro* (plantation) cashew pomiculture (Funk 1988). As a low-lying, coastal, and tropical environment, Guinea-Bissau is an ideal context to produce cashew trees. It has been argued that cashews are an ideal smallholder crop because they are (a) drought-resistant, (b) easily processed locally, (c) require low capital investment, (d) employ large portions of both the rural and urban labor force, and (e) have available and lucrative markets for the end-products (Meaney-Leckie 1991; Sahn and Sarris 1994). In Guinea-Bissau, cashews offer an alternative to the labor-intensive paddy-rice production, especially for women and children (Lundy 2009, 2012).

The above makes cashew pomiculture both a viable livelihood strategy for a large fraction of the population and at the same time an activity that increases the vulnerability of local producers due to ecosystem degradation, intensification of land-related conflicts, economic exploitation, and the erosion of rural safety nets. More than 90,000 families are involved in cashew production throughout the country (usually as an alternative to subsistence rice production), with the majority of cashew producers being smallholders with average landholdings of <3 ha each (Barry et al. 2007; Catarino et al. 2015).

As an agrarian economy, 58% of the total land area in Guinea-Bissau is dedicated to agriculture (1,630,000 ha). However, only 34% of this land was actively under production in 2017 (554,200 ha), with 50% (277,465 ha) of this agricultural land dedicated to cashews (Figure 6.2). Figure 6.2 shows that Guinea-Bissau is effectively moving toward becoming a cashew monoculture, with the global cashew economy largely determining national livelihoods, development, and wellbeing. Agriculture makes up 69% of the GDP, 90% of which comes from raw cashew nut exports, which directly affects 85% of the population (WFP 2019; see also Lundy 2012). Thus, cashew is clearly central to the national economy and rural development. But how did Guinea-Bissau become the world's sixth largest exporter of raw cashew nuts?

Cashew cultivation started and expanded during the Portuguese colonial rule throughout the 1950s, which recognized the *"potential value of the cashew tree, its hardiness and the possibility for use in intercropping or as a kind of cover for long fallow periods in order to recover soil fertility"* (Catarino et al. 2015: 461). This triggered investment in the "Cashew Development Plan in Portuguese Guinea" under the auspices of the Overseas Agronomic Research Mission (MEAU), setting up a six-year research project in the 1960s (Horta and Sardinha 1966), which eventually turned into a financial incentive known as the *taungya* system that rewarded farmers for each tree planted (Catarino et al. 2015; Horta and Sardinha 1966). These financial incentives shifted after independence to a global structural model designed to push the production of raw materials in the developing world (see Chapter 2).

However, after the 1974 independence, cashew production slowed down until a land grab in the 1980s granted property rights based on planting permanent crops. According to Catarino et al. (2015: 462), *"Cashew, due to its hardiness and quick growth, was an obvious choice."* Also, around this time, the government began to informally barter for cashew nuts at a rate of 1 kg of cashew nuts for 2 kg of rice.[3]

On 25 April 1989, the World Bank published Phase Two of its Structural Adjustment Program (SAP) for Guinea-Bissau, just two years after the first phase and six years after the government began an economic recovery program. In this 1989 report, cashews were already identified as the main export value, accounting for 70% of the total exports in 1987–1988. In order to encourage private sector growth, *"the government's strategy is to encourage diversified growth of the private sector along the lines of Guinea-Bissau's **comparative advantage**"* (World Bank 1989: 8, emphasis by the author) (see Chapter 2). When it came to the agricultural sectoral policy, the report argued that *"appropriate pricing and marketing policies"* make *"imported inputs and incentive goods in rural areas"* available to improve upon the steady growth from 1984 through 1988, and that *"in the next few years, the incentive effects of the adjustment program are expected to continue to be the main engine of agricultural growth"* (World Bank 1989: 9).

It seems counterintuitive that the government was pushing cashew pomiculture considering that the 1989 SAP report indicated the *"bleak prospects for the international price of cashews, which is expected to deteriorate again in 1989, further*

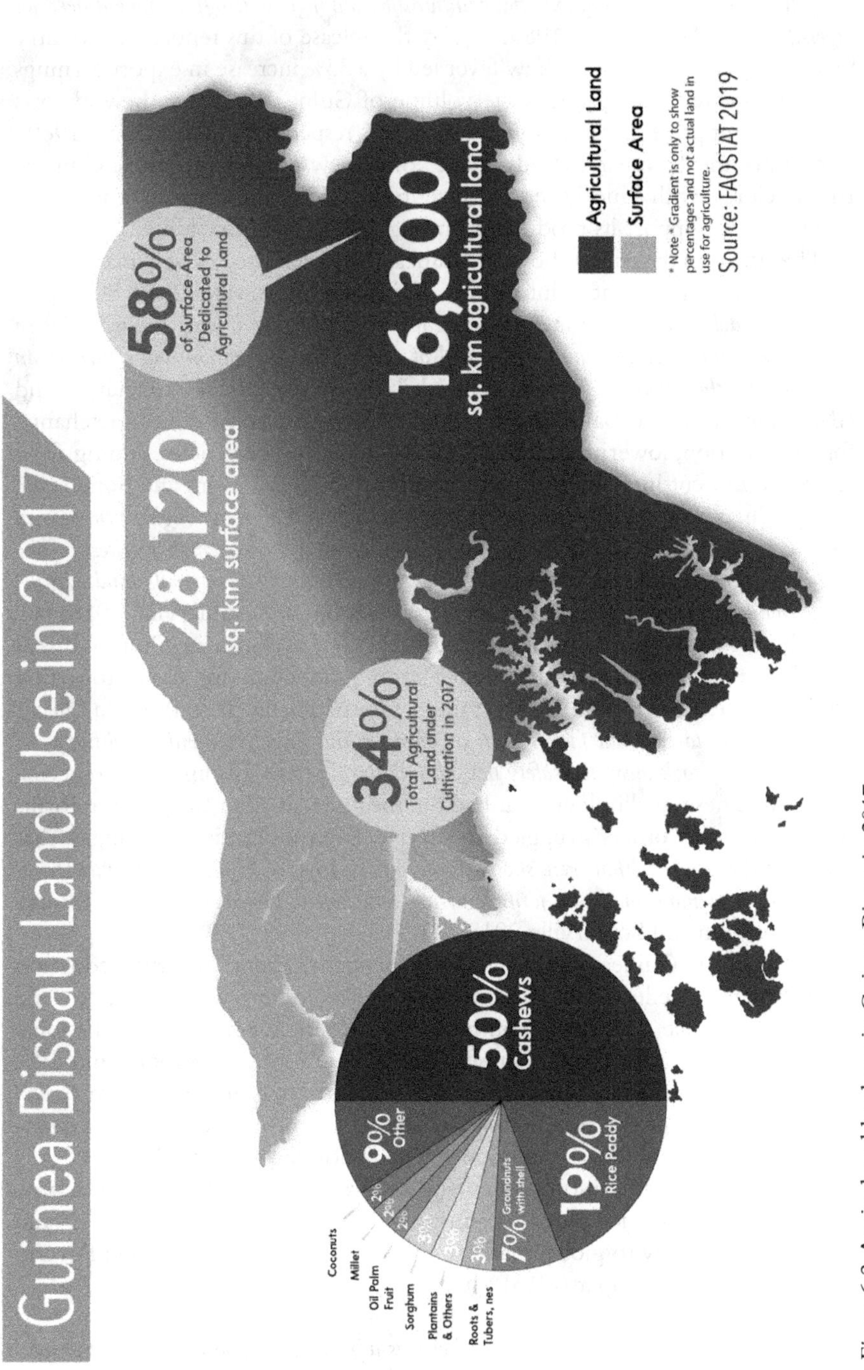

Figure 6.2 Agricultural land use in Guinea–Bissau in 2017.
Source: FAOSTAT 2019.

delay[ing] the achievement of external equilibrium, and highlight[ing] the urgent need for diversification" (World Bank 1989: 13). By the release of this report, the country was already in the grip of cashew fever led by a 33% increase in export earnings over the previous two years, as the volume of Guinea-Bissau's cashew exports and their unit price increased by 70% and 20%, respectively (World Bank 1989). For a time, the "comparative advantage" of cashew pomiculture from Guinea-Bissau felt real with a motivated producer base and the necessary environmental conditions to support its production throughout the countryside.

The situation described above in Guinea-Bissau reflected closely many other African countries at that point in time. Keim and Somerville (2018: 80) argue, *"deeply in debt and unable to repay, Africans were told that if they wanted Western help, they would have to abandon the goal of quick industrialization and return to an emphasis on the production of raw materials."* The International Monetary Fund (IMF) and the World Bank implemented SAPs by offering credit in exchange for privatization, lowering taxes and tariffs, devaluing currencies, ending food subsidies, and cutting education and healthcare budgets (Keim and Somerville 2018) (Chapter 2). Furthermore, *"farmers were allowed to charge higher prices in an effort to stimulate food and cash crop production,"* which led to *"higher prices for raw materials [that] encouraged the production of crops and supply of minerals [and agricultural products] that could be sold to Westerners"* (Keim and Somerville 2018: 80). In other words, farmers throughout Africa such as those in Guinea-Bissau abandoned labor-intensive subsistence crop production for cash crops that were more expedient and highly valued internationally, leaving them *"fragile and vulnerable to external factors such as weather and global economic conditions. … SAPs suddenly took away the safety nets of millions of people"* (Keim and Somerville 2018: 80). These smallholders that had the means to weather the policy reforms prospered, while others struggled to survive. Even more disheartening is that many of the *"SAPs [that were shown to be further triggers of underdevelopment] still exist, but in a more moderate form, and are now called Poverty Reduction Strategy Papers"*[4] (Keim and Somerville 2018: 81).

Returning to Guinea-Bissau, the IMF's "Second National Poverty Reduction Strategy Paper" reflected the country's overreliance on raw cashew nuts. The report stated that agriculture continued to account for most of the national GDP, with raw cashew nuts representing over 90% of total exports and 17% of government revenue, pointing to *"the vulnerability of the economy to cyclical fluctuations in the market price of cashews"* (IMF 2011: 16). And yet, the government and the international community have continued to tout this overreliance as a success, and furthermore, encourage cashew expansion through local processing that should add value in-country before exporting processed nuts (and byproducts) directly to global markets. Indeed, Guinea-Bissau's second Poverty Reduction Strategy Paper (PRSP) stated

> The development of this export market was a great success, particularly in the rural areas. Exports of cashew nuts, which stood at a mere 1,200 metric tons in 1970, experienced a quantitative leap to almost 100,000 metric tons in 2005, 135,500

mt in 2009, and 122,300 mt in 2010. However, only 4 percent of output is processed locally; the rest is exported unprocessed. Cashew crops occupy 47 percent of farmland and employ 80 percent of the rural farming population. **It is a strategic sector for job creation and poverty reduction.** *The possibility of increasing production is real, taking into account the new plantations and the competitive yields (500–600 kg/ha) compared with other producers, such as Brazil, India, or Vietnam. The factor influencing poverty and employment is the development of processing. The internal rate of return (IRR) on investment in cashew nut processing is high. The country's economy would benefit from locally adding value to cashew exports, which would provide major benefits in terms of growth, employment creation, and poverty reduction, compared with the current situation.*

(IMF 2011: 26, emphasis added)

In other words, cashew nuts remain a priority investment for the country, even though dependence on cashew pomiculture is seen to be creating vulnerabilities including the illegal trafficking of cashew nuts across borders and food insecurity (Pacheco de Carvalho and Mendes 2015). Furthermore, the government intends to support strong and sustained growth by "*ensuring food security and diversification of production bases in connection with the development of SMI/SME [small to medium industry/small to medium enterprise] that can serve as intermediate networks for exports of cashew nuts*" (IMF 2011: 71). The question remains: Should cashew nut production and processing further expand in Guinea-Bissau given its ability to transform ecosystems and impact overall community and household food and livelihood security?

A news story from the United Nations Integrated Peacebuilding Office in Guinea-Bissau (UNIOGBIS 2016) asked if cashew nuts were a "blessing or a curse." Quoting the Coordinator of the Rehabilitation of Private Sector and Support to Agro-industrial Development Project, funded by the World Bank, the news report stated that "*when the market is good in terms of the price paid to producers, all transformers cry, and when the price is bad for producers, everyone talks about local transformation*" (UNIOGBIS 2016: online). This contradiction between good and bad times, sets up a variable reward schedule, meaning that farmers are reluctant to change course in case the global raw cashew nut prices rebound, which they do occasionally, rewarding them for their patience with a windfall (e.g., AFP 2017). On the other hand, the government and private industry have been reluctant to invest in cashew processing.

Thus, the question of whether to promote cashew processing in Guinea-Bissau is a rather complex issue related to anticipating global demand that is sometimes shaped by unexpected circumstances (e.g., the 2006 nut allergy health scare which led to an overall decline) (IRIN 2006) and considering infrastructure development costs (Lundy 2012). For example, a 2004 report for the agricultural sector indicated that "*prices for kernels would have to be 4.4 times higher than for raw nuts if production costs were negligible*" (Thomas et al. 2004: 52). Furthermore, although processing adds great value to cashew pomiculture, it is labor-intensive, while buyers in Guinea-Bissau are only interested in raw nuts

(Lekberg 1996: 15). Additionally, due to colonial neglect, weak governance, and the damages from the armed conflict of the late 1990s, Guinea-Bissau "*does not have a railway system, and possesses only 10 percent of paved road for a network of 4,400 km. The road network in rural areas is generally unpaved and in bad condition*" (Gacitua-Mario et al. 2007: 73). These infrastructural challenges extend to the energy and water sectors as well, both of which are necessary for cashew processing. If the government earnestly pursued processing as a development strategy, according to a 2006 World Bank report, a 10% increase in export taxes to fund the endeavor would equate to a 14% loss for smallholders because "*the international sales price will not change; rather the buying price from the farmer will be reduced to maintain margins in the chain*" (World Bank 2006: 79). As argued elsewhere (Lundy 2012: 39), "*advances in processing must be pursued slowly since smallholders are unable to shoulder the cost of industrialization.*"

However, without advancing processing capabilities then, Guinea-Bissau's cashew development remains tied to increasing raw cashew nut yields. Although in 2017, Guinea-Bissau was the sixth largest global producer of cashews (155,953 metric tons of raw cashew nuts), the obtained yield was only 5.6 t/ha, ranking 24th in the world, well below the global average of 6.6 t/ha (FAOSTAT 2019). As suggested above, these below-average yields are due to a set of overlapping and complex political and ecological factors such as (a) aging trees (Catarino et al. 2015), (b) overlapping land tenure systems and ineffective conflict resolution mechanisms (Forrest 1992; Temudo et al. 2015), (c) pests and diseases (Azam-Ali and Judge 2001; Ohler 1979), (d) fire and theft (Lekberg 1996), (e) smuggling and high transaction costs, (f) indiscriminate and inferior selection of varieties and germplasm (Barry et al. 2007; cf., Sardinha et al. 1993), and (g) climate change (Africa Research Bulletin 2015; Barry et al. 2007; Catarino et al. 2015; Temudo and Bivar Abrantes 2014). In order to capture the localized effects of these overlapping and complex political and ecological factors, the sub-section "Perceptions of cashews" reports some primary ethnographic and survey data on the contemporary perceptions of cashews in Guinea-Bissau to further interrogate the appropriateness of the comparative advantage theory.

Perceptions of cashews

Linking perceptions to the political ecology of cashews

The KWIC analysis of the 12 FGDs and 6 expert semi-structured interview transcripts identified 27 direct mentions of the keyword "cashew," which, depending on the context used, were grouped into three categories: environment, livelihoods, and development based on the ways people talked about them. This follows the discussion of the cultural construction of "biodiversity," which acknowledges the "*powerful interface between nature and culture and originating a vast network of sites and actors through which concepts, policies, and ultimately cultures and ecologies are contested and negotiated*" (Escobar 1998: 75). Similarly, the perceptions of "cashews" in Guinea-Bissau are fashioned through the "interface between nature and culture" considering the plant as a biological entity, and its uses, environmental factors, economic potential and alternatives, and its real and ideal comparative advantage as a national monoculture.

We further divide the elicited perceptions into positive and negative aspects: i.e., environmental benefits or negative impacts (see "Environmental perspectives"), livelihood opportunities or barriers (see "Livelihoods perspectives"), and whether cashews are an economic development priority or misstep (see "Development perspectives").

Environmental perspectives

As already noted above, cashew production causes deforestation in Guinea-Bissau, as evidenced by the increase in woody crops (i.e., cashew and other fruit pomiculture) and a simultaneous decrease in tree-covered areas (i.e., forests) (FAOSTAT 2019). Some study participants identified that the expansion of cashew agriculture comes at the expense of forest cover, as cashew trees directly convert forest areas and divert food crop production. For example, one FGD respondent explained that "*Cashew trees [are] killing our forests. If you go outside in Guinea-Bissau, you can see only cashew trees, because of money.*"

These land use change effects were linked to biodiversity loss through the drawing analysis. In particular, when considering what respondents drew as representations of the natural environment in the past and the future, we found a statistically significant negative correlation between biodiversity and deforestation ($r = -0.206; p < 0.001$). In other words, when biodiversity (defined as "3 or more different plants or animals" represented in the drawing) was coded as present, deforestation (defined as "cut trees, clearcutting, desert-like conditions, stumps" represented in the drawing) was not, and vice versa. Furthermore, the content analysis indicated higher biodiversity in the past (i.e., more species in drawings) and that this statistically significantly decreased in drawings of the future. Conversely, the opposite effect was observed for deforestation, in other words, less deforestation in the past and more deforestation in the future, with the difference being statistically significant (Table 6.1).

Respondents touched also on the declining output of cashew agro-ecological systems due to changing environmental conditions. For example,

Table 6.1 Coding differences for biodiversity and deforestation in past and future drawings

Code	Past drawing	Future drawing	% of change	McNemar test	N	Asymptotic significance
Biodiversity	0.66	0.15	0.51	62.458	142	0.000★★★
Deforestation	0	0.30	−0.30	40.024	142	0.000★★★

Note: ★★★ $p < 0.001$.

when discussing how cashew trees seem to be less productive than in the past, one FGD participant stated that "*Cashew trees used to produce before; they no longer produce the same way, there's not many cashew nuts. Because of desertification and lack of rain. … We are lucky to have them, but we don't know how to take care of them.*" The expert interview with the agronomist further suggested,

> From an environmental point of view, we are using soils, which should be for food crops, but use them for cashew nuts. … So, we are not doing cashew in the right way. And many of these plantations have problems with productivity and so on because they are just sick, diseases, and a lot of insects and other diseases, that effect their productivity. And they [the trees] are old, 30 and 40 years old, with no real maintenance.

This reflects very well other studies in Guinea-Bissau that have argued that "*tropical deforestation, rangeland modifications, agricultural intensification and urbanization*" negatively reshape the natural environment (Temudo and Bivar Abrantes 2014: 217). The political ecology lens can help to assess these environmental vulnerabilities, as well as the resilience of local communities (Hastrup 2009), as discussed in the "Livelihoods perspectives" and "Development perspectives" sections to follow.

Livelihoods perspectives

In terms of livelihoods' perceptions, the survey data collected from college and university students and alumni point to some important intersections between cashews and livelihoods. First, only about half of the respondents had direct access to farmland (Table 6.2), which suggests that younger generations might be excluded from cashew production (especially more modernized methods that can better leverage their knowledge). This is potentially problematic for the youth as cashews are the major economic activity in the country that monopolizes available agricultural land as the need for increased yields intensifies (Lundy 2018; Sousa et al. 2014; Temudo and Bivar Abrantes 2013).

At the same time, there was a strong preference for locally grown foodstuff and for cashew juice over a common imported alternative (i.e., Coca-Cola).

Table 6.2 Perceptions related to cashew, farming, and food preferences

	Mean	Std. Dev.	N
Direct access to farmland	2.43	1.38	260
Better taste of locally grown food over expensive imported food	1.43	0.84	269
Preference of Coca-Cola to cashew juice	4.05	1.15	270
Higher importance of achieving high crop yields than protecting the natural environment	3.92	1.12	266
Diversification should be prioritized by cashew farmer	1.85	1.02	265

Note: Average age of respondents: 29.8 years old; Gender composition: 30.5% female; Responses are based on a 5-point Likert scale ranging from 1 (Highly Agree) to 5 (Highly Disagree).

These responses suggest two things, (a) cashew expansion at the expense of other agricultural land might have ramifications for national food security if agricultural land continues to be converted for cashew production and (b) there might be a national demand for cashew products from younger generations that makes an even stronger case for investing in national value addition.

Respondents also felt strongly that the natural environment should be protected at the cost of increasing agricultural outputs, while most importantly for this study, most respondents either agreed or strongly agreed that "cashew farmers should diversify what they grow." The qualitative findings from the expert interviews and FGDs support these sentiments.

For example, the importance of diversifying away from cashew production to ensure that farmer livelihoods and food security are not compromised was touched on by one of the FGD participants who stated,

> *you need to help the farmers where they plant things, how they organize their farm so that you may do the planting of the cashew tree at the same time you may work some other products instead of cashew. The reason not to depend on the cashew is because if you are dependent on the cashew, it's not good because it does not grow the same in some places than others and you won't have any other food to eat like rice.*

The main diet in Guinea-Bissau is rice-based, with rice incorporated into important rituals and life cycle events (Davidson 2016; Lundy 2009). In fact, I was told on many occasions "if you didn't have rice, it wasn't a meal." On the contrary, cashews do not hold a similar culturally valued position in society as a cash crop, although the alcoholic byproduct has become a staple of some rituals, while its availability has also increased its over-indulgence, especially among the Balanta community (Lundy 2012).

In a second FGD, one participant acknowledged the ramifications that the inability to intercrop in cashew plots can have for food security,

> *We know that when you grow corn, you can use the same place to grow, something else, like rice or peanuts. So it's possible to use the same place to grow different types [of crops]. … we don't think the cashew is sustainable. That's why we grow corn and peanuts and rice because we can use the land again. We think it is sustainable for people to grow not only cashew but to continue to produce rice and so we stop eating rice from other places and we can eat our own products.*

Here, the intersection between environment, livelihood, and development in the context of cashew pomiculture starts becoming visible (Lundy 2009). For example, agronomists suggest that to optimize cashew growth and yield, seedlings should be planted at least 10 m (35.81 feet) apart. However, in Guinea-Bissau, cashew trees are often planted much closer together and intermixed with other fruit trees, which reduces their productivity (Andrighetti et al. 1994; Lekberg 1996: 8). This further suggests that cashew cultivators in

Guinea-Bissau are not utilizing empirically derived cashew production best practices. Environmental conditions, suitable varieties, and plant and land management practices including spacing, fertilizer application, and pest control must be considered alongside farmers' capacities, communities' expectations, and overall economic demands.

Development perspectives

Despite the future development potential of cashews for foreign exchange generation or biofuel production, the interviews pointed out that these initiatives must not compete with current food production or degrade the environment. This was indicated through the expert interview with the agronomist who reasoned how the government in consultation with international development banks decided the expansion of raw cashew nut production in Guinea-Bissau for economic development:

> *This country embarked on a false economic theory. … This theory is called the theory of **comparative advantages**. Comparative advantages [theory] says, "this thing you produce for which you have advantages as compared with other countries" – "it is better," according to this theory, "to produce cashew nuts, export them, get the money, and import rice." What happens with this theory is that when you apply it this way, you will stay where you are, because you will say, "we have a comparative advantage in cashew nuts." So, just do cashew nuts, nothing else. … cashew nuts have a place in this country, in agriculture, in the economy, in export, and so on. … But not in competition with rice and other food crops. … the economy of this country, equates to cashew nuts. Can you imagine a European country that says that we are based on apples, only apples, nothing else?*
>
> (emphasis added by the author)

He then went on to discuss how the environment, livelihoods, and development must be integrated, otherwise, there is a likelihood for a system-wide breakdown.

> *Because our research and extension system in agriculture is broken, what you have, you have many cashew plantations in this country, [with] no maintenance … you don't have the best varieties. You don't have the best agricultural systems … You don't have processing. And what do we do, we export, huge exports, every single kilogram of cashew nuts to one country alone, which was India. Now we have Vietnam. … Just these two countries. If they said they are not interested in Guinea-Bissau, then the whole economy, the whole agricultural sector suffers. Because people put a lot of incentives on the side of cashew nuts, and no incentives for food crops, based on that big theory [comparative advantage] … they [international lenders] tell you [Guinea-Bissau government], "if you don't do this, we don't help you with money." Because it is called conditionality, a conditional economy. … and no innovation, no technological innovations, economic innovations, no systems. … And we talk about*

value chains. ... But value chain, is also processing. ...But how can you add value only through the supply chain?

The point made was the need to add value to exports through the in-country processing of the nuts, which could further increase farmer and government revenues. However, the lack of general infrastructure (e.g., roads, electricity, processing plants), alongside the lack of political will, suggests that related costs associated with any attempts to develop processing facilities would be shouldered by farmers (Lundy 2012) (see "Political Ecology of smallholder-based cashew pomiculture in Guinea-Bissau"). It is currently more cost-effective (monetarily, although the environmental impacts of transportation have yet to be factored into this equation) and expeditious to export raw cashew nuts for processing to China, India, or Vietnam (see "Political Ecology of smallholder-based cashew pomiculture in Guinea-Bissau").

To illustrate these development challenges further, in an earlier interview with a respected business person, he suggested that "*The government is the biggest obstacle in this country to development. I can give you evidence of this, for instance, the English took 8 billion Francs in the port of Bissau to fix in and around there ... 8 billion, I do not know how much that is in US dollars. It is about 10 million to connect to one of the fiber optic cables, so they had to decide what to do. And they made the wrong decision.*" He continued, "*To invest it in the port instead of making the connection. ... This kind of thing is keeping this country in the mess it is now.*"

This decision, which was largely a political one, prioritized the port, and hence the efficient export of cashews from the country over investment in internet infrastructure. As a highly educated tech entrepreneur, he felt that this was the wrong decision.

Cashews as an agent of ecological transformation in Guinea-Bissau

The "Perceptions of cashews" section has clearly demonstrated that cashew nuts are still viewed as "the" potential avenue toward economic development, although the overreliance on cashew production could also be Guinea-Bissau's economic and ecological downfall. These ideas have been expressed in many other studies of the cashew industry in Guinea-Bissau (Table 6.3).

The ecological transformations in Guinea-Bissau caused by cashew pomiculture have been catalyzing socioeconomic transformation (Table 6.3). According to Catarino et al. (2015: 459), most "natural" vegetation throughout Guinea-Bissau is now secondary, which is a consequence of human interventions, including shifting agriculture and the use of fire. They found that wildfires and climate change are causing deforestation rates of 1% or more per year (Catarino et al. 2015: 459). IBAP has also identified slash and burn agriculture and associated deforestation as one of the primary threats to biodiversity (IBAP 2016).

Therefore, the question remains, whether the risks and challenges associated with cashew cultivation are worth it. Some of the findings suggest that cashews are providing Guinea-Bissau with a competitive advantage internationally,

Table 6.3 Major aspects of cashew cultivation in Guinea-Bissau

Themes	Key findings	Source
- Livelihood - Development	- Cashew is the main source of livelihood for rural poor; - High transaction costs for cashew buyers and exporters decrease the farmgate price; - Absence of research and development.	Barry et al. (2007)
- Environment - Livelihood	- Land tenure practices are linked to the planting of cashew trees; - Spread of pests and diseases affecting cashew trees; - Cashew production has major impacts on living standards and food security.	Catarino et al. (2015)
- Livelihood - Development	- Increased export prices or export taxes for cashews disproportionately affect farmgate prices; - Poor rural households are more exposed to price volatility.	Cont and Porto (2014)
- Environment - Livelihood	- Some farmers plant cashews for access to rice, thus having potential to increase food security; - Risks of cashew production include fire and theft; - The value of byproducts such as alcohol are realized differently by Muslim and non-Muslim communities.	Lekberg (1996)
- Livelihood	- Cashew production is the most economical and appealing option for rural households; - Diversification as recourse for poor prices; - Cautious investment in cashew processing.	Lundy (2012)
- Environment - Livelihood	- Food crops replaced by cash crops such as cashews, compromising local livelihoods and food security; - Need of secure guidelines for sustainable production maintaining ecosystem services.	Monteiro et al. (2017)
- Development	- Added value possibilities with transformation, better markets, and improved institutional environment; - Potential for global value creation and dynamic food system; - Promote family welfare and sustainable development.	Pacheco de Carvalho and Mendes (2015)
- Environment - Livelihood	- Cashew pomiculture replaces shifting cultivation; - Response to international and national economic and conservation policies and climate instability; - Cashew production impacts on land use and forest cover change; - Cashew pomiculture degrades more sustainable and biodiverse landscapes.	Temudo and Abrantes (2013)
- Environment	- Avoiding mangrove deforestation from cashew production mitigates climate change and protects vital ecosystem services.	Vasconcelos et al. (2015)

placing it among the top ten producers in the world (see "Political Ecology of smallholder-based cashew pomiculture in Guinea-Bissau"). However, as the competition from other countries increases, and as these countries improve their market position by carefully selecting cashew tree varieties suitable for their environmental conditions and maximum high-quality outputs, Guinea-Bissau's willy-nilly approach and long-term neglect might prove disastrous. According to Barry et al. (2007: 80), a constraining factor to the cashew industry in Guinea-Bissau is "*the absence of research and development, and vulgarization activities,*" as the "*cashew trees are grown naturally without using advanced scientific methods, while in other countries research programs are carried out in order to improve the size of the nuts and their capacity to resist deadly diseases ... The sector also remains vulnerable to price shocks and a decline in export prices over time ... translating into some volatility for cashew producers*" (Barry et al. 2007: 80–81).

Local factors also drive the production and preference of cashews. For example, the cashew cycle and rice cycle in Guinea-Bissau are offset with rice being grown from July through December, while the cashew harvest takes place from March to August. "*During this period, small free-lance buyers and officers of medium sized companies travel the country acquiring cashew nuts or exchanging them for rice*" (Catarino 2015: 463). These buyers must be licensed by the government by paying a fee to the Ministry of Commerce and Chamber of Commerce, Industry and Agriculture, with currently 300-500 registered agents and less than 100 registered exporters per year. This coincides with the lean months when rice is scarce, meaning that having some extra cash from cashew sales is critical to ward off hunger (Lundy 2009: 112).

Furthermore, and somewhat obviously, cashew pomiculture is easier than labor-intensive rice production (especially paddy rice production), which requires an incredible amount of manual labor to maintain mud dikes to protect the fields from saltwater inundation (Lundy 2009, 2012). Hence, many farmers express a preference for cashew cultivation and harvest over that of rice and other subsistence activities. Additionally, Lundy (2012) found that since many Muslims do not prize the alcoholic cashew byproduct, they often exchange the cashew apple with their non-Muslim neighbors for their labor during harvesting (see also Lekberg 1996). The massive quantities of available alcohol, however, are affecting vulnerable social groups including women and youth, with a dramatic rise in alcoholism throughout the country that needs further and immediate investigation (Drift 2002).

At the international level, considering the current cultivation methods throughout Guinea-Bissau (both by smallholders and on commercial plantations), cashews would be considered organic and could potentially fetch a higher price if global hygiene standards were achieved and the nuts were marketed effectively (Catarino 2015: 463). Currently, there is very low input of artificial agro-chemicals, but some argue that by applying higher levels of irrigation, pesticides, synthetic fertilizers, and improved crop varieties (consistent with Green Revolution standards), the country could achieve a more sustainable cashew industry (Monteiro et al. 2017). Monteiro et al. (2017: 10) envision a future in which intercropping, selection of higher yield varieties, and the maximization of cashew co-products (i.e., apples, CNSL, wood) could lead to agricultural improvements (i.e., diversification), improved agroecosystems, higher and higher quality yields, and a more consistent and structured industry. These recommendations closely align with many of the study results highlighted elsewhere (Table 6.3).

Finally, Temudo and Bivar Abrantes (2014: 228) remind us of the complexity of the current situation and the need for a comprehensive political ecology approach, arguing that "*The cashew agricultural frontier in Guinea-Bissau is the outcome of complex and interconnected changes occurring at the local, national and global levels to which social, cultural, political-economic and environmental factors contributed in various degrees across different times and regional scales.*" Protecting biodiversity, improving livelihoods, developing the economy, extracting wealth, and ensuring a social safety net must all be considered when debating the next steps for the cashew industry of Guinea-Bissau.

Conclusions

Guinea-Bissau is one of the largest cashew producers in the world. National cashew production grows faster than the global industry average, but the domestic processing sector is almost non-existent, forcing the export of almost its entire cashew production in raw form. The expansion of the sector and its economic dominance has rested on the concept of competitive or comparative advantage and a series of misplaced national policies driven by international organizations.

The above factors have combined to make Guinea-Bissau one of the least diversified economies in Sub-Saharan Africa. It has been argued cashew pomiculture increases the vulnerability of the average farmer in the country because it creates a global dependency on imported foodstuffs in exchange for a volatile luxury cash crop. Increasing economic and environmental homogeneity impacts ecological adaptive capacity to natural disasters and policy crises, weakens social and economic safety nets, and limits agricultural agility and flexibility to take advantage of new opportunities as they arise. Policy priorities, financial and occupation incentives, and an ensuing land grab have all contributed to make cashew production a potent agent of ecological transformation in the country. Cashew trees practically dominate Guinea-Bissau's entire landscape.

However, a series of renewed actions are being implemented to try and reduce these negative aspects of cashew production and allow it to fulfill its socio-economic potential. For example, ongoing conservation efforts may enhance biodiversity and protect selected sensitive areas and wildlife. Small and medium industry, and the government, will also most likely continue to experiment with cashew processing options to add value in-country, including using waste as biofuel to operate processing facilities. Better-aligned and more lucrative tax and regulation structures are also being developed using real-time global data to avoid scaring off cashew buyers, as has happened through past policy failures.

For example, the African Cashew Alliance, in collaboration with the Private Sector Rehabilitation and Agribusiness Development Project (PRSPDA), has provided the National Cashew Agency of Guinea-Bissau with a market information service for cashew marketing. The National Cashew Agency of Guinea-Bissau's (ANCA) information service plans to use communication outlets such as mobile phone agencies for the daily dissemination of information on cashew market trends to farmers.

As we understand better the challenges and integrated nature of Guinea-Bissau's political ecology, coordinated actions by researchers, policymakers, and the private sector offer opportunities to improve the cashew industry, while avoiding the latent risks associated with over-specialization.

Acknowledgments

I wish to acknowledge the Fulbright Specialist Program and the US Embassy in Senegal without which the July 2018 research trip would not have been possible. I also wish to thank our local sponsors and collaborators, Raul Mendes Fernandes, vice-rector at Amilcar Cabral University; Liam Carney, a Fulbright TEFL trainer, and Mariama Elie Camara, an ECI Teacher Trainer, at Tchico Te Teacher Training College; Almamo Danfa, Public Diplomacy coordinator in the Bissau Liaison Office; and coordinator Rui Sá and assistant professor Joana Sousa and their students in the Environmental Sciences program at the University Lusofóna Guiné. I also wish to express my gratitude and appreciation for my undergraduate students from Kennesaw State University who worked so hard collecting data on this project, Rachel Langkau, Kamran Sadiq, and Sami Wilson. Their enthusiasm and positive attitude were infectious. Special thanks as well to Catholic University, and IBAP, especially General Director Justino Biai and the Director of the Rio Cacheu Mangroves Natural Park, Fernando Biag. I also want to acknowledge the assistance of agronomist José Filipe Fonseca, biology teacher Florentino A. Mango, and the organization Voz di Paz. Our heartfelt appreciation to the people of Guinea-Bissau who helped us learn together. Finally, I want to thank Nicole Connelly for designing the two figures in this chapter.

Notes

1 Since traveling to Guinea-Bissau for the first time in 2007, I have poured through processed cashew nut labels in my local US grocery stores, looking for any acknowledgement of the contents' origins. It was not until 2019 (although the label was copyrighted in 2017) that I read on a package of Back to Nature Foods Co., LLC Jumbo Cashews (verified Non-GMO): "*Cashews product of Benin, Brazil, Cambodia, Ghana, Guinea-Bissau, India, Indonesia, Ivory Coast, Kenya, Mozambique, Nigeria, South Africa, Tanzania, Vietnam.*"

2 However, critics such as James K. Galbraith (2008: 70) suggest that nations specializing in agriculture are condemned to poverty since it is dependent on finite,

non-increasing resources such as water and land, arguing that *"diversification, not specialization, is the main path out of underdevelopment, and effective diversification requires a strategic approach to trade policy."*
3 This eventually became 1:1, but by 2007 and later, it had deteriorated to 1 kg of rice for 3 kg of cashew nuts (Lundy 2009: 2012).
4 According to the IMF, PRSPs are intended to describe a country's macroeconomic, structural, and social policies and programs over a short- and long-term period in order to identify economic growth strategies and reduce poverty.

References

AFP. 2017. "Guinea-Bissau reaps reward as world goes nuts for cashews." *News24*, September 24. www.news24.com/Africa/News/guinea-bissau-reaps-reward-as-world-goes-nuts-for-cashews-20170924.

Africa Research Bulletin. 2015. "Cashews: Guinea-Bissau: Exports are up, but smuggling remains a problem." *The Daily Observer*, Banjul, April 20: 20825B.

Andrighetti, L., with G.F. Bassi, P. Capella, A.M. de Logu, A.B. Deolalikar, G. Haeusler, G.A. Malorgio, F. Mavignier Cavalcante Franca, G. Rivoira, L. Vannini, and R. Deserti. 1994. *The World Cashew Economy*. M. Procacci and J. Rees, transl. Milan, Italy: L'inchiostroblu.

Araújo, J.P.P. de, ed. 2013. *Agronegócio Caju: Práticas e Inovações*. Brasília, Brazil: Embrapa.

Atkinson, Adrian. 1991. *Principles of Political Ecology*. London: Bellhaven Press.

Azam-Ali, S.H., and E.C. Judge. 2001. "Small-scale cashew nut processing." Coventry, UK: ITDG Schumacher Centre for Technology and Development Bourton on Dunsmore. www.fao.org/3/ac306e/ac306e.pdf.

Barry, Boubacar-Sid, Edward G.E. Creppy, and Quentin Wodon. 2007. "Cashew Production, Taxation, and Poverty in Guinea-Bissau." In *Conflict, Livelihoods and Poverty in Guinea-Bissau*, edited by Boubacar-Sid Barry, Edward G.E. Creppy, Estanislao Gacitua-Mario, and Quentin Wodon, 77–88. World Bank Paper No. 88. Washington DC: The World Bank.

Blaikie, Piers. 2012. "Should Some Political Ecology Be Useful? The Inaugural Lecture for the Cultural and Political Ecology Specialty Group, Annual Meeting of the Association of American Geographers, April 2010." *Geoforum* 43(2): 231–239.

Boafo, James. 2019. "Expanding Cashew Nut Exporting from Ghana's Breadbasket: A Political Ecology of Changing Land Access and Use, and Impacts for Local Food Systems." *The International Journal of Sociology of Agriculture and Food* 25(2): 152–172.

Catarino, Luís, Yusufo Menezes, and Raul Sardinha. 2015. "Cashew Cultivation in Guinea-Bissau–risks and Challenges of the Success of a Cash Crop." *Scientia Agricola* 72(5): 459–467.

Cont, W., and G. Porto. 2014. "Measuring the Impact of Change in the Price of Cashew Received by Exporters on Farm Gate Prices and Poverty in Guinea-Bissau." Policy Research Working Paper, 7036. Washington, DC: World Bank. elibrary.worldbank.org/doi/abs/10.1596/1813-9450-7036.

Davidson, Joanna. 2016. *Sacred Rice: An Ethnography of Identity, Environment, and Development in Rural West Africa*. New York: Oxford University Press.

Drift, Roy van der. 2002. "Democracy's Heady Brew: Cashew Wine and the Authority of the Elders among the Balanta in Guinea-Bissau." In *Alcohol in Africa: Mixing Business, Pleasure, and Politics*, edited by D.F. Bryceson, 179–196. Portsmouth, NH: Heinemann.

Escobar, Arturo. 1998. "Whose Knowledge, Whose Nature? Biodiversity, Conservation, and the Political Ecology of Social Movements." *Journal of Political Ecology* 5(1): 53–82.

———. 2006. "Difference and Conflict in the Struggle over Natural Resources: A Political Ecology Framework." *Development* 49(3): 6–13.

FAOSTAT. 2019. "Cashew." FAOSTAT Data, www.fao.org/faostat/en/#search/cashew.

Forrest, Joshua B. 1992. *Guinea-Bissau: Power, Conflict, and Renewal in a West African Nation.* Boulder, CO: Westview Press.

Freire, F.C.O., J.E. Cardoso, A.A. Dos Santos, and F.M.P. Viana. 2002. "Diseases of Cashew Nut Plants (*Anacardium occidentale* L.) in Brazil." *Crop Protection* 21(6): 489–494.

Funk, Ursula. 1988. "Land Tenure, Agriculture and Gender in Guinea-Bissau." In *Agriculture, Women, and Land: The African Experience*, edited by Jean Davison, 33–58. Boulder, CO: Westview Press.

Gacitua-Mario, Estanislao, Hakon Nordang, and Quentin Wodon. 2007. "Livelihoods in Guinea-Bissau." In *Conflict, Livelihoods and Poverty in Guinea-Bissau*, edited by Boubacar-Sid Barry, Edward G.E. Creppy, Estanislao Gacitua-Mario, and Quentin Wodon, 59–76. World Bank Paper No. 88. Washington DC: The World Bank.

Galbraith, James K. 2008. *The Predator State: How Conservatives Abandoned the Free Market and Why Liberals Should Too.* New York: Simon and Schuster.

GlobeNewswire. 2019. "Cashew Nut Shell Liquid (CNSL) Market to Reach USD 450.4 Million by 2026." *GlobeNewswire*, Reports and Data, April 17. www.globenewswire.com/news-release/2019/04/17/1805582/0/en/Cashew-Nut-Shell-Liquid-CNSL-Market-To-Reach-USD-450-4-Million-By-2026-Reports-And-Data.html?culture=fr-ca.

Hastrup, Kirsten. 2009. *The Question of Resilience: Social Responses to Climate Change.* Copenhagen: Det Kongelige Danske Videnskabernes Selskab.

Horta, C., and R. Sardinha. 1966. "A indústria transformadora na Guiné Portuguesa: Problemas e perspectivas." *Boletim Cultural da Guiné Portuguesa* 21(82): 141–163.

IBAP. 2016. *The Protected Areas of Guinea-Bissau: Guide to ecotourism.* Bissau: Institute for Biodiversity and Protected Areas.

IMF. 2011. "Guinea-Bissau: Second Poverty Reduction Strategy Paper." IMF Country Report No. 11/353. Washington, DC: International Monetary Fund.

IRIN. 2006. "Guinea-Bissau: Cashew Crops Rot as Price Plummet." Integrated Regional Information Network. Dakar, www.irinnews.org/Report.aspx?ReportID=61571.

Keim, Curtis, and Carolyn Somerville. 2018. *Mistaking Africa: Curiosities and Inventions of the American Mind.* Fourth Edition. New York: Westview Press.

Lekberg, Ylva. 1996. "Cashew in Guinea-Bissau, The Small Producer's Perspective: A Minor Field Study." Working Paper 313. Uppsala: Swedish University of Agricultural Sciences, International Rural Development Centre (IRDC).

Lim, Tong Kwee. 2012. "*Anacardium occidentale*." In *Edible Medicinal and Non-Medicinal Plants*, edited by Tong Kwee Lim, 45–68. The Netherlands: Springer.

Lundy, Brandon D. 2009. "Making a Living in Kassumba, Guinea-Bissau." Ph.D. diss. SUNY at Buffalo, NY.

———. 2012. "Playing the Market: How the Cashew 'Commodityscape' Is Redefining Guinea-Bissau's Countryside." *Culture, Agriculture, Food and Environment* 34(1): 33–52.

———. 2018. "Challenging Adulthood: Changing Initiation Rites among the Balanta of Guinea-Bissau." *African Studies* 77(4): 584–606.

Maia, G.A., L.F.F. Holanda, and C.B. Martins. 1971. "Características químicas e físicas do caju *Anacardium occidentale* L." *Ciência Agronômica, Fortaleza* 1(2): 115–120.

Martinez-Alier, Joan. 2003. *The Environmentalism of the Poor: A Study of Ecological Conflicts and Valuation*. Cheltenham, UK: Edward Elgar Publishing.

————. 2014. "The Environmentalism of the Poor." *Geoforum* 54: 239–241.

Meaney-Leckie, Anne. 1991. "The Cashew Industry of Ceara, Brazil: Case Study of a Regional Development Option." *Bulletin of Latin American Research* 10(3): 315–324.

Mgaya, James, Ginena B. Shombe, Siphamandla C. Masikane, Sixberth Mlowe, Egid B. Mubofu, and Neerish Revaprasadu. 2019. "Cashew Nut Shell: A Potential Bioresource for the Production of Bio-sourced Chemicals, Materials and Fuels." *Green Chemistry* 21(6): 1186–1201.

Monteiro, Filipa, Luís Catarino, Dora Batista, Bucar Indjai, Maria Cristina Duarte, and Maria M. Romeiras. 2017. "Cashew as a High Agricultural Commodity in West Africa: Insights towards Sustainable Production in Guinea-Bissau." *Sustainability* 9(1666): 1–14. doi:10.3390/su9091666.

Mordor Intelligence. 2021. "Cashew Market - Growth, Trends, COVID-19 Impact, and Forecasts (2021–2026)." www.mordorintelligence.com/industry-reports/global-cashew-market.

Morton, Julia F. 1987. "Cashew apple, *Anacardium occidentale* L." In *Fruits of Warm Climates*, edited by Julia F. Morton, 239–240. West Lafayette, IN: Center for New Crops and Plant Products, Department of Horticulture and Landscape Architecture, Purdue University.

Moseley, William G. 2017. "The New Green Revolution for Africa: A Political Ecology Critique." *The Brown Journal of World Affairs* 23(2): 177–190.

Nguyen, Daisy. 2018. "Vietnam cashew industry: The more exports, the bigger losses." *Vietnam Insider*, July 5. vietnaminsider.vn/vietnam-cashew-industry-the-more-exports-the-bigger-losses/.

Ohler, J.G. 1979. *Cashew.* Amsterdam, Netherlands: Koninklijk Instituut voor de Tropen.

Oppong, Joseph R., and Ezekiel Kalipeni. 2005. "The Geography of Landmines and Implications for Health and Disease in Africa: A Political Ecology Approach." *Africa Today* 52(1): 3–25.

Orwa, C., A. Mutua, R. Kindt, R. Jamnadass, and A. Simons. 2009. "Agroforestry database: A tree reference and selection guide version 4.0." www.worldagroforestry.org/treedb2/AFTPDFS/Anacardium_occidentale.pdf.

Pacheco de Carvalho, Bernardo Reynolds, and Henrique Mendes. 2015. "Cashew Chain Value in Guiné-Bissau: Challenges and Contributions for Food Security: A Case Study for Guiné-Bissau." *International Journal of Food System Dynamics* 7(1): 1–13.

Paulson, Susan, and Lisa L. Gezon, eds. 2005. *Political Ecology across Spaces, Scales, and Social Groups*. New Brunswick, NJ: Rutgers University Press.

Ricardo, David. 1817. *On the Principles of Political Economic and Taxation*. London: Dent.

Sahn, David E., and Alexander Sarris. 1994. "The Evolution of States, Markets, and Civil Institutions in Rural Africa." *The Journal of Modern African Studies* 32(2): 279–303.

Sardinha, R.M.A., A.M.S. Bessa, J. Blake, D. Guyer, C. Cassamá, and P. Tambá-Bungué. 1993. "Selection of superior genotypes of cashew (*Anacardium occidentale* L.) in Guinea-Bissau and the development of in vitro techniques for their propagation." Final Summary Rep. Contract TS-2-A-0167P, Lisbon, Portugal.

Schwartz, L., D.J. Birmingham, P.C. Campbell Jr., and H.S. Mason. 1945. "Skin Hazards in the Manufacture and Use of Cashew Nut Shell Liquid Formaldehyde Resins." *Industrial Medicine and Surgery* 14(6): 500–506.

Sousa, J., A. Dabo, and A. L. Luzi. 2014. "Changing elderly and changing youth: Knowledge exchange and labour allocation in a village of southern Guinea-Bissau." Working Paper-Future Agricultures 81.

Strom, Stephanie. 2014. "Cashew Juice, the Apple of Pepsi's Eye." *The New York Times*, August 8.

Temudo, Marina Padrão, and Manuel Bivar Abrantes. 2013. "Changing Policies, Shifting Livelihoods: The Fate of Agriculture in Guinea-Bissau." *Journal of Agrarian Change* 13(4): 571–589.

———. 2014. "The Cashew Frontier in Guinea-Bissau, West Africa: Changing Landscapes and Livelihoods." *Human Ecology* 42: 217–230.

Temudo, Marina Padrão, Rui Figueira, and Manuel Bivar Abrantes. 2015. "Landscapes of Bio-cultural Diversity: Shifting Cultivation in Guinea-Bissau, West Africa." *Agroforestry Systems* 89(1): 175–191.

Tessmann, Jannes. 2018 "Governance and Upgrading in South–South Value Chains: Evidence from the Cashew Industries in India and Ivory Coast." *Global Networks* 18(2): 264–284.

Thomas, Wilbur G., John Van Dusen Lewis, and Jeff Dorsey. 2004. *Guinea Agricultural Sector Assessment (GASA) Final Report*. Arlington, VA: ARD-RAISE Consortium.

Tola, Jemal, and Yalew Mazengia. 2019. "Cashew Production Benefits and Opportunities in Ethiopia: A Review." *Journal of Agricultural and Crop Research* 7(2): 18–25.

Tullo, Alexander H. 2008. "A Nutty Chemical." *Chemical and Engineering News* 86(36): 26–27. doi:10.1021/cen-v086n033.p026.

UNDP. 2019. "Climate Change Adaptation: Guinea-Bissau." United Nations Development Programme, www.adaptation-undp.org/explore/western-africa/guinea-bissau.

UNIOGBIS. 2016. "Cashew nut central to Guinea-Bissau economy: a blessing or a curse?" United Nations Integrated Peacebuilding Office in Guinea-Bissau, April 11. uniogbis. unmissions.org/en/cashew-nut-central-guinea-bissau-economy-blessing-or-curse.

Vasconcelos, Sasha, Patrícia Rodrigues, Luís Palma, Luís F. Mendes, Agostinho Palminha, Luís Catarino, and Pedro Beja. 2015. "Through the Eye of a Butterfly: Assessing Biodiversity Impacts of Cashew Expansion in West Africa." *Biological Conservation* 191: 779–786.

WFP. 2019. *Guinea Bissau Transitional ICSP (January 2018 - June 2019)*. June 30. www. wfp.org/operations/gw01-guinea-bissau-transitional-icsp-january-2018-june-2019.

Wilson, Bee. 2015. "'Blood cashews': The toxic truth about your favourite nut." *The Telegraph*, May 4. www.telegraph.co.uk/foodanddrink/foodanddrinknews/11577928/Blood-cashews-the-toxic-truth-about-your-favourite-nut.html.

World Bank. 1989. "Report and Recommendation of the President of the International Development Association to the Executive Directors on a Proposed Development Credit of SDR 18.0 Million to the Republic of Guinea-Bissau for Phase Two of its Structural Adjustment Program." April 25, Report No. P-4980-GUB. documents. worldbank.org/curated/en/765811468029994791/pdf/multi0page.pdf.

———. 2006. *Guinea-Bissau Integrated Poverty and Social Assessment (IPSA): Transition from Post Conflict to Long Term Development: Policy Considerations for Reducing Poverty.* Vol. 1: Main Report, Report No. 34553-GW, PREM 4, Africa Region, World Bank, Washington, DC.

Part III
Agrarian transformation

7 The political ecology of genetically modified and organic cotton in India as agents of agrarian transformation

Andrew Flachs

Introduction

Indian smallholders have grown cotton for more than 5,000 years. Indian cotton was shipped across the ocean to metropoles like Zanzibar, Baghdad, London, and New York and dominated the global clothing trade for centuries. A combination of early industrial capitalism and slavery allowed American production to briefly surpass that of India until 1860 (Beckert 2014), but smallholders continue to make India the largest current cotton exporter in the world (USDA Foreign Agricultural Service 2018). Despite this longstanding dominance, the Indian cotton sector gained international renown for an agrarian crisis of pest attacks, low yields, and suicide at the end of the 20th century (Galab, Revathi, and Reddy 2009; Vaidyanathan 2006).

One potential culprit for these issues was the species of cotton itself. Indian commercial cotton since the 1970s is typically a hybrid of *Gossypium hirsutum* L., a North American cotton species susceptible to Asian insect pests. As farmers steadily increased their pesticide use, debts rose, biodiversity declined, and farmers experienced health risks associated with acute and chronic pesticide poisoning. In the 1990s, cotton accounted for nearly half of all pesticides used in India, although it was planted on only 5% of the agricultural land (Abhilash and Singh 2009).

From a political ecology perspective, which cautions that ecological risks are borne disproportionately by marginalized communities and that crises have socioeconomic and global roots, the blame for this situation rests more with neoliberal economic policies and the demands of industrialized agriculture (Chapter 2). In India, cotton farmer suicides and rural distress emerged alongside the expansion of cash-cropping, the commercialization of agricultural fertilizers and pesticides, and a withdrawal of state support (Mohanty 2005; Münster 2012), as has happened with the rise of neoliberal policies around the world (McMichael 2007). However, hundreds of millions of people depend on agriculture for their livelihoods in India, where an engaged rural citizenry makes frequent demands on the immense civil bureaucracy, and the press freely pursues such stories and challenges state responses. In this environment of need and anxiety, two potential solutions, which are mutually exclusive due to global regulatory standards, presented themselves to Indian cotton farmers: GM seeds and organic agriculture.

The appeal of these technological solutions for increasing the productivity and profitability of Indian cotton does not lay in their potential for local variation, but in their seemingly universal applicability (see Chapter 9 for similar discussion for soy in South America). The assumption in this case was that industrial agriculture or organic regulations designed in the United States could be translated to Indian farms without changes. In practice, the smooth path of agricultural development was met with stubborn friction (Tsing 2005) as Indian farmers learned to pursue particular kinds of agricultural success in this changing ecological and social landscape.

This chapter situates this friction as a historical process and uses a political ecology lens to outline how these solutions were adapted (or not) to local needs and logics in agriculture. The "Methodology" section outlines the analytical lens used in this chapter. The "Results and discussion" section begins with a historical analysis of agricultural change and development in the Indian cotton sector, and particularly how off-farm interests (including the state and private sector) have influenced farmer learning through seed technologies and how farmers have reworked these technologies to make them succeed in their own fields (see "Cotton in the colonial and postcolonial eras"). The sub-sections "Genetically modified cotton" through "Organic cotton" focus on GM and organic cotton agriculture as particular examples of solutions to agrarian distress through two ethnographic case studies in agricultural learning in Telangana (a state in South India). "Cooperative decision-making for agricultural stability" reflects on the possible role that cooperatives can play in addressing the underlying instabilities of contemporary Indian agriculture. Finally, "Genetically modified and organic cotton in India as agents of agrarian transformation" summarizes briefly how GM and organic cotton have become major agents of agrarian transformation in rural India.

Methodology

Seeds are a starting point to understanding agrarian change in India because they represent a choice that cannot be reversed. As material objects, seeds connect farmers to breeders, input retailers, and extension agents. As cultural artifacts, seeds bring agrarian aspirations into being for the coming season (Chacko 2019; Gold 2003; Nazarea 2014). Farmers who choose seeds continue to make choices that they hope will lead to yields, profits, and a good life. Asking about seed choices is a way to learn what lives are possible on these farms, and how interactions with these technologies influence rural wellbeing and sustainability.

While some scholars have blamed GM seeds or insects for the agrarian crisis of South India (Desmond 2017; Shiva et al. 2002), and others celebrate new technologies and programs as means to empower rural economies (Altenbuchner, Vogel, and Larcher 2018; Kathage and Qaim 2012), this attention presupposes technological solutions to problems that are fundamentally social. By framing these issues as technological problems that require technological solutions, stakeholders such as states, NGOs, and private sellers essentially side-step larger concerns about building an adaptive skillset during moments of rural distress. Instead, this chapter argues that the biggest impact of technologies like GM cotton and organic certification lies in the ways that they transform how farmers learn and conduct farmwork.

The information presented in this chapter is based on 16 months of ethno-graphic fieldwork in Telangana state in South India between 2012 and 2018 (Flachs 2019a). This fieldwork focused on agricultural decision-making among certified organic cotton farmers and farmers planting GM seeds. Telangana, a major cotton-producing state, was hit particularly hard by a wave of farmer suicides just as India debated legalizing GM seed technology (Rao and Suri 2006; Stone 2002). Warangal, the district containing most of the study villages, was briefly an epicenter of cotton farmer suicide as well as an early site of aggressive GM seed adoption and organic NGO intervention (Flachs 2019a; Stone 2007). To triangulate agricultural knowledge and daily practice, this chapter combines ethnography with a multi-village farmer survey (selected through a stratified random sampling approach), key informant interviews, focus group discussions, spatial analysis, and ethnobotanical transects.

When analyzed through the framework of political ecology, this approach to measuring farmer learning recognizes that the global political economy influences learning and environmental relationships through cotton grading and economic governance (Batterbury 1996; Kinchy 2012; Meek 2016; Scoones 2006). The conditions under which farmers learn, troubleshoot, perform, or make plans at this scale are all bound to cotton prices, national policies, and international governance over biotechnology and certified organic regulation. Learning is not an individual but a structural endeavor, even as farmer decision-making depends on local conditions.

At a second scale, this approach adopts the viewpoint of an analytic of per-formance to highlight agency, creativity, and opportunism among individuals and communities who have carved out new pathways for success in a chan-ging agricultural landscape (Bennike, Rasmussen, and Nielsen 2020; Flachs and Richards 2018; Fletcher, Dowd-Uribe, and Aistara 2020). Farmer learning is not pre-determined by the environmental or socioeconomic conditions that farmers find themselves in, because farmers are active agents with hopes, dreams, frustrations, and desires. By attending to both scales, it is argued that political ecologists can present a more accurate version of agrarian change.

Results and discussion

Cotton in the colonial and postcolonial eras

South Asian cotton fibers supplied large-scale cottage production across Asia in the first millennium CE. This fueled commodity taxes for Asian empires and gave Arab trader intermediaries power and wealth in deciding which merchants would gain access to Asian cloth for sale in Africa and Europe (Chaudhuri 1985). By the 15th century, European powers saw in cotton the opportunity to build a new kind of empire based on the extraction of commodities, land, and labor from colonial regions. Following the model laid during other colonial extraction regimes (Beckert 2014; Mintz 1986), English commodity traders bought cheap, raw cotton, sent it to be woven in English factories, and then sold the finished clothes at a higher price to subjects throughout the empire (Chapter 2). In the aftermath of US independence in 1776, the British turned to cotton sourcing in Africa and India, but the same extreme poverty and rural hierarchies that enabled colonizers to extract taxes and commodities also made cotton transportation and sale inefficient (Beckert 2014; Guha 2007). Facing competition from American cotton produced with enslaved labor, the East India Company invested in infrastructure and experimental farms. When the new seeds failed to increase production, the company ultimately hired American planters in 1840 to teach Indian smallholders how to farm American species for mass production (Guha 2007).

The new seeds were promising but faced a critical barrier. British-Indian cotton strains yielded 25% more fiber, but only with 200% higher cost in terms of production capital (Guha 2007). American and British growers struggled to replicate the American plantation system based on slavery, with Indian smallholders who lacked coerced and enslaved labor, reliable roads, or storage capacity that maintained cotton monocultures in the United States. Further, to retain control of the seeds, British buyers banned household seed breeding that might have produced varieties amenable to local agricultural conditions. Guha (2007) argues that input-intensive seeds were largely unattractive to Indian smallholders, who were disinterested in exchanging a flexible smallholder subsistence-cash economy for an agrocapitalist monoculture, especially since the lion's share of profits would go to colonizers. In the aftermath of the American Civil War and the decrease in US cotton exports, Indian farmers and British buyers relaxed rules governing local breeding. Farmers likely mixed local varieties with foreign cultivars to hedge their bets on the new seeds (Hazareesingh 2016). By the end of the 19th century, this resulted in a successful farmer-bred cultivar of *Gossypium hirsutum* L., a variant of the cotton that now accounts for 90% of the world's cotton clothing (Oosterhuis and Jernstedt 1999). That local strain "flourished" under the name Dharwar-American (Guha 2007). After decades of trying in vain to outcompete American long-staple cotton production, British traders ultimately relaxed planting and production restrictions to achieve their desired economic profits. Only after farmers surreptitiously bred their own varieties within their agroecological system did they come to dominate the global textile industry in the late 19th and early 20th century.

Following independence in 1947, Indian geopolitics pitted Soviet and American development interests against each other (Cullather 2013; Perkins 1997) through a nonalignment strategy that positioned capitalist and socialist development projects as a competitive field. Paranoid about the links between hunger and communism, and seeing an opportunity to steer India away from

Soviet influence, the United States initiated a series of agricultural development projects that came to be known as the Green Revolution. These entailed the use of high-yielding seed varieties, state extension services, and increased chemical inputs, irrigation, and electrification throughout the countryside (Cullather 2013; Flachs 2016b; Vasavi 2020) (Chapters 2 and 6). While 19th-century agribusiness drew farmers deeper into commodity production through new seeds and production goals, farmers themselves created the varieties that made this agriculture possible. Conversely, in the mid-20th century, wealthy farmers integrated themselves into emerging global commodity grain and textile markets, learning how to turn the higher investments and profits of state-supported cash-cropping to their advantage (Gupta 2017; Vasavi 2012). Poorer farmers did their best to emulate that success by combining fertilizer and manure to hedge bets on new seeds or voting for politicians who would keep subsidies in place to mitigate the risk of new technologies (Cullather 2013; Vasavi 2012).

In the aftermath of the Green Revolution, Indian crop breeders developed high-yielding cotton varieties and hybrid varieties that would respond to newly available pesticides, fertilizers, and water infrastructure. By 1995, 36% of the cotton area in India was planted with hybrids, about half of which were provided by private sector breeders (Basu and Paroda 1995). The rise in fertilizer and pesticide use generated new debts and insecurity, including a wave of farmer suicides that peaked in the late 1990s (Galab, Revathi, and Reddy 2009; Gruère and Sengupta 2011; Pandian 2011; Scoones 2006). For example, farmers in rural Telangana previously engaged in reciprocal finance relations with local landlords and ecological relationships with homemade inputs (Ludden 1999; Gupta 1998; Vasavi 1999) had to navigate an unfamiliar landscape of credit, labor, seeds, and inputs. At the same time, the increasing resistance of insects mirrored broader trends visible in the Green Revolution context, where insects developed pesticide tolerance in response to heavier spraying by smallholders (Brookfield 2001; Lansing 2006; Nicholls and Altieri 1997).

This combination of debt, growing disparities between rural and urban economies, and the exclusion from former socioeconomic guarantees created a sense of rural desperation, which observers linked to a persistent trend of farmer suicide resulting in more than 300,000 deaths between 1995 and 2015 (Menon and Uzramma 2018; Rao and Suri 2006; Vaidyanathan 2006). As one of the most pesticide-intensive crops, cotton offered two potential solutions to the crisis: on one hand, a GM hybrid seed that would work within the existing regime of agrochemical use, credit systems, shops, and plant scientists; on the other, a non-Bt seed that would rely upon international green marketing and farmer education. The sub-sections "Genetically modified cotton" through "Organic cotton" discuss the main elements of these cropping systems in the context of rural India.

Genetically modified cotton

In 1989, the Indian government established regulations on GM cotton that borrowed liberally from the United States, which had the world's only functioning GM regulatory framework. These rules shaped field testing, laboratory conditions, and sales (Heinemann 2012; Newell 2003) to pave the way for the commercial release of Bt cotton[1] in 2002. Bt cotton illustrates many of India's agro-environmental paradoxes, including (a) the promise of high-tech modernity as well as the threat of eroded past values (Paarlberg 2001; Pearson 2006; Scoones 2008; Shiva 1997; Stone 2002); (b) the influx of new capital and technology amid the danger of increased corporate control (Bagla and Stone 2012; Jasanoff 2005; Newell 2003; Schurman and Munro 2010; Scoones 2006); and (c) the acquisition of new farming methods at the risk of interrupting farming learning processes (Stone 2007; Stone and Flachs 2017).

Framed as a solution to an agricultural crisis characterized by yield decline, increasing pesticide use, and suicide (see "Cotton in the colonial and postcolonial eras"), GM cotton seeds (and their organic alternatives) have largely been judged on the rapidity of their adoption, their potential to decrease pesticide use, and their potential to increase yields (Flachs 2019c; Herring 2015). Undeniably, farmers rapidly adopted GM technology, with more than 95% of all cotton planted being genetically modified within 12 years of its initial introduction (Cotton Corporation of India Ltd. 2017). Yet, the simultaneous explosion in available seed choices has complicated the particular seed decisions that farmers make, in that early research showed that farmers came to plant GM seeds not for an observed yield benefit but because they achieved fad-like popularity in their villages (Stone 2007), a trend that intensified by 2013 (Stone, Flachs, and Diepenbrock 2014). Despite its rapid and enthusiastic adoption, the local cultivation and performance of agricultural knowledge are still uncertain in Indian fields.

GM cotton farmers leverage variable costs (e.g. agrochemicals, labor) against ambiguous benefits in terms of profits and yields. Given this uncertainty, they lean heavily on emulating the decisions of their neighbors or seeking the advice of experts that are external to the farming household, including pesticide shop managers and university extension agents (Flachs 2019a). Over time, many have planted particular seed brands *en masse*, to the point that the government issued permits to control sales in the face of rampant black markets (Flachs 2016c).[2] Despite this enthusiasm, farmers during 2012–2016 overwhelmingly abandoned these seeds the year after planting them, justifying their choices with *manci digubadi*, a Telugu phrase that translates literally to "good yield."

Unfortunately, in this highly speculative environment, the desire to obtain good yields can be very far from actually taking concrete steps to achieve them. Data since 2012 show that (a) there is no yield rationale to choosing particular seeds, (b) that farmers gamble by switching seeds frequently, (c) that farmers themselves do not know very much about the seeds they are planting, and (d) that the market is increasingly confusing (Flachs 2019a; 2019b). As a result, local and experiential knowledge has been eroded in favor of social or didactic knowledge, a process that arguably destabilizes local control over decision-making and the resources necessary for agricultural production (Stone 2016).

When faced with an overload of choices and no usable metrics to differentiate between them, consumers in many contexts feel anxious and overwhelmed and defer to experts who may offer a more informed choice (Chernev, Böckenholt, and Goodman 2015; Iyengar, Huberman, and Jang 2004; Iyengar and Lepper 2000). Despite not knowing which seeds they should plant, farmers must still make a choice. In this way, seed choice is paradoxically crucial yet uncertain. This suggests that *manci digubadi* is an important benchmark for agricultural decision-making, not because it reveals an agronomic truth, but because it is a script performed on this neoliberal agrarian stage (Flachs 2019b). Farmers choose seeds amidst a deluge of marketing, competition, consumption, and the persistent erosion of experiential knowledge.[3] Yet, a farm is a public stage that invites the gazes of neighbors and visiting scientists. To enhance their well-being under the conditions of a closely watched uncertainty, farmers fall back on seeking higher yields, mirroring the language of agribusiness as discussed in "Cotton in the colonial and postcolonial eras" (see Chapter 9). This hope for higher yield, of course, glosses a range of aspirations to stay on land, be successful, make good choices, and ultimately live well. But as part of the lives of cotton farmers, these seeds limit the futures imagined by growers. This new normal of the farm masks a deep ambivalence about what it means to farm well.

Organic cotton

As with GM seeds, the proponents of organic agriculture tout its potential to mitigate high agrochemical use, reverse nutritional deficiency, combat poverty, and bring domestic products to new markets (Altenbuchner, Vogel, and Larcher 2018; Panneerselvam et al. 2012). Similarly, by aligning themselves with American regulation, organic cotton producers marked their product as legally exclusive with GM technology (Schmid 2007),[4] as no GM cotton can be legally sold in organic markets, regardless of how it is produced. Indian organic standards (APEDA 2012) monitor organic cotton production practices including chemical inputs, crop rotation, and potential contamination. Indian farmers currently provide most of the organic cotton sold in global markets (Willer and Lernoud 2016), by building trust that their products adhere to standards through intensive documentation (Galvin 2018). Inspectors of farms and processing facilities are, at a minimum, accredited by state boards and monitored through TraceNet, an electronic repository of quality assurance data collected by operators and producers within the Indian organic supply chain. Some of the Telangana organic supply chains considered in this study have their own, additional monitoring systems that trace cotton to individual villages and even particular farm plots.

Even when organic farming is not technically certified, non-GM cotton farming relies on regular audits to add value and achieve sales in alternative commodities markets. Some field trials have shown that organic cotton can deliver profits that are comparable to the GM cotton cultivated with agrochemical inputs (Forster et al. 2013). However, the yields of organic cotton farmers in Telangana are only about a third of those growing GM cotton in the same areas, as documented by this ethnographic research (Flachs 2016c). As the rules are strict and the yields are low, many programs offer farmers incentives to adopt organic agriculture practices. For example, organic cotton projects buffer low yields by providing infrastructure and subsidies, including free seeds, seasonal jobs, equipment, loans, or easier access to government programs. However, there are social benefits as well, as these farmers earn celebrity in the news and on social media, they travel to farms or conferences in other states, and they discover a new platform to speak to neighbors and visitors. In essence, organic programs provide an alternative stage on which to achieve rural wellbeing.

Some farmers get especially involved in such activities, lending their likeness and stories to sell organic commodities. Some programs rely on these charismatic "show farmers" (Flachs 2017b; Stone 2014) to both extoll the virtues of organic farming and to ground-truth agricultural suggestions for future implementation. In fact, model farms and farmers have a long and important history in South Asian agricultural development, as British and American entrepreneurs employed similar model agriculture platforms to compel Indian farmers to grow high-yielding crops in the mid-19th century (Taylor and Bhasme 2018). In this case, charismatic interlocutors are one-part celebrities modeling benefits to villagers and documenters, and one-part local enforcers, who adapt agricultural management to local ecological conditions and economic needs. Sometimes, show farmer media gives the impression that organic regulation (and by proxy organic clothing consumption) is helping these farmers. This elision hides the crucial efforts of NGOs and organic companies who reduce vulnerability and build social capital by creating stages for show farmers. It also hides the efforts of ambitious, charismatic, opportunistic, and earnest farmers who take it upon themselves to implement organic farming. Furthermore, it ignores the iterative labor and knowledge that farmers and organic programs create together, including learning how to navigate new markets, creating value in commercially marginal crops, or solving farm-specific pest and soil fertility issues. In this way, social relationships between key farmers and NGO project managers reshape the landscape in ways that have positive ecological consequences in terms of soil quality, biodiversity conservation, and exposure to agrochemicals (Flachs 2016d). Such institutions are critical for helping these communities build resilient environmental relationships.

Yet, it would be a mistake to conclude that these spaces are performative and thus illegitimate. Rather, it is through the social work of reducing vulnerability, creating social capital, and incentivizing new ways of farming that organic programs support sustainable agrarian livelihoods. Organic agriculture groups that do not recognize the value that these performances provide for participating farmers can frustrate or alienate participants (Flachs 2018). In other studies (Panneerselvam et al. 2012), farmers have listed their suspicion that organic programs will not deliver on their promises as the main deterrent to joining up with organic agriculture groups. Farmers have learned that they can reap short-term benefits from projects that need short-term deliverables, such as a one-year study or a mural on village empowerment. Yet, by recognizing show farmers as partners and stakeholders looking to take advantage of new opportunities, development programs (including those related to organic cotton cultivation) have been seen to foster social relationships over short-term input incentives.

Just as GM seeds became faddish commodities in the hands of seed companies and black-market brokers, organic farmers also pursue sociocultural goals tangential to agricultural production during organic marketing. In practice, many farmers learn to perform opportunistically for visiting donors or the media, entering into a mutually beneficial relationship wherein the sponsoring program "exhibits" these farmers to visitors and potentially interested farmer converts. In exchange, these performing farmers reap the best material and social rewards for working with organic programs. On the organic cotton farms where I have conducted research, even those farmers peripheral to show farming and NGO decision-making see increased opportunities for diversified livelihoods. These include access to new jobs, markets, loans, or social networks that provide parallel values and ways to aspire to continue farming in the context of a larger moment of agrarian uncertainty (Flachs 2019b; Flachs and Panuganti 2020). While critically investigating the dynamics that guide learning on organic farms, it can be argued that they provide a future for rural wellbeing outside of *manci digubadi*. Organic cotton agriculture is less a technological panacea to the larger issues of agrarian distress and more a new kind of performance on a new kind of agricultural stage. The lack of an engaged audience has left other Telangana cotton farmers exposed to the free market and bereft of social or economic support systems.

Cooperative decision-making for agricultural stability

Political ecologists understand sustainability as depending on a particular historical, political, ecological, and economic context rather than an inevitable outcome of a technological intervention to solve a specific problem (Nightingale et al. 2020). One of the key problems of GM cotton outlined in "Genetically modified cotton" is the confusing market structure and the unreliability of much of the first-hand information used by farmers in their iterative learning process. Conversely, the advantage of organic agriculture lies in its collective decision-making and risk-management approach. This suggests that GM seeds sold and managed in a more collective system may help alleviate much of this stress and risk. For example, cotton farmers in one key village from the Warangal district of Telangana that I will call Srigonda are locally famous for a core of high-caste coastal migrants who settled large swaths of more fertile dark earth, ideal for cotton growing. But perhaps more important to the success of cotton production in this village than good soil quality or high caste has been its cooperative shop.

First, this cooperative shop is welcoming. Unlike many rural shops in the region, its cement foundation extends for several feet beyond the storefront, creating a porch. Most days, the cooperative's manager sets out chairs and extends a cloth awning over the space. As soon as the shop opens, farmers linger on the porch, drinking tea and debating newspaper and television reports. The cooperative is the first place that I go when arriving in Srigonda because it is a central node in this local social network. Like me, research scientists and corporate brokers also visit the cooperative first when coming to the area. This gives the manager and cooperative members access to a wide range of centralized expertise and resources.

During a period of seed shortage in 2012, the manager called a meeting to explain why some popular seeds were unavailable. Farmers wanted to know the best alternative seeds, and so the manager called a local extension scientist with family ties to the village, who in turn consulted with his colleagues at the extension service and at corporate breeders. This lengthy and collective discussion on seeds and their alternatives would be unfeasible for any normal village shop. A lengthy and democratic discussion is unthinkable in the melee of the Warangal seed and input sellers, who have no time for such back-and-forths with farmers because they have their own problems with inventory, thin profit margins, and stiff competition. The manager grimaces when I tell him about farmers' experiences in the Warangal shops. "We're all part of the cooperative," he explains, gesturing to fellow farmers on the porch. "There wouldn't be any point in cheating them." This grain of trust in an anarchic GM cotton market may explain why he sold half of the village's cottonseeds from 2012 to 2014, and why he dwarfed the market share of any other individual vendor named by farmers interviewed in this research.

Such community-led initiatives can have major lasting consequences, unlike the short-term interventions that are frequent in agricultural development programs. For example, a key Warangal crop scientist stayed in the area after an integrated pest management workshop and worked with a prominent family to start this cooperative. Through a long and slow process, farmers adapted these suggestions to their own land, keeping some suggestions, such as bird perches and the strategic application of chicken fertilizer, and abandoning others, such as time-intensive homemade neem pesticides. Over time, the cooperative and the scientist tweaked interventions that worked for the local community, discussing the complexities of the market and ways to diversify local agricultural production to move away from cotton. The cooperative provided not only seeds and trust but also access to small loans and collectively available machinery to mill rice and maize. This added value to local agricultural activities. With local financial instability being a major factor throughout cotton's agrarian crisis (Mohanty 2005) and long-distance micro-credit schemes falling short of assuaging farmer debts (Taylor 2011, 2013), village cooperatives that offer diverse loans alongside local and kin obligations can provide additional security for smallholders (Krishnamurthy 2018; Ramprasad 2019) by situating that risk within an existing network.

Cooperatives are not a perfect social institution. The manager described above belongs to the village's most prominent high-caste family. Vargas-Cetina (2005) notes that cooperative organizations with unstable sizes and political affiliations can be ephemeral and subvert an initial solidarity by consolidating resources within key actors. Similarly, research on collective soy buying and market negotiations has shown that barriers related to caste and education can be exacerbated in new social and technological organizations of agricultural decision-making (Kumar 2015, 2016). In this case, some lower-caste and Adivasi farmers complained during household interviews that the manager's family essentially controls access to agricultural resources and the flows of information from the university extension officers. As wealthier and predominantly high-caste farmers invest more money in the cooperative, they have a greater say in how money is to be spent than the poorer lower-caste and Adivasi farmers. For example, the manager's friends and social circle are often the first to participate in new and subsidized development schemes. And yet, the institution as a whole strengthens the farmers' ability to learn about their seeds and apply local management knowledge. Yields in the area are higher, no farmer affiliated with this cooperative has committed suicide, local biodiversity has increased, and agro-chemical use is more judicious. This is not because the cooperative's technology is better, but because the cooperative provides stability and alternatives.

Every organic program that I have seen works through cooperatives, offering financial, social, environmental, and biodiversity benefits. The agro-diversity in organic farms is striking with these farms often containing twice the number of actively managed plants compared to conventional farms (Flachs 2016d). This is not an inherent feature of organic agriculture but is achieved through explicit design in response to program demands for diversified agriculture. Farmers do not wash out fertilizer bags and pesticide bottles in the streams where farmers and sweating anthropologists swim and fish because they do not use them. By design, organic programs recognize that farmers will not succeed if they have to do this alone, and they instead provide social and material incentives for diversified work and farm products.

Genetically modified and organic cotton in India as agents of agrarian transformation

The contemporary conventional agriculture system in India is extremely effective at selling seeds and associated inputs. In the past 18 years, the private seed market has exploded, while India is now the top cotton producer in the world (USDA Foreign Agricultural Service 2018). However, the current system is far less effective at addressing the underlying problem of rural distress and suicides, low yields, and pesticide overuse that demanded GM and organic technologies in the first place. The stubborn persistence of farmer suicides appears to have virtually nothing to do with GM cotton (Gruère and Sengupta 2011; Plewis 2014). Actually, suicides peaked in the late 1990s, before GM cotton was available to farmers, and have plateaued since its introduction. While both pro-GM and anti-GM voices claim that suicides will support their narrative, a recent exhaustive study reveals an obvious, if painful, truth (Gutierrez et al. 2015): (a) farmers are still committing suicide, with historically disenfranchised communities facing the greatest risks (Vasavi 2012); (b) rural poor communities continue to lack access to credit or water infrastructure that would make them more productive (Taylor 2013); and (c) the risks that farmers assume are part of a larger story of aspiration and desire (Münster 2012, 2015a; Ramamurthy 2011). This disconnect between aspirations and possibilities in the cotton sector can seem devastating.

While much is made of how GM seeds have raised cotton yields in India (Herring 2013; Kathage and Qaim 2012; Qaim 2003), the story becomes more complicated in individual fields (Stone 2011). As more farmers planted Bt cotton across the country, yields plateaued. In fact, while Indian cotton production experienced a large increase in yields just after 2001, this came when fewer than 10% of farmers were planting GM seeds. These yield gains had stagnated by 2008, when most farmers had switched to Bt cotton (Kranthi and Stone 2020; Stone 2011).

Although Bt cotton was supposed to reduce pesticide use for Lepidopteran pests, pesticide application plummeted when Bt cotton was adopted in 2002. Yet, since 2008, when farmers adopted Bt cotton on a larger scale, pesticide use for sucking pests unaffected by Bt proteins has increased, bringing total pesticide use to even higher levels than previously observed (Flachs 2017a; Kranthi 2014). It does not always matter what the pest species are in a given field, because farmers have to be seen by neighbors to be spraying and caretaking. This is what it means to be a good and responsible farmer. One farmer explained this as: "if your neighbor sprays four times you need to spray five." This does not mean that farmers are incapable of rationally calculating their costs, but that agriculture is social. Those spraying pesticides are exposed to harmful substances while working in their plots. This affects disproportionately the largely young and female labor force of weeders and pickers who are exposed to these harmful substances during crop management. Furthermore, the agrochemicals enter groundwater and wells, finding their way to the local food systems. Paradoxically, all this is the outcome of a technology whose sole purpose is to reduce pesticide use. Conversely, organic farmers avoid this issue because they do not use these chemicals.

Considering the above, both GM and organic cotton have arguably catalyzed agrarian transformation in the Indian cotton sector. Returning to the two political ecology scales discussed in "Methodology," both technologies transform farmer relationships with the global political economy through new international and state governance, different interactions with markets, new risk exposures, and changes in access to agricultural inputs. Locally, GM and organic cotton have redefined how farmers view themselves as successful, and thus how they make decisions during quotidian farm labor. Table 7.1 summarizes some of the key impacts of GM and organic cotton in Indian agrarian life.

Ultimately, neither technology acts in a vacuum. Some of the most promising aspects of organic agriculture or GM cotton for socioeconomic development or ecological restoration come not from these technologies themselves, but from the ways that cooperatives and institutions associated with these technologies work with farmers to create networks of knowledge-sharing,

Table 7.1 Influences of genetically modified and organic cotton on agrarian transformation in India

Dimensions	Bt cotton	Organic agriculture
Global governance	- GMOs developed by multinational companies; - Cotton sold on global commodities markets governed by World Trade Organization (WTO).	- International certification and value-added organic markets facilitated through TraceNet and Agricultural and Processed Food Export Development Authority (APEDA).
National and local governance	- Varieties approved by federal regulators; - Varieties released in partnership with local extension and domestic breeding companies.	- Project managers, NGOs, and show farmers enforce guidelines and document compliance.
Agricultural inputs	- Purchased at maximum retail price* in shops or from traveling brokers.	- Heavily subsidized, household-made, or directly provided by sponsoring programs.
Seed decision-making	- Choices made in overloaded context, as part of "herds" or "fads"; - Little first-hand knowledge.	- Choices pre-determined by sponsoring organic programs.
Knowledge emphasis	- Follows regional seed trends; - Focus on yield maximization within a given season.	- Follows and troubleshoots project directives; - Presentation of self during field audits or project meetings; - Prioritizes diversified agriculture at the expense of cotton yields.
Key risks	- Insect resistance; - Global price fluctuations; - Exposure to agrochemicals; - Chronic indebtedness.	- Low cotton yields; - Dependence on project support.
Market sales	- Negotiated between buyers and farmers at open-air urban markets.	- Mediated through sponsoring organic projects.
Pathways to success	- High yields; - High farm productivity.	- Socioeconomic safety nets through sponsoring programs; - Employment generation; - Improved social standing through regional media.
Supporting institutions	- Cooperative seed and chemical shops	- Cooperative buying, selling, and regulating inherent to organic program structure.
Agricultural and agrarian impacts	- Near universal adoption of GM cotton; - Yield increases in the first decade and plateau in the second decade; - Increases in pesticide use since 2008; - Increased pest resistance.	- Production of most organic cotton sold globally; - Relegation to niche markets based in Japan, North America, and Europe; - Extensive and supportive international community.

Note: * In India's semi-controlled economy, agricultural inputs have state-mandated maximum retail prices at which they can be purchased in shops.

collective buying, and collective selling. This conclusion aligns with other social science perspectives on India's agrarian crisis (Gupta 2017; Gutierrez et al. 2015; Münster 2015b; Vasavi 2012) in focusing more attention on the ways in which farming communities experience desperation, lack of resources, and exclusion from comprehensive socioeconomic support.

Conclusions

The chapter examined the social, botanical, and historical context of cotton in India, with a specific focus on the Telangana region. Since the colonial period, cotton-related policies and technologies have largely been translated from an American to an Indian context. The most recent iteration of this process has manifested in the push to foster a sustainable and resilient cotton sector through the adoption of GM or organic cotton production. A history of agrocapitalism that views technology adoption as inevitable might argue that the high adoption rates of GM cotton in India prove the inevitable success of GM seeds, or at least their inevitable integration into cotton smallholder agriculture in the country. However, such an uncomplicated progression has never been true throughout the history of Indian cotton capitalism and is a misleading narrative of agricultural change in other parts of the world.

This chapter has argued that organic and GM cotton production in India has developed in ways that were unintended by their proponents. For example, while GM seeds were intended to be an uncomplicated agricultural intervention, they have destabilized farmer knowledge and led to a process where farmers scramble to buy specific seeds for one or two years, abandon them the following year, and then repeat the process. Conversely, while organic agriculture intended to provide farmers with new (and arguably more sustainable) cotton production methods, it often teaches farmers how to benefit from foreign buyers willing to underwrite their costs. However, by changing the definitions of success in agriculture, both these technological options also create new avenues for gaining social or economic benefits.

Although both GM and organic cotton are heavily mobilized in international debates about the future of agriculture, farmers are responding to the risks and rewards of these technological packages, situating themselves as students, wise practitioners, producers, and consumers. Cooperatives in both settings have invested substantial time to work with farmers and incentivize a different pathway to success in an agricultural sector where GM cotton farmers internalize their failures and commit suicide. In doing so, they address fundamental causes of agrarian distress where GM seeds or mandatory organic regulations only scratch the surface. In both cases, sustainability is not a question of technology, but of social institutions and daily practices.

Acknowledgment

Arguments in this chapter are adapted and updated from Flachs (2016b).

Notes

1 Bt cotton has been engineered to contain genes from *Bacillus thuringiensis*, which produce a toxin deadly to Lepidopteran. Bt cotton (and other GM crops using similar approaches) were designed to offer protection against major pests like the American bollworm (*Helicoverpa armigera*) and pink bollworm (*Pectinophora gossypiella*).

2 In 2012, smugglers circumvented permit requirements in Andhra Pradesh by bringing desired seeds from other states and charging extra for them. In other years, brokers bought popular seeds on wholesale and sold them off the backs of motorcycles. Brokers sold seeds to farmers across the region, but such seeds were disproportionally more to Scheduled Tribe Banjara farmers who live outside village propers, in comparatively poorer *thanda* communities with comparatively less infrastructure (Flachs 2016a).

3 For ethnographic descriptions of this process, see (Flachs 2019a).

4 In practice, organic farmers sometimes also grow Bt cotton on separate, uncertified land. This potential for cross-contamination, along with the persistent concerns of gene drift and cotton-mixing in gins is a frequent concern for organic auditors. For example, a 2010 study found that large quantities of Indian "organic" cotton contained Bt genes (Chua 2010).

References

Abhilash, P.C., and N. Singh. 2009. "Pesticide Use and Application: An Indian Scenario." *Journal of Hazardous Materials* 165 (1–3): 1–12. doi.org/10.1016/j.jhazmat.2008.10.061.

Altenbuchner, C., S. Vogel, and M. Larcher. 2018. "Social, Economic and Environmental Impacts of Organic Cotton Production on the Livelihood of Smallholder Farmers in Odisha, India." *Renewable Agriculture and Food Systems* 33 (4): 373–385. doi.org/10.1017/S174217051700014X.

APEDA. 2012. "National Programme for Organic Production (NPOP)." 2012. www.apeda.gov.in/apedawebsite/organic/index.htm.

Bagla, P., and R. Stone. 2012. "India's Scholar-Prime Minister Aims for Inclusive Development." *Science* 335 (6071): 907–908. doi.org/10.1126/science.335.6071.907.

Basu, A.K., and R.S. Paroda. 1995. "Hybrid Cotton in India: A Success Story." 1995/1. Bangkok: Asia-Pacific Association of Agricultural Research Institutions.

Batterbury, S.P.J. 1996. "Planners or Performers? Reflections on Indigenous Dryland Farming in Northern Burkina Faso." *Agriculture and Human Values* 13 (3): 12–22. doi.org/10.1007/BF01538223.

Beckert, S. 2014. *Empire of Cotton: A Global History.* New York: Knopf.

Bennike, R.B., M.B. Rasmussen, and K.B. Nielsen. 2020. "Agrarian Crossroads: Rural Aspirations and Capitalist Transformation." *Canadian Journal of Development Studies / Revue Canadienne d'études Du Développement* 41 (1): 40–56. doi.org/10.1080/02255189.2020.1710116.

Brookfield, H.C. 2001. *Exploring Agrodiversity.* New York: Columbia University Press.

Chacko, X.S. 2019. "Creative Practices of Care: The Subjectivity, Agency, and Affective Labor of Preparing Seeds for Long-Term Banking." *Culture, Agriculture, Food and Environment* 41 (2): 97–106. doi.org/10.1111/cuag.12237.

Chaudhuri, K.N. 1985. *Trade and Civilisation in the Indian Ocean: An Economic History from the Rise of Islam to 1750.* New York: Cambridge University Press.

Chernev, A., U. Böckenholt, and J. Goodman. 2015. "Choice Overload: A Conceptual Review and Meta-Analysis." *Journal of Consumer Psychology* 25 (2): 333–358. doi.org/ 10.1016/j.jcps.2014.08.002.

Chua, J.M. 2010. "H&M, Other Brands Guilty of 'Organic Cotton Fraud'?" *Ecouterre*, January 25, 2010. www.ecouterre.com/hm-other-brands-guilty-of-organic-cotton-fraud/.

Cotton Corporation of India Ltd. 2017. "47th Annual Report 2016–2017." Annual Report 47. Mumbai: Cotton Corporation of India Ltd.

Cullather, N. 2013. *The Hungry World: America's Cold War Battle against Poverty in Asia.* Reprint edition. Cambridge: Harvard University Press.

Desmond, E. 2017. "Risk Definition and the Struggle for Legitimation: A Case Study of Bt Cotton in Andhra Pradesh, India." *Journal of Risk Research* 20 (1): 135–150. doi. org/10.1080/13669877.2015.1042504.

Flachs, A. 2016a. "Cultivating Knowledge: The Production and Adaptation of Knowledge on Organic and GM Cotton Farms in Telangana, India." Doctoral Dissertation, St. Louis: Washington University in St. Louis.

———. 2016b. "Green Revolution." In *Encyclopedia of Food and Agricultural Ethics*, edited by Paul B. Thompson and David M. Kaplan, 1–7. Dordrecht: Springer Netherlands. link.springer.com/10.1007/978-94-007-6167-4_567-1.

———. 2016c. "Redefining Success: The Political Ecology of Genetically Modified and Organic Cotton as Solutions to Agrarian Crisis." *Journal of Political Ecology* 23 (1): 49–70.

———. 2016d. "The Economic Botany of Organic Cotton Farms In Telangana, India." *Journal of Ethnobiology* 36 (3): 683–713. doi.org/10.2993/0278-0771-36.3.683.

———. 2017a. "Transgenic Cotton: High Hopes and Farming Reality." *Nature Plants* 3 (January): 16212. doi.org/10.1038/nplants.2016.212.

———. 2017b. "'Show Farmers': Transformation and Performance in Telangana, India." *Culture, Agriculture, Food and Environment* 39 (1): 25–34. doi.org/10.1111/cuag.12085.

———. 2018. "Development Roles: Contingency and Performance in Alternative Agriculture in Telangana, India." *Journal of Political Ecology* 25 (1): 716–731.

———. 2019a. *Cultivating Knowledge: Biotechnology, Sustainability, and the Human Cost of Cotton Capitalism in India.* Global Change/Global Health. Tucson, AZ: University of Arizona Press.

———. 2019b. "Planting and Performing: Anxiety, Aspiration, and 'Scripts' in Telangana Cotton Farming." *American Anthropologist* 121 (1): 48–61. doi.org/10.1111/ aman.13175.

———. 2019c. "The Factish in the Field: An Anthropological Inquiry of Genetically Modified Seeds and Yields as Beings." *Science and Technology Studies* 32 (3): 26–43.

Flachs, A., and S. Panuganti. 2020. "Organic Aspirations in South India." *Economic Anthropology* 7 (1): 38–50. doi.org/10.1002/sea2.12158.

Flachs, A., and P. Richards. 2018. "Playing Development Roles: The Political Ecology of Performance in Agricultural Development." *Journal of Political Ecology* 25 (1): 638–646.

Fletcher, R., B. Dowd-Uribe, and G.A. Aistara, eds. 2020. *The Ecolaboratory: Environmental Governance and Economic Development in Costa Rica.* Tucson: University of Arizona Press.

Forster, D., C. Andres, R. Verma, C. Zundel, M.M. Messmer, and P. Mäder. 2013. "Yield and Economic Performance of Organic and Conventional Cotton-Based Farming Systems – Results from a Field Trial in India." *PLoS ONE* 8 (12): e81039. doi.org/ 10.1371/journal.pone.0081039.

Galab, S., E. Revathi, and P.P. Reddy. 2009. "Farmers' Suicides and Unfolding Agrarian Crisis in Andhra Pradesh." In *Agrarian Crisis in India*, edited by D. Narasimha Reddy and S. Mishra. New Delhi, New York: Oxford University Press.

Galvin, S.S. 2018. "The Farming of Trust: Organic Certification and the Limits of Transparency in Uttarakhand, India." *American Ethnologist* 45 (4): 495–507. doi.org/10.1111/amet.12704.

Gold, A.G. 2003. "Vanishing: Seeds' Cyclicality." *Journal of Material Culture* 8 (3): 255–272. doi.org/10.1177/13591835030083002.

Gruère, G., and D. Sengupta. 2011. "Bt Cotton and Farmer Suicides in India: An Evidence-Based Assessment." *Journal of Development Studies* 47 (2): 316–337. doi.org/10.1080/00220388.2010.492863.

Guha, S. 2007. "Genetic Change and Colonial Cotton Improvement in Nineteenth and Twentieth Century India." In *Situating Environmental History*, edited by R. Chakrabarti, 307–322. New Delhi: Manohar.

Gupta, A. 1998. *Postcolonial Developments: Agriculture in the Making of Modern India*. Durham: Duke University Press Books.

———. 2017. "Farming as Speculative Activity: The Ecological Basis of Farmers' Suicides in India." In *The Routledge Companion to the Environmental Humanities*, edited by Ursula K. Heise, Jon Christensen, and Michelle Niemann, 1st edition, 185–193. London; New York: Routledge.

Gutierrez, A.P., L. Ponti, H.R. Herren, J. Baumgärtner, and P.E. Kenmore. 2015. "Deconstructing Indian Cotton: Weather, Yields, and Suicides." *Environmental Sciences Europe* 27 (1): 12. doi.org/10.1186/s12302-015-0043-8.

Hazareesingh, S. 2016. "'Your Foreign Plants Are Very Delicate': Peasant Crop Ecologies and the Subversion of Colonial Cotton Designs in Dharwar, Western India, 1830–1880." In *Local Subversions of Colonial Cultures: Commodities and Anti-Commodities in Global History*, 1–10. Cambridge Imperial and Post-Colonial Studies Series. Hampshire, UK: Palgrave MacMillan.

Heinemann, J. 2012. "Suggestions on How to Apply International Safety Testing Guidelines for Genetically Modified Organisms." Christchurch, New Zealand: Centre for Integrated Research in Biosafety.

Herring, R.J. 2013. "Reconstructing Facts with Bt Cotton: Why Scepticism Fails." *Economic and Political Weekly* 48 (33): 63–66.

———. 2015. "State Science, Risk and Agricultural Biotechnology: Bt Cotton to Bt Brinjal in India." *The Journal of Peasant Studies* 42 (1): 159–186. doi.org/10.1080/03066150.2014.951835.

Iyengar, S.S., G. Huberman, and W. Jang. 2004. "How Much Choice Is Too Much? Contributions to 401(k) Retirement Plans." In *Pension Design and Structure: New Lessons from Behavioral Finance*, edited by Olivia S. Mitchell and Steve Utkus, 83–95. Oxford: Oxford University Press.

Iyengar, S.S., and M.R. Lepper. 2000. "When Choice Is Demotivating: Can One Desire Too Much of a Good Thing?" *Journal of Personality and Social Psychology* 79 (6): 995–1006. doi.org/10.1037/0022-3514.79.6.995.

Jasanoff, S. 2005. *Designs on Nature: Science and Democracy in Europe and the United States*. Princeton, NJ: Princeton University Press.

Kathage, J., and M. Qaim. 2012. "Economic Impacts and Impact Dynamics of Bt (Bacillus Thuringiensis) Cotton in India." *Proceedings of the National Academy of Sciences* 109 (29): 11652–11656. doi.org/10.1073/pnas.1203647109.

Kinchy, A. 2012. *Seeds, Science, and Struggle: The Global Politics of Transgenic Crops.* Cambridge: MIT Press.

Kranthi, K.R. 2014. "Cotton Production Systems – Need for a Change in India." *Cotton Statistics & News*, December 16, 2014: 4–7.

Kranthi, K.R., and G.D. Stone. 2020. "Long-Term Impacts of Bt Cotton in India." *Nature Plants* 6 (3): 188–196. doi.org/10.1038/s41477-020-0615-5.

Krishnamurthy, M. 2018. "Fields, Markets and Agricultural Commodities." In *Critical Themes in Indian Sociology*, edited by S. Srivastava, Y. Arif, and J. Abraham. New Delhi: SAGE Publications India.

Kumar, R. 2015. *Rethinking Revolutions: Soyabean, Choupals, and the Changing Countryside in Central India.* 1st edition. New Delhi, India: Oxford University Press.

———. 2016. "The Perils of Productivity: Making 'Good Farmers' in Malwa, India." *Journal of Agrarian Change* 16 (1): 70–93. doi.org/10.1111/joac.12084.

Lansing, J.S. 2006. *Perfect Order: Recognizing Complexity in Bali.* Princeton Studies in Complexity. Princeton, NJ: Princeton University Press.

Ludden, D. 1999. *An Agrarian History of South Asia.* Cambridge: Cambridge University Press.

McMichael, P.D. 2007. "Globalization and the Agrarian World." In *The Blackwell Companion to Globalization*, edited by G. Ritter, 216–238. Malden, MA: Blackwell Publishing.

Meek, D. 2016. "The Cultural Politics of the Agroecological Transition." *Agriculture and Human Values* 33 (2): 275–290. doi.org/10.1007/s10460-015-9605-z.

Menon, M., and Uzramma. 2018. *A Frayed History: The Journey of Cotton in India.* Oxford, New York: Oxford University Press.

Mintz, S.W. 1986. *Sweetness and Power: The Place of Sugar in Modern History.* New York: Penguin Books.

Mohanty, B.B. 2005. "'We Are Like the Living Dead': Farmer Suicides in Maharashtra, Western India." *The Journal of Peasant Studies* 32 (2): 243–276. doi.org/10.1080/03066150500094485.

Münster, D. 2012. "Farmers' Suicides and the State in India: Conceptual and Ethnographic Notes from Wayanad, Kerala." *Contributions to Indian Sociology* 46 (1–2): 181–208. doi.org/10.1177/006996671104600208.

———. 2015a. "'Ginger Is a Gamble': Crop Booms, Rural Uncertainty, and the Neoliberalization of Agriculture in South India." *Focaal* 2015 (71): 100–113. doi.org/10.3167/fcl.2015.710109.

———. 2015b. "Farmers' Suicides as Public Death: Politics, Agency and Statistics in a Suicide-Prone District (South India)." *Modern Asian Studies* 49 (5): 1580–1605. doi.org/10.1017/S0026749X14000225.

Nazarea, V.D. 2014. *Heirloom Seeds and Their Keepers: Marginality and Memory in the Conservation of Biological Diversity.* Reprint edition. Tucson: University of Arizona Press.

Newell, P. 2003. "Biotech Firms, Biotech Politics: Negotiating GMOs in India." IDS Working Paper 201. Brighton: Institute of Development Studies.

Nicholls, C.I., and M.A. Altieri. 1997. "Conventional Agricultural Development Models and the Persistence of the Pesticide Treadmill in Latin America." *International Journal of Sustainable Development & World Ecology* 4 (2): 93–111. doi.org/10.1080/13504509709469946.

Nightingale, A.J., S. Eriksen, M. Taylor, T. Forsyth, M. Pelling, A. Newsham, E. Boyd, et al. 2020. "Beyond Technical Fixes: Climate Solutions and the Great Derangement." *Climate and Development* 12 (4): 343–352. doi.org/10.1080/17565529.2019.1624495.

Oosterhuis, D.M., and J. Jernstedt. 1999. "The Origin and Domestication of Cotton." In *Cotton: Origin, History, Technology, and Production*, edited by C. Wayne Smith and J. Tom Cothren, 175–206. New York: John Wiley & Sons.

Paarlberg, R.L. 2001. *The Politics of Precaution: Genetically Modified Crops in Developing Countries*. Baltimore: The Johns Hopkins University Press.

Pandian, A. 2011. "Ripening with the Earth: On Maturity and Modernity in South India." In *Modern Makeovers: A Handbook of Modernity in South Asia*, edited by S. Dube, 1st ed., 157–169. New Delhi: Oxford University Press.

Panneerselvam, P., N. Halberg, M. Vaarst, and J.E. Hermansen. 2012. "Indian Farmers' Experience with and Perceptions of Organic Farming." *Renewable Agriculture and Food Systems* 27 (02): 157–169. doi.org/10.1017/S1742170511000238.

Pearson, M. 2006. "'Science', Representation and Resistance: The Bt Cotton Debate in Andhra Pradesh, India." *The Geographical Journal* 172 (4): 306–317.

Perkins, J.H. 1997. *Geopolitics and the Green Revolution: Wheat, Genes, and the Cold War*. New York: Oxford University Press.

Plewis, I. 2014. "Indian Farmer Suicides: Is GM Cotton to Blame?" *Significance* 11 (1): 14–18. doi.org/10.1111/j.1740-9713.2014.00719.x.

Qaim, M. 2003. "Bt Cotton in India: Field Trial Results and Economic Projections." *World Development* 31 (12): 2115–2127. doi.org/10.1016/j.worlddev.2003.04.005.

Ramamurthy, P. 2011. "Rearticulating Caste: The Global Cottonseed Commodity Chain and the Paradox of Smallholder Capitalism in South India." *Environment and Planning A: Economy and Space* 43 (5): 1035–1056. doi.org/10.1068/a43215.

Ramprasad, V. 2019. "Debt and Vulnerability: Indebtedness, Institutions and Smallholder Agriculture in South India." *The Journal of Peasant Studies* 46 (6): 1286–1307. doi.org/10.1080/03066150.2018.1460597.

Rao, P.N., and K.C. Suri. 2006. "Dimensions of Agrarian Distress in Andhra Pradesh. Economic and Political Weekly." *Economic and Political Weekly* 41 (16): 1546–1552.

Schmid, O. 2007. "Development of Standards for Organic Farming." In *Organic Farming: An International History*, edited by William Lockeretz, 152–174. Cambridge: CABI.

Schurman, R., and W.A. Munro. 2010. *Fighting for the Future of Food: Activists versus Agribusiness in the Struggle over Biotechnology*. Minneapolis: University of Minnesota Press.

Scoones, I. 2006. *Science, Agriculture and the Politics of Policy: The Case of Biotechnology in India*. New Delhi: Orient Blackswan.

———. 2008. "Mobilizing Against GM Crops in India, South Africa, and Brazil." *Journal of Agrarian Change* 8 (2 and 3): 315–344.

Shiva, V. 1997. *Biopiracy: The Plunder of Nature and Knowledge*. Boston: South End Press Collective.

Shiva, V., A.H. Jafri, A. Emani, and M. Pande. 2002. *Seeds of Suicide*. 2nd ed. New Delhi: Research Foundation for Science, Technology, and Ecology.

Stone, G.D. 2002. "Both Sides Now: Fallacies in the Genetic-Modification Wars, Implications for Developing Countries, and Anthropological Perspectives." *Current Anthropology* 43 (4): 611–630.

———. 2007. "Agricultural Deskilling and the Spread of Genetically Modified Cotton in Warangal." *Current Anthropology* 48 (1): 67–103.

———. 2011. "Field versus Farm in Warangal: Bt Cotton, Higher Yields, and Larger Questions." *World Development* 39 (3): 387–398.

———. 2014. "Theme Park Farming in Japan." Blog. *Fieldquestions* (blog). June 5, 2014. fieldquestions.com/2014/06/05/theme-park-farming-in-japan/.

————. 2016. "Towards a General Theory of Agricultural Knowledge Production: Environmental, Social, and Didactic Learning." *Culture, Agriculture, Food and Environment* 38 (1): 5–17.

Stone, G.D., and A. Flachs. 2017. "The Ox Fall down: Path-Breaking and Technology Treadmills in Indian Cotton Agriculture." *The Journal of Peasant Studies* 45 (7): 1272–1296. doi.org/10.1080/03066150.2017.1291505.

Stone, G.D., A. Flachs, and C. Diepenbrock. 2014. "Rhythms of the Herd: Long Term Dynamics in Seed Choice by Indian Farmers." *Technology in Society* 36 (1): 26–38. doi.org/10.1016/j.techsoc.2013.10.003.

Taylor, M. 2011. "'Freedom from Poverty Is Not for Free': Rural Development and the Microfinance Crisis in Andhra Pradesh, India." *Journal of Agrarian Change* 11 (4): 484–504. doi.org/10.1111/j.1471-0366.2011.00330.x.

————. 2013. "Liquid Debts: Credit, Groundwater and the Social Ecology of Agrarian Distress in Andhra Pradesh, India." *Third World Quarterly* 34 (4): 691–709. doi.org/10.1080/01436597.2013.786291.

Taylor, M., and S. Bhasme. 2018. "Model Farmers, Extension Networks and the Politics of Agricultural Knowledge Transfer." *Journal of Rural Studies* 64 (November): 1–10. doi.org/10.1016/j.jrurstud.2018.09.015.

Tsing, A.L. 2005. *Friction: An Ethnography of Global Connection.* Princeton, NJ, Oxford: Princeton University Press.

USDA Foreign Agricultural Service. 2018. "Cotton: World Markets and Trade." Trade Report. World Production, Markets, and Trade Reports. Washington DC: United States Department of Agriculture.

Vaidyanathan, A. 2006. "Farmers' Suicides and the Agrarian Crisis." *Economic and Political Weekly* 41 (38): 4009–4013.

Vargas-Cetina, G. 2005. "Anthropology and Cooperatives: From the Community Paradigm to the Ephemeral Association in Chiapas, Mexico." *Critique of Anthropology* 25 (3): 229–251. doi.org/10.1177/0308275X05055210.

Vasavi, A.R. 1999. *Harbingers of Rain: Land and Life in South Asia.* New Delhi: Oxford University Press.

————. 2012. *Shadow Space: Suicides and the Predicament of Rural India.* Gurgaon: Three Essays Collective.

————. 2020. "The Tiger and the Tube Well: Malevolence in Rural India." *Critical Asian Studies* 52 (3): 429–445. doi.org/10.1080/14672715.2020.1764855.

Willer, H., and J. Lernoud, eds. 2016. *The World of Organic Agriculture: Statistics and Emerging Trends 2016.* Bonn: Research Institute of Organic Agriculture (FiBL), Frick, and IFOAM – Organics International.

8 Changing agrarian dynamics in oil palm and jatropha production areas of Ghana

A feminist political ecology perspective

Abubakari Ahmed and Alexandros Gasparatos

Introduction

Agricultural modernization and its linkages to economic growth and rural development, and the transformation of agrarian livelihoods has been a major theme in industrial crop research (Bernstein, 2004, 2010) (Chapters 1 and 3). As discussed throughout this edited volume, industrial crops are major agents of agrarian transformation, shaping profoundly agricultural systems at different scales through various context-specific mechanisms (Chapters 3, 7, 10, and 11).

Much of the current literature on industrial crops has focussed on their environmental or socioeconomic impacts in terms of land use change, employment/income generation, food security, and poverty alleviation (Chapters 1 and 3). However, many of the studies on the impacts of industrial crops have been criticized as being depoliticized. This is mainly due to their failure to consider the underlying power relations in contexts of industrial crop expansion, and the strong linkages between the political, socioeconomic, and environmental processes shaping the dynamics of expansion (Darkoh and Ould-Mey, 1992; Masanjala, 2006). Many scholars have pointed to the clear need to document better the politico-socio-ecological transformations catalyzed by industrial crop production, including changes in agrarian structures, class struggles, and the restructuring of labour relations (Moyo, 2008; O'laughlin, 2016). Indeed, an emerging body of literature has been framing industrial crop production at the intersection of interrelated socio-political and ecological processes (Ariza-Montobbio et al., 2010; Borras et al., 2010; Dietz et al., 2014; White and Dasgupta, 2010).

Gender has featured saliently within this broad context of agrarian transformation, power relations, and interrelated processes that characterize industrial crop expansion (Dao, 2018; Dietz et al., 2014; Elmhirst et al., 2017) (Chapter 3). For example, many studies have shown the very different ways that men and women engage in (and are affected by) industrial crop systems (Behrman et al., 2012; Elmhirst et al., 2017), not the least via differentiated tasks and production responsibilities, knowledge, and tenure rights (Gyapong, 2019; Manley and Van Leynseele, 2019; Zakaria, 2017). For example, in plantation settings, men and women are often assigned very different tasks, which is "justified" in terms of physical labour requirements and generally leads to more employment opportunities for men (Hill and Vigneri, 2014; Li, 2015) (Chapter 3). Furthermore, in smallholder settings, most industrial crops are perceived to be "men's crops" due to their income generation potential (i.e. cash crops), while the prevailing land politics and insecure land tenure rights tend to prevent most women from engaging in their production (Carr, 2008; Elmhirst et al., 2017; Fonjong, 2017; Zakaria, 2017). Furthermore, in many agrarian settings of the global South, men tend to have better access to knowledge for industrial crop production than women (Quisumbing et al., 2014; Ragsdale et al., 2018; Yeboah et al., 2020).

Ghana is one of the countries in Sub-Saharan Africa (SSA) that has experienced both the forceful expansion of industrial crops during the colonial period, as well as a continued reliance after its independence in 1957. Indeed, some industrial crops such as cocoa and oil palm have played a critical role in national economic growth, foreign exchange generation, and rural development, having been widely promoted and received government support (Manley and Van Leynseele, 2019; Ministry of Food and Agriculture, 2015) (Chapter 1). Other industrial crops such as cotton, sugarcane, and jatropha have experienced boom-and-bust cycles characterized by large initial expansions followed by almost total collapses (Ahmed et al., 2017; Boafo et al., 2018; Francois et al., 2015). Similar to many other SSA countries, Ghana experienced in the mid-2000s a wave of large-scale land acquisitions (LSLAs), mainly for large-scale jatropha production (Ahmed et al., 2017, 2019a, 2019b, 2019c) (Chapter 3). Even though the jatropha sector eventually collapsed (Ahmed et al., 2019b), it reinvigorated the interest in the large-scale production of other industrial crops such as sugarcane and oil palm (Manley and Van Leynseele, 2019; Ministry of Trade and Industry, 2016).

Considering the above, this chapter aims to analyze how industrial crop systems have become agents of agrarian transformation in Ghana. In particular, we draw from feminist political ecology to explore these phenomena in oil palm and jatropha production sites in Kwae and Yeji, respectively. "Methodology" briefly outlines the conceptual framing of feminist political ecology (see "Research approach"), study sites (see "Study sites"), and the data collection and analysis methods (see "Data collection and analysis"). Sections "Changes in agrarian structure" through "Gender differentiation" contain the main results and discussion, in terms of how oil palm and jatropha production have changed the agrarian structure (see "Changes in agrarian structure") and caused social differentiation (see "Creation and differentiation of social groups") and gender differentiation (see "Gender differentiation"). "Mechanisms underpinning agrarian transformation" synthesizes the main mechanisms through which oil palm and jatropha production have catalyzed agrarian transformation in the two study sites, as well as the main implications of these mechanisms for Ghana and beyond.

Methodology

Research approach

As mentioned throughout this edited volume, industrial crop production is a major cause of ecological, agrarian, and socioeconomic transformation. Political ecology offers a very powerful lens to unravel the different factors shaping their production and their impacts (Adams et al., 2019; Ariza-Montobbio et al., 2010; Dell'Angelo, et al., 2017; Dietz et al., 2014) (Chapters 1–3). However, whereas in "standard" political ecology the focus is on the general political and institutional factors shaping environmental change (Peet et al., 2011; Perreault et al., 2015), a feminist political ecology lens delves more into how everyday practices intersect with gender (Harcourt, 2019; Lamb et al., 2017; Sato and Soto Alarcón, 2019; Sundberg, 2017). Thus, in feminist political ecology, the emphasis is placed on what people do (i.e. practices) and say (i.e. discourse), as well as how these can be disaggregated (Elmhirst et al., 2017).

However, it is important to note that, in feminist political ecology, gender is linked closely to power and other intersectional categories that shape everyday practices such as race, ethnicity, and poverty (Harris et al., 2017). The underlying logic is that different social groups are not homogenous in their everyday practices and discourses, but they vary due to factors such as the ones outlined above. As discussed throughout this book, this is very evident in industrial crop settings as well (see "Introduction") (Chapters 3, 7, and 11).

In this sense, feminist political ecology offers an ideal lens in understanding how industrial crop systems (re)produce, situate, and unevenly distribute inequalities on the basis of gender, class, race, and ethnicity (Elmhirst et al., 2017; Lamb et al., 2017). Feminist political ecology views gendered relations in the context of industrial crop expansion as the product of the interconnection of gendered responsibilities and tasks during their production, as well as of differentiation in knowledge, rights, politics, and access to resources (Awanyo, 2001; Dao, 2018; Elmhirst et al., 2017; Gunewardena, 2010; Lamb et al., 2017; Mollett and Faria, 2013; Vadjunec et al., 2011).

In this chapter, we also employ the concept of agrarian structure (see "Changes in agrarian structure"), which is defined as the distribution of assets and rights tied to land in areas where agrarian activities are the basis of livelihoods (Kevane, 1996; Lehman, 1977). By appreciating changes in the agrarian structure before and after the introduction of industrial crops, it is possible to obtain a better perspective on how their production has shaped local agrarian dynamics and gender relations.

Study sites

This chapter focusses on the Smart Oil jatropha plantation (in Yeji) and the GOPDC oil palm plantation and its surrounding smallholders (in Kwae)

Table 8.1 Key characteristics of the study sites

	Oil palm (Kwae)	Jatropha (Yeji)
Location	6°14′40.82″ N	8°13′34.46″ N
	0°58′12.43″ W	0°39′12.93″ W
Investor/operator	GOPDC	Smart Oil
Land acquired (ha)	14,000	6,750
Land cultivated (ha)	8,200	720
Company employees	4,500	250
Smallholders	7,200 GOPDC outgrowers Unknown number of independent growers	NA

Source: (Ahmed et al., 2019a, 2019c).

(Table 8.1). The two study sites are associated with different periods of industrial crop expansion in Ghana. The oil palm site reflects a state-led compulsory LSLA that sought to boost export-oriented development during the colonial and early parts of the post-colonial cash crop revolution in western Africa (Tosh, 1980) (see also Chapters 1 and 11). Conversely, the LSLA Yeji occurred during the global land rush of the early 2000s and was facilitated through neoliberal policies to attract private investors through foreign direct investments (FDIs) (Chapters 1, 3, and 5). In both areas, industrial crop production entailed large-scale landscape transformation and intensified crop production in monocultures, shaping in the process prevailing agrarian structures (see "Results and discussion"). Below, we introduce briefly the two study sites, but the interested reader is diverted to Ahmed et al. (2019a, 2019c) for further details about the sites and the land acquisition processes.

The oil palm site is located in Kwae and contains a large core plantation and mill currently operated by the Ghana Oil Palm Plantation Development Company (GOPDC), surrounded by thousands of individual outgrowers and independent smallholders (Ahmed et al., 2019a, 2019c) (Chapter 1). Oil palm production in the area started in the 1970s and experienced many changes over the years (Ahmed et al., 2019a). The core GOPDC plantation and mill spans 8,200 ha and employs over 4,500 staff members. More than 70% of these employees engage in agricultural activities, such as harvesting, fertilizer application, and weeding, and are paid monthly salaries. An additional 10,000 ha approximately are under smallholder-based oil palm production. The GOPDC works closely with about 7,200 outgrowers, providing them agricultural knowledge and inputs for oil palm cultivation. In turn, these outgrowers are contractually required to sell their oil palm harvest solely to the GOPDC mill. Furthermore, many thousands of independent smallholders are engaged in oil palm production in the broader area, but in contrast to the GOPDC outgrowers, they can decide depending on price signals whether to sell their harvest to the GOPDC mill, other licensed oil palm buyers (e.g. Oboama Company), or small independent processors depending on price signals (Ahmed et al., 2019a, 2019c). Poverty is fairly low in the study area compared to the rest of the country, with 16.6% of the local population characterized as poor.

The Smart Oil plantation in Yeji was incorporated in 2006 and started large-scale jatropha cultivation in 2011. Considering the lack of a functional biodiesel programme in Ghana (Ahmed et al., 2017), the initial plan was to export jatropha seeds to Burkina Faso and some European Union (EU) countries (Chapter 3). The land concession that accommodates jatropha production spans 6,750 ha and was created through the consolidation of land parcels from the Kadue, Agentriwa, Kwaese communities (Ahmed et al., 2019a). Most permanent employees are involved in nursery and agricultural activities such as harvesting, and fewer in managerial or professional tasks (e.g. machinery operators). Seasonal workers pick jatropha seeds during harvesting and transplant trees from the nursery to the plantation (Ahmed et al., 2019a, 2019c). The plantation is located in the Pru district, which is one of the poorest districts in Ghana, registering a 43.1% poverty incidence rate.

Data collection and analysis

The present study is part of a broader research project on the impacts of industrial crop production in Ghana (Ahmed et al., 2017, 2018, 2019a, 2019c). This chapter synthesizes selected qualitative and quantitative data collected through two fieldwork campaigns.

During the first fieldwork campaign, we conducted 400 household surveys in the GOPDC area (December 2016–January 2017) and 250 in the Smart Oil area (August 2017–September 2017). These surveys were conducted with households having different types of involvement in oil palm and jatropha production, including (a) workers in the GOPDC and Smart Oil plantations, (b) oil palm smallholders (i.e. outgrowers, independent growers), and (c) households not involved in oil palm/jatropha production but engaged in subsistence agriculture (i.e. control groups). More detailed information about the survey and sampling design can be found elsewhere (Gasparatos et al., 2018b). The data reported in this chapter include information about (a) land ownership, (b) income, (c) dependence on ecosystem services, and (d) perceptions of change in the access to ecosystem services following landscape conversion for oil palm and jatropha production.

During the second fieldwork campaign (August–September 2017), we conducted in-depth follow-up interviews with a subset of the original household survey respondents. We selected randomly 53 respondents in the oil palm area and 32 respondents in the jatropha area (Table 8.2). These interviews extracted information about (a) changes in rural agrarian structures, (b) labour practices (i.e. what respondents do), and (c) discourses (i.e. what respondents say about what they do) that shape agrarian dynamics.

During the second fieldwork campaign, we also conducted four Focus Group Discussions (FGDs) in each study area. In the GOPDC area, we conducted four

Table 8.2 Data collection mechanisms

Mechanism	Oil palm (Kwae)	Jatropha (Yeji)
Household surveys	- 100 GOPDC workers - 100 oil palm out-growers - 100 independent oil palm growers - 100 subsistence farmers (control)	- 100 permanent plantation workers - 50 seasonal plantation workers - 100 subsistence farmers (control)
Household interviews	- 20 GOPDC workers - 9 oil palm out-growers - 12 independent oil palm growers - 12 subsistence farmers (control)	- 20 permanent plantation workers - 12 subsistence farmers (control)
FGDs	4	4
Expert interviews	3	2

FGDs, two with oil palm growers and two with control groups. For each of these groups, we conducted one FGD only consisting of women and one only consisting of men. We used a similar FGD strategy in the jatropha study area. Each FGD involved 7–10 respondents. This division across gender sought to elicit better their unique experiences, practices, and discourses in engaging with industrial crop production. Finally, we conducted three expert interviews in Kwae and two in Yeji with local chiefs.

Household surveys were analyzed through basic statistics; see more details in Ahmed et al. (2019c). Household interviews, FGDs, and expert interviews were transcribed and coded manually to extract the main themes and repeated statements, especially those related to gendered labour practices and discourses; see more details in Ahmed et al. (2019a).

Results and discussion

Changes in agrarian structure

When understanding the changes that industrial crops have brought to agrarian structure, it is first important to appreciate the context within which their expansion has occurred. Much like other parts of the global South, industrial crop expansion (and associated LSLAs) in Ghana occurred in two major waves: during the colonial and post-colonial periods, and the global land rush of the 2000s (Chapters 1 and 3). During the former wave, LSLAs reflected the concept of eminent domain in the public interest and entailed government-led compulsory land acquisitions with compensation to local communities. During the latter wave, LSLAs were mainly undertaken by the private sector through foreign direct investments (FDIs) that were facilitated by neoliberal development policies (Chapter 3).

In more detail, the first wave entailed 835 colonial and 501 post-colonial large compulsory land acquisitions that covered 53,123 ha and 103,720 ha, respectively, before 2005 (Larbi et al., 2004). The second wave entailed the acquisition of over 1 Mha between 2005 and 2011, which were on average much larger in size compared to the government-led compulsory land acquisitions of the post-colonial period (Schoneveld, 2014). Essentially, the oil palm site was developed during the first wave and the jatropha site during the second wave.

In Kwae, prior to the introduction of oil palm, the land was predominantly used for the smallholder-based production of cocoa and food crops (for subsistence). Traditionally, the land tenure system in the area has been stool land ownership, whereby the local chief holds the land in trust for the subjects, namely the local community (Ahmed et al., 2019a). Community members have usufructuary rights to use the land for agricultural production and other livelihood activities such as hunting and gold mining (Amanor, 2001). Land inheritance occurs matrilineally, whereby male heirs inherit the land from their mother's side, with women occasionally inheriting the land to cement marriage or business relations (Amanor, 2001).

On the 11th of March 1976, over 14,000 ha in the broader Kwae area was "acquired for the public interest" under the 1962 State Lands Act signed by IK Acheampong (Ahmed et al., 2019a). Most of this land was eventually used for oil palm cultivation in an intensive plantation-based commercial model, catering mainly to the domestic market. Eventually, the GOPDC was privatized in 1992 following the mass divestiture programme of the Ghanaian government during the 1990s. Currently, GOPDC is solely owned by the SIAT Group, a Belgian agro-industrial group. However, the introduction of oil palm and the development of the GOPDC plantation had profound ramifications for the local agrarian system, affecting significantly the landscape, livelihoods, and demography of the wider area.

In more detail, the GOPDC plantation created thousands of formal jobs in an area that had very few such opportunities beforehand. These jobs did not only attract members of the local communities but also migrants from other parts of Ghana. The subsequent development of the outgrower scheme and the emergence of independent oil palm smallholders meant that a substantial fraction of the local farmers started growing oil palm as a cash crop (see "Study sites"). This was facilitated by the availability of different stable market streams due to the steady national/international palm oil demand. The expansion of GOPDC plantations and the smallholders engaged in the oil palm sector caused significant land and labour shifts in the broader area (Ahmed et al., 2019a), having important ramifications in terms of class formation and gender-differentiated outcomes (see sub-sections "Creation and differentiation of social groups" through "Gender differentiation").

In Yeji, before the introduction of jatropha, the local landscape was predominantly covered with patches of farmland and natural vegetation, primarily savanna grasses and woody species such as shea (Adams et al., 2019). Subsistence farming and livestock rearing formed traditionally (and overwhelmingly) the main livelihoods for the local rural communities. The land tenure system is mixed, with some plots belonging to individuals, families, and clans while other areas being under the control of the chiefs (i.e. stool land) (Ahmed et al., 2019a).

On the 4th of November 2011, approximately 6,540 ha of the stool land was given to Smart Oil to develop a large-scale commercial jatropha plantation (Ahmed et al., 2019a). Over time, the company employed both permanent and seasonal workers for plantation activities. However, in contrast to the GOPDC, these jobs were fewer in number and mainly targeted the local communities (Ahmed et al., 2019a). However, due to the landscape conversion and the very different nature of plantation compared to the other livelihood options in the region, Smart Oil also caused changes in the agrarian structure of the community (see sub-sections "Creation and differentiation of social groups" through "Gender differentiation").

Creation and differentiation of social groups

Table 8.3 contains information about the land ownership and income levels of groups with different involvement in oil palm and jatropha production in the two sites. Table 8.3 suggests the very different characteristics of groups with different types of engagement in industrial crop production, with some groups being better off than their respective controls, and others not. We argue below that industrial crop production in both sites has not only influenced changes in agrarian structure (see "Changes in agrarian structure") but also has further created new social groups, leading or reinforcing their differentiation.

In Kwae, the land acquisition and compensation modalities followed by the state (Ahmed et al., 2019a) caused land dispossession for many local households (Amanor, 2001). In particular, the land was acquired through the right to eminent domain and a compulsory acquisition under the 1962 State Lands Act (Act 125) (Ahmed et al., 2019a). When the GOPDC started operating, the community members that provided land for the development of the plantation received cash compensation instead of land resettlement, which could have allowed them to have land in an alternative location (Ahmed et al., 2019a). Due to the matrilineal system of land inheritance in the wider Kwae area, women were more severely affected by land dispossession (Amanor, 2001). The interview with the chief of Kwae indicated that:

> We have people here who do not have land now, although they used to own land before the coming of GOPDC. As compensation, most of them work with GOPDC. In the early 1980s, there are many migrant farmers here too. Most of them came for a season works in the plantation, but some ended up not going back.
>
> [pers. comm., Local Chief in Kwae]

The above statement also points to the influx of migrants drawn from employment opportunities in the plantation since the very early stages of GOPDC development. Many of these migrant workers are still employed permanently

Table 8.3 Income and land ownership across study groups

Site	Group	Land ownership			Total income (GHC)			Income per capita (GHC/cap)		
		Total (ha)	Cropland (ha)	Unused land (ha)	Entire sample	Men	Women	Entire sample	Men	Women
Kwae	Plantation workers	0.6	0.5	0.1	5,834.5	6,195.8	5,172.1	2,585.1	2,849.2	2,031.4
	Oil palm outgrowers	7.2	4.1	3.1	12,915.2	13,864.6	10,438.3	3,331.6	3,577.6	2,666.6
	Oil palm independent growers	7.1	3.5	3.6	13,429.9	12,185.8	13,890.1	3,474.6	3,058.5	3,628.5
	Food crop farmers (control)	4.9	1.9	3.0	9,092.5	8,998.1	9,360.8	2,714.5	2,760.6	2,583.2
Yeji	Permanent plantation workers	3.6	1.4	2.2	5,086.8	5,367.9	4,866.0	1,142.0	1,355.1	822.8
	Seasonal plantation workers	2.5	2.0	0.5	4,275.4	4,103.5	4,364.0	1,043.8	720.2	1,210.5
	Food crop farmers (control)	3.5	3.1	0.4	5,907.3	6,372.5	5,209.5	1,254.9	1,486.7	907.1

Source: (Ahmed et al., 2019a; 2019c).

or seasonally (Ahmed et al., 2019a). This migrant influx changed to some extent the local demographic profile, as the migrant men either brought their wives or married local women from the Kwae area. What is important to note is that the traditional land tenure system in Ghana makes it very difficult to outsiders to gain access to land, including migrants (Amanor, 2001).

Table 8.3 echoes the above, in that the GOPDC plantation workers have both the lowest land ownership and household incomes compared to other study groups in the region. All these differences are statistically significant for all variables and comparisons (significant at $p < 0.01$ or $p < 0.05$), with the only exception being the lack of statistical significance for income per capita between plantation workers and the control group (Ahmed et al., 2019c). Although the actual reasons for landlessness vary between migrant and local labourers as outlined above, the fact remains that both have specialized in off-farm employment, usually undertaking low-skill and manual tasks for low wages (Amanor, 2001). Their significantly lower land ownership and incomes compared to other groups (and particularly the control group) suggests that it is the poorer community members that engage in plantation employment. More importantly, the low salaries for most of these labourers suggest that they might be stuck in a poverty trap considering both the general lack of other employment opportunities in the area and their lack of economic resources to pursue other livelihood options. Arguably, the above suggest the formation of a landless social group that is compelled to work for the GOPDC plantation undertaking manual tasks for little payment, and with possibly few opportunities to escape poverty.

Outgrowers and independent oil palm growers have by far the highest land ownership and household incomes in Kwae (Table 8.3). Both groups have significantly higher land ownership compared to the control groups (both significant at $p < 0.01$), which suggests that only the most well-endowed households in the area tend to (or are able to) engage in oil palm cultivation (Ahmed et al., 2019c). It is interesting to point that the independent oil palm growers have higher household incomes compared to the outgrowers, though the difference is not statistically significant. This suggests that they may benefit more through their flexibility to choose oil palm markets depending on price signals (see "Study sites") (Ahmed et al., 2019c).

However, involvement in smallholder oil palm production also comes with major shifts in intra-household allocation of labour, and essentially their livelihoods. This is captured in an interview with an outgrower that stated:

> *Previously, I was only into food crop farming for just my family. Nevertheless, when the GOPDC started the outgrower schemes, I joined them. Since then, my family and I have reduced our land area for food crop and also the time we spend on a food crop farming. We rather invest more time in the oil palm, and the income we get is used to buy food. It is better than just cultivating food crops from hand to mouth. Many outgrowers like me do not invest too much time and money on cultivating food crops.*

[pers. comm, Outgrower in Kwae]

This statement also reflects the similar experiences among independent oil palm growers. It further suggests that smallholder-based oil palm cultivation has changed the agrarian structures in Kwae, by reducing for a large proportion of local households the investment of time and resources in subsistence farming. This agrees with a plethora of studies identifying industrial crop smallholders as rather differentiated social groups in many agrarian settings of the global South (Gatto et al., 2017; von Maltitz et al., 2019; Yaro et al., 2017) (Chapter 11). When considering both the significantly higher land ownership and incomes of oil palm outgrowers and individual growers compared to the control group (significant at $p < 0.01$ or $p < 0.05$), it can be argued that the already better-off households involved in oil palm production would also benefit more from higher levels of income accumulation, possibly increasing disproportionally their wealth over time compared to other groups in the wider area (Ahmed et al., 2019c). This possibly points to the gradual emergence of a rather distinct social group involved in market-oriented agriculture and characterized by comparatively larger land ownership and wealth (Cousins, 2010).

In Yeji, the introduction of jatropha did not catalyze the emergence of a "landless labourer" social group to the same extent as in Kwae. Both permanent and seasonal workers in the Smart Oil plantation own substantial amounts of land, which for permanent workers are on par with control groups (Table 8.3). This is because, in contrast to Kwae, most of the Smart Oil workers come from the local communities, which had access to local land through the prevailing traditional land tenure system in this part of Ghana. However, when compared to the other study groups, the permanent workers do not use a large proportion of their land for food crop production (Table 8.3), possibly due to the extensive women labour diversion from agriculture to waged plantation employment as discussed below.

Still, many households lost their entire land (or a substantial portion of it) during the land consolidation process for the jatropha LSLA. These households were offered employment as part of a compensation package, which, however, has raised many concerns about its fairness (Ahmed et al., 2019a). For example, an interview with a permanent worker in Yeji indicated that:

> I opted to work in the jatropha plantation because I lost my land to the company. The only option for making a living was to accept the job offer as compensation. For now, I do not have land, but I work permanently with the company.
>
> [pers. comm, Jatropha Worker in Yeji]

Conversely, some households engage in seasonal plantation employment during periods of low activity on their family farms. For example, women FGDs alluded to the fact that seasonal employment is essentially a measure for supplementing household income:

> I only go for seasonal work with the company when our farm work is minimal, and my husband permits to do so.
>
> [pers. comm, Participant in women FGD in Yeji]

Instead, the FGDs with men provide a rather different perspective over seasonal employment. Firstly, men respondents believe that most seasonal employees are women who divert labour and time from their household and family farm activities, to engage in paid employment. Though seasonal, for most of these women, this is their first-ever formal employment. Secondly, men respondents perceive that Smart Oil uses the seasonal employment as a means of obtaining cheap labour, considering that the wages for most women are generally low, due to the low skill requirement of their tasks and the fact that women lack strong bargaining power (see also "Gender differentiation"). Throughout interviews and FGDs, men respondents believe that seasonal employment is an exploitative labour mechanism, as for example indicated by a male FGD respondent:

> [o]ur wives rather leave the farm work and household chores but work for the company. I think it is because every woman here wants to receive a letter of employment from the company. However, their salary is meager, as they do not have a common voice to negotiate a higher wage. The company is rather exploiting our women.
>
> [pers. comm, Participant in Men FGD in Yeji]

Thus, despite some counter-perspectives, the introduction of jatropha in Yeji seems to have catalyzed the emergence of a social sub-group of labourers characterized by seasonal and precarious employment, in which only the least well-off in the community are willing to engage. The above are very well reflected by the fact that seasonal workers have significantly lower land ownership and total household incomes compared to all other study groups in Yeji (Table 8.3) ($p < 0.01$) (Ahmed et al., 2019c). Interestingly, this also points to some interesting gender dynamics that are further unpacked in the "Gender differentiation" sub-section.

Gender differentiation

Differentiation in work tasks and economic gains

Labour tasks in industrial crop plantations are often gender-differentiated depending on the workload and physical requirements, with this differentiation being sometimes discretionary (Gyapong, 2019) (Chapter 3). However, as a general rule, men tend to have more opportunities than women, especially for the more economically rewarding tasks (see "Introduction"). The above are also rather evident in the two study plantations, as the FGDs clearly indicate gender differentiation in both the type of tasks and perceived economic returns (Tables 8.4–8.5).

In the oil palm plantation, it seems that more types of tasks are available to men including harvesting, pruning, loading, spraying, and irrigation that are available to men (Table 8.4). This is because such tasks tend to require more intense labour, and are thus often associated with slightly higher perceived economic returns. Conversely, tasks assigned to women, such as picking of loose

Table 8.4 Gendered tasks and perceived economic gains in the Kwae oil palm plantation

Task	Gender	Target	Economic gains
Harvesting	Men	90 bunches	High
Pruning	Men	30–40 bunches	Average
Loose picking	Women	5 bags	Average
Round weeding	Both	30 trees	Average
Fertilizer application	Women	200 trees	Average
Slashing	Both	10 m²	Low
Security	Men	–	Low
Technical support	Men	–	High
Operations	Men	–	High
Loading	Men	2 trips/day	Average
Spraying	Men	20 litres	High
Irrigation of nursery	Men	–	Average
Carrying	Women	Per bunch	Flat rate

Table 8.5 Gendered tasks and perceived economic gains in the Yeji jatropha plantation

Task	Gender	Target	Economic gains
Harvesting	Women	Fixed rate per	Low
Pruning	Men	hour/day of	High
Weeding	Both	work	High
Fertilizer application	Women		Average
Security	Men		Low
Loading	Men		Average
Spraying	Both		High
Shelling	Men		Average
Drying	Women		Low

fruit, fertilizer application, and carrying are associated with less intense labour requirement and lower economic rewards. Similar patterns are also observed in the jatropha plantation, in that women have access to fewer and less economically rewarding tasks, under the perception that they do not require an intense labour effort. Such tasks include harvesting, fertilizer application, and drying of jatropha seeds (Table 8.5). Conversely, men have access to more employment options, with some of these tasks such as pruning having relatively high economic returns (Table 8.5). These findings reflect well-existing studies that have identified a gender divide in terms of labour opportunities, tasks, and remuneration in plantations in Ghana (Gyapong, 2019), and other developing countries (Chapter 3).

However, the participants in the women FGDs in both sites have contested this labour status quo, asserting that women can perform equally well some of the tasks assigned solely to men. For example, women employees in the oil palm plantation argued that they could undertake nursery irrigation, security, and some other operational tasks if they receive proper instructions and training (Table 8.4). Similarly, women employees in the jatropha plantation have argued the same for security tasks and sheller operations (Table 8.5). These were encapsulated in some of the FGDs. In particular, participants in the women FGDs argued as follow:

> *I think most of us here can do the work of the security [guard] and also the watering of the oil palm seedlings at the nursery. So I do not know why the company reserves such tasks to only men.*
>
> [pers. comm, Participant in Women FGD in Kwae]

> *For me, I can operate the sheller if they train me on the instructions. The male operator was given training and he does not have formal education like me.*
>
> [pers. comm, Participant in Women FGD in Yeji]

Some women have also contested the gender divide in work tasks based on the fact many have undefined tasks and that task assignment by managers and supervisors is usually discretionary. For example, women FGD participants claimed that some of the "men only" tasks are sometimes given to women when male labour is in short supply. This implies that women are capable of undertaking "male-only" tasks effectively despite their "labour intensiveness", and that there is a high degree of discretionary thinking in task assignment. For example, in a woman FGD, a participant in Yeji recalled that:

> *Here, there are more women employees than men. Most men do not want to work for the company because of the low wages. Sometimes when the company is not getting enough men to assign to some tasks, we, the women, are given those tasks but are advised to work harder. Some women were involved in weeding and pruning of jatropha, and they did well. As women, sometimes, we are asked to do different things. Today you are assigned to harvest the next day you are asked to do drying.*
>
> [pers. comm, Participant in Women FGD in Yeji]

Many women have pointed to pay gaps for the same tasks. For example, when both women and men undertake slashing, men tend to be paid more than women on the basis that they cover comparatively larger areas per person. Similarly, some of the tasks assigned to women, such as fertilizer and agrochemical application, expose them to disproportionally higher health risks. However, many women workers are not insured, and especially those on seasonal employment. For example, a participant in the women FGD in Kwae indicated that:

> *Fertilizer application does expose us to health risks, yet the company does not have any long-term insurance for us. Also when I do slashing with men, they get a higher wage than me simply because they cover a larger area than me.*
>
> [pers. comm, Participant in Women FGD in Kwae]

Most of the above points are reflected very well in the differentiated income levels between the male- and female-headed households of plantation workers. Male-headed households of permanent workers in both plantations tend to have higher incomes compared to female-headed households (Table 8.3). However, when it comes to seasonal jatropha plantation workers, female-headed households tend to have higher incomes (Table 8.3). Even though the income difference is rather small, it is not clear why this happens. For example, it might be due to the more extensive amounts of time that these households are willing to invest in plantation employment to supplement their income.

Differentiation in access to ecosystem services

A mix of agricultural land and natural vegetation was converted in each site for oil palm and jatropha production (see sub-sections "Data collection and analysis" and "Changes in agrarian structure"). As in other parts of Ghana and SSA, such land conversion can cause the loss of ecosystem services related to food and materials in the converted areas and their scarcity in surrounding areas (Gasparatos et al., 2018a). As in other parts of SSA, such ecosystem services contribute manifold to rural livelihoods (IPBES, 2018). However, considering that men and women access differently such ecosystem services both in type and magnitude, it is possible that the ecological transformation accompanying the two LSLAs might affect disproportionally the female-headed households.

Indeed, the results from both sites clearly show that for most groups, a comparatively higher proportion of female-headed households report a "significant" to "moderate" decline in aggregate access to ecosystem services (Table 8.6). Furthermore, on average, female-headed households access more ecosystem services compared to male-headed households in their respective groups (Table 8.6).

Beyond gender, the observed differences in ecosystem services access can shed further light on changes in agrarian structure and class differentiation in the study sites. In both sites, permanent worker households (whether male- or female-headed) tend to access fewer ecosystem services compared to other groups, and especially their respective controls. This further suggests the gradual disconnect of plantation workers from traditional agrarian livelihoods and their ever-constant specialization to waged employment.

Access to other off-farm livelihoods

In both study sites, many women engage in off-farm livelihood activities that were generated one way or another from the plantations. The most notable examples include petty trading such as the sale of food, water, beverages, and alcohol to plantation workers, within and around the plantations. Such indirect income generation potential has been observed in some plantation settings across SSA, through different spillover effects (Hall et al., 2017).

Table 8.6 Change in access to ecosystem services by study group and gender in the two sites

Study site	Group	Male-Headed households						Female-Headed households					
		Access to ecosystem services	Decreased significantly (%)	Decreased moderately (%)	Remained the same (%)	Increased moderately (%)	Increased significantly (%)	Access to ecosystem services	Decreased significantly (%)	Decreased moderately (%)	Remained the same (%)	Increased moderately (%)	Increased significantly (%)
Kwae	Plantation workers	0.3	1	4	93	0	1	0.3	11	4	70	7	7
	Oil palm outgrowers	3.0	9	10	75	6	0	4.6	41	22	28	0	9
	Oil palm independent growers	2.9	11	19	66	1	3	2.5	44	37	7	7	4
	Food crop farmers (control)	2.1	3	16	77	3	1	4.2	33	48	15	4	0
Yeji	Permanent plantation workers	3.5	23	48	18	0	11	6.0	20	54	14	13	0
	Seasonal plantation workers	4.2	24	24	18	0	35	5.8	27	45	12	15	0
	Food crop farmers (control)	6.2	23	17	50	2	8	8	33	15	40	5	8

Note: Access to ecosystem services denotes the average of the number of ecosystem services accessed by the households in each group.

Many of these petty traders are women plantation workers that use their access to plantation grounds to reach clients. According to the FGDs, these engage in such off-farm activities mainly to supplement plantation wages and diversify their sources of income. However, many of these women petty traders are not plantation workers, but target workers before and/or after their daily shifts outside the plantations. For example, a participant in the women FGD in Yeji indicated that:

> *In the morning I sell rice along the street mainly to plantation workers. When the bus is ready to take us to the plantation, I leave the rice selling to my daughter. But other women sell drinks and water. I use the income from the sales of rice to support the wage I get from the plantation and help my family.*
>
> [pers. comm, Participant in Women FGD in Yeji]

Mechanisms underpinning agrarian transformation

Our observations suggest that oil palm and jatropha production has catalyzed agrarian transformation in both study sites through multiple mechanisms. According to Table 8.7, the more prominent of these mechanisms relate to (a) the creation and differentiation of social groups with different types of engagement in industrial crop production, and (b) the emergence of some new gender dynamics.

Table 8.7 Main mechanisms of agrarian transformation related to oil palm and jatropha production in the two study sites

Mechanism	Description	Oil palm	Jatropha
Permanent plantation employment	Households specialize in waged plantation employment, diverting *almost completely* their labour from subsistence farming	✓	✓
Seasonal plantation employment	Households engage in waged plantation employment, diverting *partly* their labour from subsistence farming	✓	✓
Commercial farming	Households specialize in the production of industrial crops for commercial crop markets, diverting significantly their land and labour	✓	-
Gendered labour tasks	Women have fewer and less well-paid job opportunities in plantations, with often arbitrary task assignment	✓	✓
Off-farm income	Women divert *partly* their labour from subsistence farming and other household activities to engage in petty trading	✓	✓
Loss of ecosystem services	Female-headed households experience higher loss of ecosystem services due to landscape conversion for industrial crop production.	✓	✓

In terms of social differentiation, we see marked differences between plantation workers and control groups in Yeji and Kwae. Plantation workers divert most or part of their labour in waged employment, with little-to-no access to land (in Kwae) or the ability to fully cultivate it (in Yeji) (see "Creation and differentiation of social groups"). Furthermore, such groups exhibit markedly lower reliance on ecosystem services for their livelihoods. Such shifts from subsistence-based livelihoods to waged plantation employment have been associated with profound restructuring and transformation of the peasantry in many rural SSA contexts (Bryceson, 2019; Manley and Van Leynseele, 2019).

In Kwae, well-endowed households have the opportunity to engage directly in the cultivation of oil palm as outgrowers or independent growers. These households have diverted a major fraction of both their land and labour from subsistence to cash crop farming, becoming in the process dependent on national and international commodity markets for income generation. Many scholars have pointed to the emergence of this type of commercial farming as another major mechanism of agrarian transformation (Byerlee, 2014; Dao, 2018; Lehman, 1977).

Wealth and labour allocation are central elements of class dynamics in contexts of agrarian change (Bernstein, 2010, 2004; O'laughlin, 2016). Arguably, their significant differentiation between study groups suggests that the groups involved in industrial crop production seem to gradually coalesce into distinct social classes (Cousins, 2010). Following loosely the class-analytic typology for SSA agrarian contexts proposed by Cousins (2010: 100) and acknowledging the fluidity between the different categories, the study groups can be classified as:

- Control groups: "petty commodity producers" that are able to *"reproduce themselves from farming alone (or with only minor additional forms of income)"*;
- Workers in oil palm plantation: "allotment holding wage workers" that *"work small plots or gardens but are primarily dependent on wages for their simple reproduction"*;
- Workers in jatropha plantation (both permanent and seasonal): "worker-peasants" that farm but *"are also engaged in wage labour, and combine these in their simple reproduction"*;
- Oil palm smallholders: "small-scale capitalist farmer" that can *"begin to engage in expanded reproduction and capital accumulation"*.

The above processes have had significant gender-related undertones, both by facilitating changes in gender roles and reinforcing gender inequality. In order to fully appreciate these effects, it is important to first understand the traditional women roles in the study areas. Traditionally, women in Ghana are responsible for taking care of homes and children (Dery and Bawa, 2019; Dery and Ganle, 2020), as well as in the production, harvesting, storage, and marketing of food crops (Bosak et al., 2018). Many women also engage in some minor income generation activities though these are secondary and opportunistic in order to supplement household income, e.g. charcoal and fuelwood sales, petty trading (Andersson Djurfeldt et al., 2013; Canagarajah et al., 2001).

Industrial crop production can alter to a considerable extent such traditional gender roles in Ghanaian agrarian contexts by offering women some more stable off-farm income opportunities. For example, women constitute a major fraction of the permanent and seasonal workforce in both plantations. Furthermore, by boosting the local economy, plantations also create fertile conditions for other off-farm income opportunities for women such as petty trading in and around plantations. Such off-farm income generation options are generally scarcer for women in most rural SSA contexts (Van den Broeck and Kilic, 2018) and were rather uncommon in the original agrarian structure of the study communities (see "Changes in agrarian structure").

However, these otherwise positive opportunities for gender empowerment come with some challenges. First, there is high gender inequality in plantation employment, in that women tend to have fewer and less well-paid opportunities than men (see "Differentiation in work tasks and economic gain"). This type of gender divide in terms of available tasks and economic returns is common in plantation settings around the world and has been noted to further reinforce agrarian transformation by both changing gender roles and also creating new types of inequalities (Elmhirst et al., 2017; Fonjong and Gyapong, 2021). Second, off-farm employment can divert women's time from food crop production, household tasks, and personal time, having important ramifications for food security and their quality of life (Fonjong and Gyapong, 2021; Jarzebski et al., 2020; Mingorría et al., 2014). While such time diversion might be limited early on, it can become permanent if these off-farm activities become stable, further reinforcing agrarian transformation (Barrett et al., 2001; Matshe and Young, 2004; Pfeiffer et al., 2009).

It is worth noting that the processes of agrarian transformation outlined above have created some serious discontents in both Kwae and Yeji. Various social conflicts and struggles of different severities can be observed, with the more notable including (a) conflicts between the local communities and the industrial crop companies, (b) conflicts between different categories of plantation employees and between employees and companies, and (c) conflicts between men and women at the household level.

The first type of conflict between the local communities and the private companies operating the plantations has been anchored to the discontent over the past and ongoing land dispossession. For example, recently in Kwae, some of the local youth undertook protests to reclaim land under oil palm and stop GOPDC from further expanding the plantation. The youth perceives this as forceful eviction and the company as unfounded requests considering that from the company's perspective compensation has been paid in the past (Ahmed et al., 2019a) (see "Creation and differentiation of social groups"). For example, according to an independent oil palm grower,

> GOPDC is expanding its plantation to new land areas. The youth asked the company to pay compensation, which the company claims it has done several years back. The youth whose family lands were taken by GOPDC, went last year on a demonstration to stop the company from expanding its cultivated area into the

existing farming lands. Some of the youth also go into the plantation to steal palm fruits and even destroy palm trees. As a measure, the company has dug a big deep ditch around the whole plantation to prevent people from entering the area unless through the approved secured entrances. Since the ditch was done, several people have lost their lives.

[pers. comm, oil palm grower in Kwae]

The second type of conflict is anchored to the competition and re-alignment of the interests of plantation workers. This reflects many studies showing both the conflicts between different types of labourers due to diverging interests, as well as their alliances to resist corporate exploitation (Deininger, 1999; Gyapong, 2019). This type of conflict is mainly witnessed in Yeji, whereby permanent workers receive in a sense more "benefits" that creates some discontent with seasonal workers, but in some cases, their interests align when agitating against Smart Oil. For example, as a seasonal worker recalled:

When the company has food items to share such as rice or oil during Christmas break, permanent workers are given priority. But when the permanent workers want to resist poor working condition, they often want our support. Last year, all the workers agitated for an increment in the daily wage.

[pers. comm , Seasonal worker in Yeji]

The third type of conflict manifests at the household level and is anchored on the diversion of women's time from household duties to plantation employment, as well as the allocation of the benefits. For example, in the jatropha site, many of the male respondents complained about exactly this, considering the high proportion of women in the plantation labour force (see sub-sections "Gender differentiation" through "Mechanisms underpinning agrarian transformation"). According to a male respondent,

I divorced last year because ever since she [his ex-wife] *started working there, she refused to perform her home duties. She goes early to work, comes late and cooks late at night, does not take care of the children and does not have time for me. Also the money she gets from the plantation, she keeps it alone without sharing with me or using it for the family. Several times we had to fight at home.*

[pers. comm, Male Respondent in Yeji]

Conclusions

This chapter employed a feminist political ecology lens to unpack how oil palm and jatropha production in Ghana has catalyzed agrarian transformation. In particular, the chapter highlights how industrial crops created new livelihood opportunities and gendered tasks, transforming in the process the agrarian systems of the study communities. Arguably, engagement in industrial crop production can form new social groups that divert their labour from subsistence agriculture to paid plantation employment or commercial farming. These groups have very differentiated characteristics and gradually seem to coalesce into distinct social classes. At the same time, these processes have had significant gender-related undertones, both by facilitating changes in gender roles and reinforcing gender inequality. Only time will tell whether the observed social differentiation and gender inequalities will consolidate in radically different social classes and gender dynamics. We argue that rural development policies promoting industrial crops should anticipate such processes and should seek to adopt gender-sensitive practices and prevent/mitigate social inequalities and disparities.

Acknowledgements

We acknowledge the support of the Asahi Glass Foundation through a Research Grant for Young Scientists and a Continuation Grant and the Japan Science and Technology Agency (JST) for the Belmont Forum project FICESSA. AA acknowledges the support of a DAAD ClimapAfrica Postdoc Fellowship.

References

Adams, E.A., Kuusaana, E.D., Ahmed, A., Campion, B.B., 2019. Land dispossessions and water appropriations: Political ecology of land and water grabs in Ghana. Land use policy 87. doi.org/10.1016/j.landusepol.2019.104068

Ahmed, A., Campion, B.B., Gasparatos, A., 2017. Biofuel development in Ghana: Policies of expansion and drivers of failure in the jatropha sector. Renew. Sustain. Energy Rev. 70, 133–149. doi.org/10.1016/j.rser.2016.11.216

Ahmed, A., Abubakari, Z., Gasparatos, A., 2019a. Labelling large-scale land acquisitions as land grabs: Procedural and distributional considerations from two cases in Ghana. Geoforum 105, 1–15. doi.org/10.1016/j.geoforum.2019.05.022

Ahmed, A., Campion, B.B., Gasparatos, A., 2019b. Towards a classification of the drivers of jatropha collapse in Ghana elicited from the perceptions of multiple stakeholders. Sustain. Sci. 14, 315–339. doi.org/10.1007/s11625-018-0568-z

Ahmed, A., Dompreh, E., Gasparatos, A., 2019c. Human wellbeing outcomes of involvement in industrial crop production: Evidence from sugarcane, oil palm and jatropha sites in Ghana. PLoS One 14, e0215433. doi.org/10.1371/journal.pone.0215433

Ahmed, A., Jarzebski, M.P., Gasparatos, A., 2018. Using the ecosystem service approach to determine whether jatropha projects were located in marginal lands in Ghana: Implications for site selection. Biomass Bioenerg. 114, 112–124.

Amanor, K.S., 2001. Land, Labour and the Family in Southern Ghana: A Critique of Land Policy Under Neo-Liberalisation. Nordiska Afrikainstitutet, Uppsala.

Andersson Djurfeldt, A., Djurfeldt, G., Bergman Lodin, J., 2013. Geography of gender gaps: Regional patterns of income and farm-nonfarm interaction among male- and female-headed households in eight African countries. World Dev. 48, 32–47. doi.org/10.1016/j.worlddev.2013.03.011

Ariza-Montobbio, P., Lele, S., Kallis, G., Martinez-Alier, J., 2010. The political ecology of *Jatropha* plantations for biodiesel in Tamil Nadu, India. J. Peasant Stud. 37, 875–897. doi.org/10.1080/03066150.2010.512462

Awanyo, L., 2001. Labor, ecology, and a failed agenda of market incentives: The political ecology of agrarian reforms in Ghana. Ann. Assoc. Am. Geogr. 91, 92–121. doi.org/10.1111/0004-5608.00235

Barrett, C., Reardon, T., Webb, P., 2001. Nonfarm income diversification and household livelihood strategies in rural Africa: Concepts, dynamics, and policy implications. Food Policy 26, 315–331. doi.org/10.1016/S0306-9192(01)00014-8

Behrman, J., Meinzen-Dick, R., Quisumbing, A., 2012. The gender implications of large-scale land deals. J. Peasant Stud. 39, 49–79. doi.org/10.1080/03066150.2011.652621

Bernstein, H., 2004. "Changing before our very eyes": Agrarian questions and the politics of land in capitalism today. J. Agrar. Chang. 4, 190–225. doi.org/10.1111/j.1471-0366.2004.00078.x

Bernstein, H., 2010. Class Dynamics of Agrarian Change. Kumarian, Fernwood, MA.

Boafo, Y.A., Balde, B.S., Saito, O., Gasparatos, A., Lam, R.D., Ouedraogo, N., Chamba, E., Moussa, Z.P., 2018. Stakeholder perceptions of the outcomes of reforms on the performance and sustainability of the cotton sector in Ghana and Burkina Faso: A tale of two countries. Cogent Food Agric. 4. doi.org/10.1080/23311932.2018.1477541

Borras, S.M., McMichael, P., Scoones, I., 2010. The politics of biofuels, land and agrarian change: Editors' introduction. J. Peasant Stud. 37, 575–592. doi.org/10.1080/03066150.2010.512448

Bosak, J., Eagly, A., Diekman, A., Sczesny, S., 2018. Women and men of the past, present, and future: Evidence of dynamic gender stereotypes in Ghana. J. Cross. Cult. Psychol. 49, 115–129. doi.org/10.1177/0022022117738750

Bryceson, D.F., 2019. Gender and generational patterns of African deagrarianization: Evolving labour and land allocation in smallholder peasant household farming, 1980–2015. World Dev. 113, 60–72. doi.org/10.1016/j.worlddev.2018.08.021

Byerlee, D., 2014. The fall and rise again of plantations in tropical Asia: History repeated? Land 3, 574–597. doi.org/10.3390/land3030574

Canagarajah, S., Newman, C., Bhattamishra, R., 2001. Non-farm income, gender, and inequality: Evidence from rural Ghana and Uganda. Food Policy 26, 405–420. doi.org/10.1016/S0306-9192(01)00011-2

Carr, E.R., 2008. Men's crops and women's crops: The importance of gender to the understanding of agricultural and development outcomes in Ghana's central region. World Dev. 36, 900–915. doi.org/10.1016/j.worlddev.2007.05.009

Cousins, B., 2010. What is a "smallholder"?: Class-analytic perspectives on small-scale farming and agrarian reform in South Africa, in: Hebinck, P., Shackleton, C. (Eds.), Reforming Land and Resource Use in South Africa: Impact on Livelihoods. Routledge, London, pp. 86–109. doi.org/10.4324/9780203839645

Dao, N., 2018. Rubber plantations and their implications on gender roles and relations in northern uplands Vietnam. Gender, Place Cult. 25, 1579–1600. doi.org/10.1080/0966369X.2018.1553851

Darkoh, M.B.K., Ould-Mey, M., 1992. Cash crops versus food crops in Africa: A conflict between dependency and autonomy. Transafrican J. Hist. 21, 36–50. doi.org/10.2307/24520419

Deininger, K., 1999. Making negotiated land reform work: Initial experience from Colombia, Brazil and South Africa. World Dev. 27, 651–672. doi.org/10.1016/S0305-750X(99)00023-6

Dell'Angelo, J., D'Odorico, P., Rulli, M.C., Marchand, P., 2017. The tragedy of the grabbed commons: Coercion and dispossession in the global land rush. World Dev. 92, 1–12. doi.org/10.1016/J.WORLDDEV.2016.11.005

Dery, I., Bawa, S., 2019. Agency, social status and performing marriage in postcolonial societies. J. Asian Afr. Stud. 54, 980–994. doi.org/10.1177/0021909619851148

Dery, I., Ganle, J.K., 2020. "Who knows, you may overpower him": Narratives and experiences of masculinities among the Dagaaba youth of Northwestern Ghana. J. Men's Stud. 28, 82–100. doi.org/10.1177/1060826519846932

Dietz, K., Engels, B., Pye, O., Brunnengräber, A., 2014. The Political Ecology of Agrofuels, Routledge ISS Studies in Rural Livelihoods. Routledge, London.

Elmhirst, R., Siscawati, M., Sijapati Basnett, B., Ekowati, D., 2017. Gender and generation in engagements with oil palm in East Kalimantan, Indonesia: Insights from feminist political ecology. J. Peasant Stud. 44, 1135–1157. doi.org/10.1080/03066150.2017.1337002

Fonjong, L., 2017. Interrogating Large-Scale Land Acquisition and Its Implications for Women's Land Rights in Cameroon, Ghana and Uganda. International Development Research Centre (IDRC), Ottawa.

Fonjong, L.N., Gyapong, A.Y., 2021. Plantations, women, and food security in Africa: Interrogating the investment pathway towards zero hunger in Cameroon and Ghana. World Dev. 138, 105293. doi.org/10.1016/j.worlddev.2020.105293

Francois, M., Kofi, A., Saviour, A.W., Harvim, P., 2015. Time series and factor analysis of sugar cane production: Case study at Tosukpo, Volta Region, Ghana. Int. J. Innov. Res. Dev. 4, 367–379.

Gasparatos, A., Romeu-Dalmau, C., von Maltitz, G.P., Johnson, F.X., Shackleton, C., Jarzebski, M.P., Jumbe, C., Ochieng, C., Mudombi, S., Nyambane, A., Willis, K.J., 2018a. Mechanisms and indicators for assessing the impact of biofuel feedstock production on ecosystem services. Biomass Bioenerg. 114, 157–173.

Gasparatos, A., von Maltitz, G.P., Johnson, F.X., Romeu-Dalmau, C., Jumbe, C.B.L., Ochieng, C., Mudombi, S., Balde, B.S., Luhanga, D., Lopes, P., Nyambane, A., Jarzebski, M.P., Willis, K.J., 2018b. Survey of local impacts of biofuel crop production and adoption of ethanol stoves in southern Africa. Sci. Data 5, 180186. doi.org/10.1038/sdata.2018.186

Gatto, M., Wollni, M., Asnawi, R., Qaim, M., 2017. Oil palm boom, contract farming, and rural economic development: Village-level evidence from Indonesia. World Dev. 95, 127–140. doi.org/10.1016/J.WORLDDEV.2017.02.013

Gunewardena, N., 2010. Bitter cane: Gendered fields of power in Sri Lanka's sugar economy. Signs 35, 371–396. doi.org/10.1086/605481

Gyapong, A.Y., 2019. Land deals, wage labour, and everyday politics. Land 8, 94. doi.org/10.3390/land8060094

Hall, R., Scoones, I., Tsikata, D., 2017. Plantations, outgrowers and commercial farming in Africa: Agricultural commercialisation and implications for agrarian change. J. Peasant Stud. 44, 515–537. doi.org/10.1080/03066150.2016.1263187

Harcourt, W., 2019. Feminist political ecology practices of worlding: Art, commoning and the politics of hope in the class room. Int. J. Commons 13, 153. doi.org/10.18352/ijc.929

Harris, L., Kleiber, D., Goldin, J., Darkwah, A., Morinville, C., 2017. Intersections of gender and water: Comparative approaches to everyday gendered negotiations of water access in underserved areas of Accra, Ghana and Cape Town, South Africa. J. Gend. Stud. 26, 561–582. doi.org/10.1080/09589236.2016.1150819

Hill, R.V., Vigneri, M., 2014. Mainstreaming gender sensitivity in cash crop market supply chains, in: Gender in agriculture: Closing the knowledge gap. Springer, Berlin. pp. 315–342. doi.org/10.1007/978-94-017-8616-4_13

IPBES, 2018. Regional Assessment Report on Biodiversity and Ecosystem Services for Africa. Intergovernmental Science-Policy Platform on Biodiversity and Ecosystem Services (IPBES), Bonn.

Jarzebski, M.P., Ahmed, A., Boafo, Y.A., Balde, B.S., Chinangwa, L., Saito, O., von Maltitz, G., Gasparatos, A., 2020. Food security impacts of industrial crop production in sub-Saharan Africa: A systematic review of the impact mechanisms. Food Secur. 12, 105–135. doi.org/10.1007/s12571-019-00988-x

Kevane, M., 1996. Agrarian structure and agricultural practice: Typology and application to Western Sudan. Am. J. Agric. Econ. 78, 236–245. doi.org/10.2307/1243794

Lamb, V., Schoenberger, L., Middleton, C., Un, B., 2017. Gendered eviction, protest and recovery: A feminist political ecology engagement with land grabbing in rural Cambodia. J. Peasant Stud. 44, 1217–1236. doi.org/10.1080/03066150.2017.1311868

Larbi, W.O., Antwi, A., Olomolaiye, P. (2004). Compulsory land acquisition in Ghana—policy and praxis. *Land Use Policy*, 21, 115–127. https://doi.org/10.1016/j.landusepol.2003.09.004

Lehman, D., 1977. Agrarian structures and paths of transformation. J. Contemp. Asia 7, 79–91. doi.org/10.1080/00472337785390071

Li, T.M., 2015. Social Impacts of Oil Palm in Indonesia: A Gendered Perspective from West Kalimantan (No. 124), Social Impacts of Oil Palm in Indonesia: A Gendered Perspective from West Kalimantan. Center for International Forestry Research (CIFOR), Bogor. doi.org/10.17528/cifor/005579

Locke, A., Henley, G., 2013. A review of the literature on biofuels and food security at a local level assessing the state of the evidence. Overseas Development Institute (ODI), London.

Manley, R., Van Leynseele, Y., 2019. Peasant agency in Ghana's oil palm sector: The impact of multiple markets on food sovereignty. J. Agrar. Chang. 19, 654–670. doi.org/10.1111/joac.12323

Masanjala, W.H., 2006. Cash crop liberalization and poverty alleviation in Africa: Evidence from Malawi. Agric. Econ. 35, 231–240. doi.org/10.1111/j.1574-0862.2006.00156.x

Matshe, I., Young, T., 2004. Off-farm labour allocation decisions in small-scale rural households in Zimbabwe. Agric. Econ. 30, 175–186. doi.org/10.1111/j.1574-0862.2004.tb00186.x

Mingorría, S., Gamboa, G., Martín-López, B., Corbera, E., 2014. The oil palm boom: Socio-economic implications for Q'eqchi' households in the Polochic valley, Guatemala. Environ. Dev. Sustain. 16, 841–871. doi.org/10.1007/s10668-014-9530-0

Ministry of Food and Agriculture, 2015. Agriculture in Ghana: Facts and Figures. Ministry of Food and Agriculture, Accra.

Ministry of Trade and Industry, 2016. Ghana Sugar Policy: Intent and Development Process. Ministry of Trade and Industry, Accra.

Mollett, S., Faria, C., 2013. Messing with gender in feminist political ecology. Geoforum 45, 116–125. doi.org/10.1016/j.geoforum.2012.10.009

Moyo, S., 2008. African Land Questions, Agrarian Transitions and the State: Contradictions of Neo-liberal Land Reforms. Council for the Development of Social Science Research in Africa, Dakar.

O'laughlin, B., 2016. Bernstein's puzzle: Peasants, accumulation and class alliances in Africa. J. Agrar. Chang. 16, 390–409. doi.org/10.1111/joac.12177

Peet, R., Robbins, P., Watts, M., 2011. Global Political Ecology. Routledge, London.

Perreault, T.A., Bridge, G., McCarthy, J., James P., 2015. The Routledge Handbook of Political Ecology. Routledge, London.

Pfeiffer, L., López-Feldman, A., Taylor, J.E., 2009. Is off-farm income reforming the farm? Evidence from Mexico. Agric. Econ. 40, 125–138. doi.org/10.1111/j.1574-0862.2009.00365.x

Quisumbing, A.R., Meinzen-Dick, R., Raney, T.L., Croppenstedt, A., Behrman, J.A., Peterman, A. (2014) Closing the Knowledge Gap on Gender in Agriculture, in: Quisumbing, A., Meinzen-Dick, R., Raney, T., Croppenstedt, A., Behrman, J., Peterman, A. (Eds.) Gender in Agriculture. Springer, Dordrecht. https://doi.org/10.1007/978-94-017-8616-4_1.

Ragsdale, K., Read-Wahidi, M.R., Wei, T., Martey, E., Goldsmith, P., 2018. Using the WEAI+ to explore gender equity and agricultural empowerment: Baseline evidence among men and women smallholder farmers in Ghana's Northern Region. J. Rural Stud. 64, 123–134. doi.org/10.1016/j.jrurstud.2018.09.013

Sato, C., Soto Alarcón, J.M., 2019. Toward a postcapitalist feminist political ecology' approach to the commons and commoning. Int. J. Commons 13, 36. doi.org/10.18352/ijc.933

Schoneveld, G., 2014. The geographic and sectoral patterns of large-scale farmland investments in sub-Saharan Africa. Food Policy 48, 34–50. doi.org/10.1016/j.foodpol.2014.03.007

Sundberg, J., 2017. Feminist political ecology, in: Richardson, D., Castree, N., Goodchild, M.F., Kobayashi, A., Liu W., Marston, R.A. (Eds.) International Encyclopedia of Geography: People, the Earth, Environment and Technology. John Wiley & Sons, Ltd, Oxford, pp. 1–12. doi.org/10.1002/9781118786352.wbieg0804

Tosh, J., 1980. The cash-crop revolution in Tropical Africa: An agricultural reappraisal. Afr. Aff. 79, 79–94. doi.org/10.2307/721633

Vadjunec, J.M., Schmink, M., Gomes, C.V.A., 2011. Rubber tapper citizens: Emerging places, policies, and shifting rural-urban identities in Acre, Brazil. J. Cult. Geogr. 28, 73–98. doi.org/10.1080/08873631.2011.548481

Van den Broeck, G., Kilic, T., 2018. Dynamics of Off-Farm Employment in Sub-Saharan Africa: A Gender Perspective, Dynamics of Off-Farm Employment in Sub-Saharan Africa: A Gender Perspective. World Bank, Washington DC. doi.org/10.1596/1813-9450-8540

von Maltitz, G.P., Henley, G., Ogg, M., Samboko, P.C., Gasparatos, A., Read, M., Engelbrecht, F., Ahmed, A., 2019. Institutional arrangements of outgrower sugarcane production in Southern Africa. Dev. South. Afr. 36, 175–197. doi.org/10.1080/0376835X.2018.1527215

White, B., Dasgupta, A., 2010. Agrofuels capitalism: A view from political economy. J. Peasant Stud. 37, 593–607. doi.org/10.1080/03066150.2010.512449

Yaro, J.A., Teye, J.K., Torvikey, G.D., 2017. Agricultural commercialisation models, agrarian dynamics and local development in Ghana. J. Peasant Stud. 44, 538–554. doi.org/10.1080/03066150.2016.1259222

Yeboah, T., Chigumira, E., John, I., Anyidoho, N.A., Manyong, V., Flynn, J., Sumberg, J., 2020. Hard work and hazard: Young people and agricultural commercialisation in Africa. J. Rural Stud. 76, 142–151. doi.org/10.1016/j.jrurstud.2020.04.027

Zakaria, H., 2017. The drivers of women farmers' participation in cash crop production: The case of women smallholder farmers in Northern Ghana. J. Agric. Educ. Ext. 23, 141–158. doi.org/10.1080/1389224X.2016.1259115

Part IV

Socioeconomic and institutional transformation

9 Political ecology of soybeans in South America

Gustavo de L. T. Oliveira

Introduction

Deforestation rates in the Amazon are once again rising, while the extent of dryland forests such as the Chiquitania in eastern Bolivia and the Chaco across northern Argentina and Paraguay decline at alarming rates due to the expansion of agriculture and logging (Fair 2019). The intensification of soy production across the grasslands of the Cerrados of Brazil and the Pampas of Argentina is now extending into the edges of these ecosystems, displacing cattle ranching deeper into neighboring forests (Fair 2019). Soybean agribusinesses are sacrificing entire landscapes in South America in the name of agroindustrial modernization and capital accumulation. Indeed, this is a highly mechanized, modernist production system relying on packages of standardized seeds and agrochemicals. It leverages political and economic forces that reposition whole countries in international trade and geopolitical affairs and restructures class configurations in their countryside. This reconfiguration of property and labor relations occurs through flexible production relations that interconnect small-scale independent producers, fragmented contract farming and land-leasing operations, and transnational vertically integrated corporations managing mega-farms and entire agroindustrial commodity chains (Oliveira 2016; Oliveira and Hecht 2016). Flexible production also intertwines various firms in the financial, industrial, biotechnological, and energy sectors through the integration of livestock feed, vegetable oil, biodiesel, and other industrial markets (Oliveira and Schneider 2016) (Chapter 5). Precisely because of this "flex crop" framework (Oliveira and Schneider 2016), soy has expanded dramatically over remarkably distinct ecologies, interlinking and homogenizing highly diverse socio-ecological relations.

This expansion undergirds the creation of a new rural middle class, elite agribusiness professionals, and leading transnational firms from South America. Certain benefits have accrued primarily to these rising elites and South American states that balance international trade and consolidate rural development and geopolitical strategies. Meanwhile, rural areas subsumed by soy production have hollowed out in terms of traditional populations and agrobiodiversity. Indigenous peoples and other traditional populations have been dispossessed, small-scale farmers are squeezed out or adversely integrated into soy systems, and landless rural workers are exploited (Eloy et al. 2016; Goldfarb and van der

Haar 2016) (Chapter 5). Moreover, the environmental effects of agroindustrial soy expansion are not limited to deforestation alone, but encompass a complex set of interlinked ecological calamities, as entire biomes risk collapse (Chapter 1). Some of the main environmental impacts include loss of ecosystem connectivity, soil erosion and compaction, water and soil pollution, changes in water cycling and availability (both locally and regionally), increasing prevalence of pest outbreaks due to monoculture expansion, and loss of biodiversity due to habitat loss and agrochemical use (Altieri and Pengue 2006, Pacheco 2012; Elgert 2016) (Chapters 1 and 3). These compounding environmental impacts in turn are responsible for many negative socioeconomic outcomes. Some of these impacts include public health risks (sometimes leading to cancer and increased mortality) through exposure to agrochemicals, as well as increased vulnerability through the displacement of traditional food crops, reduced rural employment opportunities, slavery and labor rights violations, land and wealth concentration, and violence against indigenous, traditional, and peasant communities (Rulli 2007; McKay and Colque 2016; Leguizamón 2016) (Chapters 1 and 3).

This chapter draws upon empirical narratives from history, geography, and critical agrarian studies and the rich tradition of Marxist political economy to produce a political-ecological analysis of soybean production, use, and trade in South America. By employing the concept of agroindustrial flexing from the literature on flex crops, this chapter discusses the political economy of "who benefits" and "who does not benefit" from the expansion of soy across South America. Through the concept of neo-nature, the chapter examines not simply soy's environmental impacts but more fundamentally the production of nature itself through agroindustrialization. This reveals the blind spots and contradictions of the eco-modernist discourse that attempts to frame soybean agribusiness as the preeminent mechanism to conjoin economic development and nature conservation through agroindustrial intensification. Instead, the chapter seeks to demonstrate how these socio-ecological dynamics are ultimately a constitutive process of capital accumulation and concentration that has soy at its core and necessarily reproduces the social exclusion and ecological exploitation of capitalism.

The next section outlines the methodology and theoretical approach to the historical data used in this chapter, and the subsequent section describes the historical geography of soy production in South America. The fourth section analyzes the dialectic of socio-economic and ecological transformation of soybeans, discussing the production of neo-natures through commodification and uniformity, and the concentration of wealth and power among agribusiness elites through agroindustrial flexing. The penultimate section synthesizes the mechanisms and trajectories through which soy has become an agent of ecological and socioeconomic transformation in South America.

Theory and methods

Political ecology, as an academic field, hinges on the theorization of human–environment relations as always and everywhere political (Chapter 2). Rather than identifying "purely" economic, technological, demographic, or cultural interpretations of environmental problems (and similarly for designing proposals for environmental policy and management), political ecology insists on analyzing *power relations* as essential for comprehending environmental problems and formulating solutions (Watts et al. 2010; Svarstad et al. 2018). The pioneering and most well-established practice in political ecology entails the use of Marxist political economy for analyzing agrarian and ecological change (e.g. Watts 1983/2013; Hecht and Cockburn 1989; Peluso 1992, 2012; Castree 2000). Such studies focus on how historically produced social structures (i.e. land distribution, labor relations, and other class articulations with legacies of colonialism and ongoing forms of imperialism) condition the agency of individual smallholders, elite landowners, government officials, and other actors. Thus, this theoretical framework calls for examining a broad set of empirical evidence, which can often be gleaned from historical and geographical narratives that inform critical agrarian studies.

Given the large scope of actors, the multiplicity of relevant sites of interaction and multi-scalar relations of production, commercialization, investment, and political struggle that structure soy's expansion and impacts in South America, this chapter draws primarily upon empirical narratives from the secondary literature on the history and geography of soybeans in South America. This material is supplemented with in-person interviews with soy farmers, agrochemical and seed salesmen, and soybean agribusiness officials and sectoral representatives undertaken between 2012 and 2015 across multiple locations (primarily in the Cerrado ecosystem of central Brazil). This methodological engagement enables the theoretical analysis of the central guiding questions of Marxist political economy, namely "who owns what," "who does what," and "who gets what," which Bernstein (2010) has popularized as the central pillar of critical agrarian studies more generally. This analysis is undertaken in relation to two concepts from the literature of critical agrarian studies and political ecology, namely agroindustrial flexing and neo-nature.

Agroindustrial flexing reflects the point that agroindustrial production has become increasingly dynamic in terms of the variety of agricultural inputs that are incorporated by processing firms, and their increasingly malleable operations to attend the growing number and diversity of markets (e.g. food, livestock feed, fuel). This is what Borras et al. (2016: 94) have called the rising "flexible-ness and multiple-ness" of "flex crops and commodities." For example, biofuel producers around the world may opportunistically shift the composition of their inputs between sugarcane, soybeans, maize, palm oil, rapeseed, jatropha, and animal fats according to shifting price signals and government policies. Similarly, agroindustrial trading companies juggle supplies not only between North and South American soy exporters, Southeast Asian oil palm giants, and European rapeseed processors, but also bring them into competition with each other as almost any flex crop can often substitute for another in the production of meal for livestock feed, vegetable oil, and the multiple derivative industrial products that can be synthesized from basic biochemical components

(Gasparatos et al. 2015; Oliveira and Schneider 2016; McKay et al. 2016; Borras et al. 2016; Oliveira, McKay, and Plank 2017) (Chapters 5 and 10–11). Thus, agroindustrial flexing must be examined as a central element in the shifting political economy of soy in South America, which is reshaping relations of property, production, and consequent distribution of both economic goods and ecological impacts.

Neo-nature, as a concept, emerges from the literature in the fields of geography and Marxist political ecology on the "production of nature" (Castree 2000). Elsewhere, it has been theorized that the biotechnological mutations and ecosystem-wide transformations undertaken by South American soybean agribusiness essentially produce neo-nature (Oliveira and Hecht, 2016; Oliveira, 2020). Yet this does not refer simply to the innovations in biotechnology that produce novel varieties of soybeans, nor merely to the anthropogenic landscapes of soybean monocultures that replace rainforests and grasslands alike, but conceptualizes new sorts of socio-natures that emerge through the *commodification* of "nature's products, places, and processes" (Peluso, 2012: 79; Castree, 2000). This neo-nature, in other words, is the ecological expression of the neoliberal "commodification of everything" (Watts et al. 2010), which renders even ancient varieties of soybeans and other crops valuable now for the genetic material they may provide for future applications (Kloppenburg 2010). It can be argued that this casts soy farms that consistently lose money as financial assets that are still valuable, and can be dubbed as "developed land" that can be profitably leveraged in the portfolios of pension funds and hedge fund managers (Fairbairn 2014; Ouma 2014; Ducastel and Anseew 2015; Pitta and Mendonça 2015).

Therefore, attending to the political economy of soybean agribusiness as the production of a neo-nature extends the analysis of power relations from the structural transformations of agroindustrial flexing itself to post-structural and discursive forms of power that imbue socio-natures with multiple forms of value (Watts et al. 2010; Svarstad et al. 2018). Thus, this chapter sets out to analyze the interlinking struggles over the economic value of capital accumulation, the socio-cultural values of traditional livelihoods and modernization, and the political values mobilized to cast these conflicts as crises or opportunities for South Americans across the continent.

Results and discussion

Historical geography of soybeans in South America

It is possible to identify four main periods in the history of soy production in South America (Table 9.1), spanning its early introduction, small-scale establishment, expansion and shift to large-scale production, and accelerated growth and concentration. The following four sub-sections discuss in greater detail these four periods.

Table 9.1 Chronology and characteristics of soy production in South America

Period	Duration	Areas	Characteristics
Introduction	1880–1930	– Throughout South America	– Experimental introductions – Non-commercial production by Japanese immigrants for food consumption
Small-scale establishment	1940–1960	– Argentinean Pampas – Brazilian Pampas and southern pine forests	– Limited commercial production in rotation with wheat and/or maize for green fertilizer or edible oil production, mostly by migrant smallholders from Japan and Europe (<100 ha farms) – Commercial introduction in Paraguay and Bolivia
Expansion and shift to large-scale production	1970–1995	– Argentinean Pampas – Southern Brazilian Cerrados – Paraguayan and Bolivian forests on Paraná Basin	– Government support for geographical expansion and growth in the number and size of soybean farms – Government-developed non-GM varieties adapted for sub-tropical climates and acidic soils – Establishment of domestic soybean agribusiness companies, and expansion of average farm sizes (100–1,000 ha) – Entrance of transnational trading companies – Increased use of soy as input for livestock feed
Accelerated growth and concentration	1995–present	– Argentinean Pampas and Chaco – All Brazilian Cerrados and southern Amazon – Paraguayan forests on Paraná Basin – Bolivian lowland forests – Uruguay	– Introduction of GM varieties and privatization of R&D – Removal of taxes on whole-bean exports from Brazil – Concentration of the seed and input industries in the hands of transnational companies – Government support for expansion into marginal and vulnerable ecosystems – Intensification in former pasture areas and displacement of other crops – Increase in cross-regional and global investments – Land concentration and expansion of very large farms in Brazil (>1,000 ha) – Establishment of large-scale *pools de siembra* in Argentinean Pampas and beyond – Promotion of soybeans for biodiesel and other industrial uses – Vertical integration of production systems

Source: Author's own elaboration based on (Shurtleff and Aoyagi, 2009; Oliveira and Hecht, 2016; Oliveira, 2016).

Introduction and small-scale establishment

Soybeans were first introduced in South America for agronomic experiments in Brazil in 1882, and subsequently for similar reasons in French Guyana (1893), Suriname (1905), Argentina (1908), Uruguay (1911), Venezuela (1913), Paraguay (1921), Peru (1928), Colombia (1929), Chile (1934), and Ecuador (1936) (Shurtleff and Aoyagi 2009). All these introductions in tropical climates failed to advance beyond agronomic experiments (Chapter 2), and steady soybean production only took place in temperate climates by Japanese immigrants until the 1920s. The earliest documented instance of Japanese migrants bringing soybean seeds to Brazil was in 1908. However, South American commercial farmers did not adopt these pre-industrial soybean varieties, which had been selected for their nutritional qualities and characteristics and ability to be manually harvested and incorporated into diversified farming systems.

The earliest efforts to establish soybeans as a commercial crop in South America took place in 1924, when the Argentinean government imported 15,000 kg of soybean seeds (from 15 varieties) from the United States, and distributed them to 8,000 wheat and maize farmers (Shurtleff and Aoyagi 2009). These US varieties had straighter stems and pods that opened more easily to facilitate mechanized harvesting and were co-developed with the machinery itself. This suggests that the expansion of soy production in South America has been intimately related not simply to soy as a "natural" resource, but more specifically to a particular neo-nature characterized by chemical-intensive and mechanized agricultural production. Their underlying economic driver was the incipient agroindustrial flexing of multiple grains for the production of vegetable oil and animal feed.

The first known exports of soybeans from South America took place in 1929, when Brazil exported 800 tons of soybeans, with Argentina not exporting soybeans until 1962 (Shurtleff and Aoyagi 2009). Thus, until the 1960s, soybeans were not planted in South America for export to international markets but were sporadically introduced in agricultural experiments throughout the continent, as an instrument of producing neo-nature.

Eventually, production expanded slowly in the Argentinian and Brazilian pampas, where small-scale (<100 ha) commercial farmers, many recent emigrants from Europe, began integrating soybeans in rotation with wheat and/ or maize. They did so primarily as green manure or fodder, and secondarily for the emerging domestic vegetable oil and animal feed industries. In 1946–1947, soybeans were grown on 1,650 ha in Argentina, of which 460 ha yielded a total of 572 tons of grain/seed, with the remainder used as fodder and green manure. Argentinian production remained stable around an average of 1,000 tons per year throughout the 1950s, but there was a much faster growth in Brazil during this period. During the 1940s, the state of Rio Grande do Sul contained almost all soy production in Brazil at around 10,000 tons per year. As soy consolidated and expanded to Paraná state, production grew tenfold to over 100,000 tons per year by the 1950s (with 10,000 tons exported to Japan in 1957 alone). By 1969, soy production in Brazil grew tenfold once again, reaching 1 million tons. Until then, production remained geographically limited to southern Brazil and the Argentinian pampas, with soybean uses gradually shifting from green manure, fodder, and self-consumption among smallholders, to the emerging vegetable oil and livestock-feed markets in South America and abroad (Shurtleff and Aoyagi 2009).

This initial stage of agroindustrial flexing and soybean neo-natures was essentially a modernist compromise between small-scale European colonists in the Brazilian and Argentinian pampas, agroindustrial traders and emerging entrepreneurs, and urban consumers. Soy farmers gained a new instrument that delayed soil fertility loss and diversified their incomes to cope with the declining rate of return on wheat/maize rotations due to technological treadmills and international competition (Mazoyer and Roudart 2006). Agroindustrialists gained a new and cheap input for flexing innovation, and urban consumers gained a relatively low-cost source of food and oil.

Yet, the early adoption of soy production in South America also consolidated a neo-nature that highly excludes indigenous people and landless rural workers, and set the foundation for technological treadmills of the mechanization and chemicalization of agricultural production that would ultimately result in the dramatic ecological degradation currently observed throughout the continent (Chapters 1, 3 and 7).

Expansion and shift to large-scale production

The major turning point of soybean production in South America was in the 1970s. At that point, the US government imposed moratoria on soybean exports due to shortages of domestic production, while the collapse of Peruvian anchovy production (until then the most important protein input for animal feed) triggered a rush for alternative protein sources for the fast-growing confined animal feeding operations (CAFOs) of the United States, Soviet Union, Europe, and Japan (Morgan 2000). Japan then began collaborating with South American governments to expand soybean production by providing cheap credit and infrastructure. At the same time, Brazilian and Argentinian governments implemented scholar exchange programs and sponsored South American agronomists for training at Midwestern US universities, who upon return began adapting soybean varieties to South American conditions. Given that the bulk of South American soybean expansion has occurred in Brazil, the role of the Brazilian state-owned agricultural research company Embrapa is especially noteworthy in adapting soybean to the sub-tropical climates and acidic soils of the Cerrado (Nehring 2016).

In addition to this public investment in technology, soybean production and consumption were included in multiple agricultural and development programs in Brazil and Argentina. These included the expansion of transport and grain storage infrastructure, price support mechanisms, cheap credit, and other fiscal and financial incentives for purchasing seeds, agrochemicals, and large-scale planting and harvesting machinery (Jepson et al. 2010; Gordillo 2014). Commercial farmers seized these government incentives and responded avidly to the high prevailing prices on international markets, reinvesting profits into more intensive soy monocultures and converting additional land into production, even if it meant moving deeper into the Cerrados of central Brazil or even across borders into Paraguay to find sufficiently cheap land (Hecht 2005; Oliveira 2016). By 1978, South America had surpassed Asia to become the second largest soybean-producing region in the world. At the same time, soybeans became one of the major feedstocks for the vegetable oil industry and the emerging CAFOs across Europe, Asia, and South America itself (Oliveira and Schneider 2016).

Thus, it can be argued that agroindustrial flexing and the production of a neo-nature in South America placed soybean agribusiness at the vortex of political economic and agrarian transformations across the continent. This rendered the production of soybean neo-natures as some of the most important mechanisms for capital accumulation, discursive formation of modernity, and political-ecological articulation of agroindustrial intensification as the main mechanism for "sparing land for nature" (Oliveira and Hecht 2016; Thaler 2017). By this point, however, the consolidation of commercial agriculture over the Brazilian and Argentinian pampas, and its expansion into the Cerrados of central Brazil and the eastern lowlands of Paraguay, was propelled by much more than economic interests in capital accumulation and modernist desires for development. In particular, peasant uprisings and communist organizers demanding land redistribution and agrarian reform were questioning the political economic foundation that sustained agroindustrial flexing and soybean neo-natures (Patel 2013; Oliveira 2016). And so, the consolidation of soybean neo-natures across the southern cone of the South American continent was also a political project, implemented through military coups and brutal dictatorships that imposed capitalist relations of property and production at the expense of peasants, workers, and any landscapes that would not conform to agroindustrial neo-natures.

Accelerated growth and concentration

The military dictatorships of South America embarked on an international debt-fueled developmentalist and modernizing project, which collapsed with the Volcker Shock of 1979 and the imposition of neoliberal structural adjustment and austerity measures in the decades that followed (Harvey 2005). Particularly since the mid-1990s, Brazilian, Argentinian, and Paraguayan governments encouraged commercial farmers and transnational corporations to drastically accelerate the expansion of soy production in order to generate revenues in US dollars to repay mounting international loans. A few soy farmers that had previously organized themselves into cooperatives to facilitate input purchase and harvest sale established their own soybean export companies and crushing facilities (Chase 2003, Fajardo 2005, Jepson et al. 2010). But the transnational trading corporations that had an oligopoly over US soy exports (i.e. Bunge, Cargill, Louis Dreyfus, and later ADM) also began investing in soybean storage, processing, and trade in South America, which have since made them dominant exporters from the region (Goldsmith et al. 2004; HighQuest Partners and Soyatech 2011; Turzi 2011). Until 1995, these companies only owned about 10% of soybean crushing capacity in South America, but by 2002 they controlled about 50% of crushing capacity and 85% of whole bean exports from the region (Wesz Jr. 2016).

Similar concentrations of transnational companies have also occurred in the seed and agrochemical input sectors. During the 1970s, Brazilian and Argentinian state-owned agricultural research companies were undertaking most agronomic research and development of soybean varieties adapted to the South American landscapes and sub-tropical climatic conditions. Domestic seed companies played leading roles until the 1980s, when Brazil and Argentina served as "incubators" for sub-tropical and tropical soybean production technologies that were then transferred to Paraguay, Uruguay, and Bolivia (Craviotti 2016). But the development of transgenic technologies by US and European chemical companies has since displaced these public and domestic enterprises, even though they continue to play key roles in developing soybean varieties that contain patented transgenic material from transnational companies, as well as multiplying and commercializing seeds with traits from transnational companies (Kloppenburg 2010; Craviotti 2016). Transgenic varieties tolerant to herbicides (particularly glyphosate) were approved in Argentina in 1995 and smuggled into Brazil, Paraguay, and Bolivia until the respective governments also approved their use between 2003 and 2005. During this period, a handful of transnational chemical companies from the Global North have come to dominate soybean seed and associated agrochemicals markets (see similar discussion for the cotton sector in India, Chapter 7). Currently, the top three companies, Bayer (with Monsanto), ChemChina (with Syngenta), and Dow-DuPont, control over 55% of global soybean seed markets, with this concentration being even greater in South America where genetically modified (GM) varieties predominate. The top four companies (those listed above and BASF) control 69.5% of global agrochemical markets, and the first three alone control over 49.1% of the USD 11.5 billion agrochemical market in Brazil (EcoNexus and Berne Declaration 2013, Silva and Costa 2012).

Globally, the restructuring of the soybean agroindustrial complex since the 1990s is characterized by the simultaneous concentration of input and trading markets by transnational corporations and the deconcentration of soybean production and processing from North America and Western Europe. Between 1990 and 2014, the US share of global soybean production declined from 50% to 31%, while Brazil's share increased from 18% to 31% and Argentina's share from 10% to 19%. Soybean processing industries also shifted from the United States (where the share of global soybean crush declined between 1990 and 2014 from 37% to 19%), while they increased in Brazil (from 15% to 16%), Argentina (from 8% to 16%), and China (from 4% to 29%) (Oliveira and Schneider 2016).

As South America surpassed the United States in soy production, therefore, the processing industry shifted radically, particularly from North America and Western Europe to the newest and fastest growing market for soybean exports in China (Chapter 10).

In short, the neoliberal reforms that accelerated the growth of soy production in South America catalyzed a relative loss of market share for US farmers and South American state-owned seed companies. However, US agroindustrial companies increased both their market share and profits through their expansion into South America. Conversely, smaller-scale farmers have been squeezed out by larger private entities that are able to reinvest a larger amount of capital for the expansion of farmland expansion, the acquisition of larger machinery, and the constant upgrading of agrochemical and biotechnological seed packages.

Although this period has often been characterized as a neoliberal "corporate food regime" (McMichael 2012), there has been continuous state support for cross-regional and global investments (Chapter 3). These essentially push soybean producers to expand horizontally onto increasingly more marginal land and vulnerable ecosystems, such as the Argentinean Chaco, the Brazilian Amazon, and the semi-deciduous forests of Paraguay and Bolivia (Chapters 4–5), and replace degraded pastures and less profitable crops with intensive soybean monocultures in Uruguay, the Argentinian Pampas, and the central Cerrados of Brazil (Oliveira 2016, McKay and Colque 2016, Goldfarb and van der Haar 2016, Elgert 2016). These increasingly large-scale soy monocultures in South America concentrate farmland, wealth, and power in the hands of fewer and fewer large-scale farmers and transnational agroindustrial companies that benefit from the multiple and increasingly flexible markets. Thereby, they also consolidate a soybean neo-nature in which cleared lands and soils drenched with agrochemicals become increasingly barren substrates for the reproduction of a particular sort of agroindustrial capital and modernity.

Flexing and neo-natures: Dialectic of socio-economic and ecological transformation of soybean

Soybean neo-natures: Commodification, concentration, and uniformity

The high costs of specialized harvesting equipment imported to South America meant that only the well-capitalized commercial farmers could adopt soy cultivation. Furthermore, once the fertility gains from the nitrogen-fixing bacteria in soybeans were exhausted, the increasing use of chemical fertilizers and pesticides further increased the production costs of wheat-soy production systems in southern Brazil and the Argentinian pampas. Wherever such "green revolution" production systems were adopted, farmers became entrapped in a technological treadmill that requires continuous reinvestment for chemical inputs, larger production areas, and bigger machinery. This constant commodification of production inputs and farmland concentration expels the smaller and least productive farmers and favors land and wealth concentration among the larger and more capitalized farmers (Mazoyer and Roudart 2006, Oliveira 2009, VanderVennet et al. 2016, Mier y Teran 2016; Leguizamón 2016). Consequently, by the 1970s, it also became increasingly difficult for the youth in these families to inherit enough land for an economically viable farm in southern Brazil or the Argentinian Pampas. It was this crisis of social reproduction that triggered a process of outmigration that conditioned the trajectories of soy expansion across South America. As explained by a soy farmer whose family planted soy in southern Brazil in his youth, "at that time the government wanted us to move to the Cerrado, *clear everything*, and plant soy. And my father's farm [in southern Brazil] wasn't enough for me and my three brothers, so we didn't have any choice but to come here where land was cheap enough" (Personal Communication: soy farmer in Goiás, Brazil, April 2014).

While a substantial number of young people from soy-farming families in the Brazilian and Argentinian pampas migrated to cities, equally many of these (such as the farmer interviewed above) moved to agricultural frontier zones where land was cheaper and state policies encouraged the expansion of commercial farming (e.g. the Cerrados of Brazil, the Argentinian Chaco, and eastern Paraguay) (Desconsi 2011, Galeano 2012, Goldfarb and Van der Haar 2016). Those who refused to migrate to cities but lacked the capital to purchase farms elsewhere became landless, and some even self-organized into social movements for agrarian reform, as illustrated most famously in the origins of the Landless Rural Workers' Movement (MST) in southern Brazil (Stédile and Fernandes 1999). Although soy production is not as labor-intensive as other industrial crops, its massive expansion into relatively unpopulated areas also created a demand for migrant and temporary farm labor, often originating in poverty-stricken areas (e.g. northeastern Brazil) where the agroindustrial sector did not expand (Rumstain 2012). Consequently, most areas of soybean expansion areas are produced by complex migration patterns, which are marked by sharp cultural differences, tension, and even conflicts between migrants and locals (Gordillo 2014), or between distinct groups of migrants such as *gaúchos* (southerners) and northeasterners in central Brazil (Desconsi 2011, Rumstain 2012).

Simultaneously, the expansion of soybean production to frontier zones should not be understood as a simple linear trajectory leading solely toward concentration to large-scale farms. This is a long-term *trend* that has been visible and documented across the continent. However, multiple regional factors and historical particularities allow for a relatively broad spectrum of farm sizes and their associated production practices. In the Argentinean Pampas, for example, most soy is grown on farm units that range from 150 to 1,000 ha. However, the figure of the independent "soy farmer" has been splintered into multiple different categories, as smaller-scale landowners rent their farms to companies that hire agronomists, farm managers, machinery operators, and other specialized laborers to run the soy production system. These companies, known as *pools de siembra*, collect their capital from many different investors, ranging from rural and urban individuals to institutional investors and finance corporations. Consequently, while each farm unit might remain "medium-sized," these might ultimately be operated by companies that manage hundreds of thousands of hectares across Argentina and elsewhere in South America, while small farmers themselves become proletarianized (Oliveira and Hecht 2016).

Some of the leading *pools de siembra* and soybean production companies are expanding into Paraguay and Bolivia as well, but local conditions have not allowed for a similar consolidation of this system as in the Argentinian pampas. Yet, similar forms of farmland concentration through the technological treadmill and land-leases have also been taking place in Bolivia and Paraguay through very different social dynamics. In these geographical contexts, many of the farmers who previously obtained small (approximately 50 ha) plots under colonization schemes found themselves without any other option but to lease their land to neighboring soy farmers that gradually increased their land to a few hundred hectares (McKay and Colque 2016; Correia 2019).

In some areas of southern Brazil, relatively small soy farms (<300 ha) still exist. Yet, they are only economically viable insofar as they are able to capture price premiums from niche markets or integrate soybeans as part of a more diversified farming system (Vander Vennet et al. 2016). In the Cerrado region of central Brazil, most soy farms are medium in terms of size (300–1,000 ha), but it is the large farms (1,000–30,000 ha) that account for most of the production. In some areas, for example, farms larger than 1,000 ha account for over 90% of the cultivated area (Mier y Teran 2016). Moreover, large-scale farm management firms are increasing rapidly their operations in the Cerrado region, with multiple farms reaching sizes between 10,000 and 30,000 ha. Soy farms in the Amazon are predominantly large (>3,000 ha) and the expansion of soy production has been very fast, but still accounts for a very small portion of the overall production in Brazil (Sauer 2018).

However, across all these varied landscapes and scales of soybean production in Brazil, it is possible to observe the same technological package of transgenic seeds and their associated chemical products and production practices. Furthermore, in all these settings the pressure for soybean farmers is to "get big or get out." In other words, across the entire southern cone of South America, there is a remarkable homogeneity in the agronomic techniques and technologies of production, from the smallest to the largest farms. In this sense, the multiplicity of socio-economic actors and the flexible variations of production and commercial relations that interconnect them contrasts powerfully with the highly simplified and homogenous production of soybean neo-natures across the continent.

The long-term tendency toward farmland expansion and concentration for soy production has been aggravated with the incorporation of transgenic seeds since the late 1990s, enabling intensification through no-till practices that facilitate double-cropping. No-till techniques are certainly better in terms of soil erosion control and more economically efficient from a farm-management perspective, but the eco-modernist discourses that this intensification "spares land for nature" are highly disingenuous (Oliveira and Hecht 2016; Elgert 2016; Thaler 2017).

In short, the claim is that intensification happens in land that is already cultivated, thereby curtailing expansion into forested and marginal landscapes. Yet, while agronomic research in experimental fields demonstrates some productivity gains and reduction of agrochemical use in small areas and over short periods (1–2 years), the larger-scale and longer-term observations challenge all these optimistic claims still made by these agribusiness companies (Cotacora-Vargas et al. 2018). What in fact ensues is the aggravation of technological treadmills due to rising production costs, increasing pest/weed resistance, and pressure for farmers to reinvest in larger-scale production as a means of increasing their incomes and compensating for declining profit (Altieri and Pengue 2006; Mazoyer and Roudart 2006; Binimelis et al. 2009; Cotacora-Vargas et al. 2018).

Finally, the challenges imposed by "super weeds" that are resistant to glyphosate herbicides associated with transgenic soy and pest outbreaks that require more frequent application of increasingly toxic pesticides coalesce in the dialectical socio-economic contradictions of the production of this neo-nature. While the Brazilian Soybean Producers Association (APROSOJA) and the national landowners association (Confederação Nacional da Agricultura, CNA) publicly defend the use of transgenic soy varieties and push the Brazilian government to expedite the approval of transgenic varieties and agrochemicals, many soy farmers (including those in leadership positions in these associations) complain that "the chemical companies do not have our interests at heart," because "they are posting record profits while our production costs keep going up" (based on interviews and field notes, Goiás, Brazil, March 2014). Some have even stated during an official meeting of soy producers at a major agribusiness fair that "we are being held hostage" by the agrochemical companies (based on interviews and field notes, Bahia, Brazil, July 2015). Indeed, agroindustrial inputs such as agrochemicals and seeds already accounted for 37–47% of production costs for soybean farmers in Brazil in 2011(Silva and Costa 2012). Production costs have continued to increase, with perceived benefits to agrochemical and seed companies and negative effects to soy farmers and local ecosystems (Personal Communication: Soy farmers in Goiás, Tocantins, and Bahia, Brazil, 2014, 2015, 2020).

Agroindustrial flexing: Concentration of wealth and power among agribusiness elites

Given the direct role of farmers in the development of soy neo-natures, it is easy to imagine that they are the fundamental drivers and beneficiaries of this agroindustrial production system. However, this would ignore not only the class differences and ongoing process of differentiation between small-scale farmers (who are effectively trapped by technological treadmills until they are forced out of soy production altogether), and large-scale farmers (who incorporate their properties and their services as employees or independent contractors along this process). But soy production itself is only one element of the broader global political economy of agribusiness. As already discussed, the leading seed and agrochemical input manufacturers (alongside major soybean trading companies), effectively control the inputs and farming practices of most soy farmers across South America and *lock in* prices and access to significant portions of their harvests through the prearranged supply of fertilizer, pesticides, herbicides, and seeds (West Jr. 2016; Craviotti 2016). Smaller and less-capitalized farmers might commit as much as two-thirds of their harvest to input providers and/or trading companies before the planting season. Even large and well-capitalized farmers frequently contract around a quarter of their harvest in exchange for fertilizers and other inputs (Personal Communication: Soy farmers in Goiás, Tocantins, and Bahia, Brazil, 2014, 2015 and 20120.

Through these virtual seed and agrochemical monopolies, and soybean commercialization monopsonies, a handful of transnational agribusiness companies have come to control soybean production systems, prices, and commodity flows throughout the continent. The four largest agroindustrial trading companies (i.e. ADM, Bunge, Cargill, and Louis Dreyfus, collectively known as the ABCDs) are estimated to manage about 80% of international soybean trade and control 50% of installed crushing capacity and 85% of soybean exports in South America (Wesz Jr. 2016). Their power and market share have since been challenged by massive investments across the continent by the leading Chinese and Japanese trading companies, mainly COFCO and Marubeni (Oliveira 2017). In both instances, it is the possibility of agroindustrial flexing through vertical integration or strategic partnerships across multiple markets and production chains that enable these soybean trading and processing companies to make such gains (Murphy et al. 2012, HighQuest Partners and Soyatech 2011).

Agroindustrial suppliers and traders establish this power over soy producers the moment farmers take their harvest to their local warehouse, controlling much of the storage capacity, port terminals, cargo ships, and processing facilities that ultimately crush the soybeans into meal and vegetable oil (Morgan 2000, Goldsmith et al. 2004; Wesz Jr. 2016). Their global reach in terms of sourcing soybeans and other flex crops are controlling the chokepoints of transnational trading logistics combines with an oligopolistic share of agroindustrial processing capacity to provide them incomparable flexibility when sourcing from (and redirecting sales) to multiple markets around the world (Oliveira and Schneider 2016). They are even able to reroute cargo ships mid-ocean to gain marginal profits on large volumes, and speculate on futures markets with the privileged information gained through controlling significant shares of non-transparent markets (Salerno 2017). Ultimately, this brings them many benefits such as increasing control over the soybean production chain and its price setting-mechanisms, increasing profit margins, reducing production and transaction costs, minimizing risks, profiting from future-market hedging and speculation, and generating complementarities and synergies between the different sectors of the soybean production and processing complex (Oliveira and Schneider 2016; Salerno 2017).

Against the efforts of agribusiness companies (located both up- and downstream from framing activities) to control soy production processes and profits, soy farmers have often organized themselves into cooperatives to pool capital to purchase inputs, build storage and processing facilities, and increase their bargaining power vis-à-vis trading companies (Chase 2003, Fajardo 2005). Some of the largest scale farmers, such as the Maggi family in Brazil and the Grobocopatel family in Argentina (Oliveira and Hecht 2016), have even been able to expand vertically into the construction and operation of their own trading operations. Yet, all farmers are still price takers in domestic and international soybean markets, which are controlled by a virtual monopsony of transnational trading companies. Thus, many soy farmers have sought to cope with their weak position relative to soybean crushers and trading companies, predominately through increasing production and controlling the "quality" of their harvests through agrochemicals. Despite the failure of biotechnology to increase productivity and reduce production costs beyond the short-term, most soy farmers have avidly embraced glyphosate-resistant transgenic varieties to facilitate the management of their extensive soybean monocultures (Binimelis et al. 2009, Cravioto 2016; Cotacora-Vargas et al. 2018).

Moreover, when farmers deliver their harvest to warehouses, they receive a price deduction based upon the percentage of non-grain material in a sample. This is the only aspect of price-setting over which farmers have any control. Since the intensive use of glyphosate and other herbicides drastically reduces the "contamination" of fields and harvests with weeds, leaves, and stems, farmers now consider glyphosate-resistant GM varieties necessary to achieve profitable farming operations, even if they must continuously increase the use of agrochemicals due to growing weed and pest resistance (Binimelis et al. 2009; Cravioto 2016; Cotacora-Vargas et al. 2018). The result of this political economic pressure of agroindustrial flexing is the aggravation of technological treadmills, which further simplify agroecosystems and drive farmers to clear more land, increase agrochemical use, and reinvest (whenever possible) in the expansion of soy production.

Soybeans as a driver of socioeconomic transformation in South America

Table 9.2 synthesizes the mechanisms and trajectories through which soy has become an agent of socioeconomic transformation in South America. The production of neo-natures and agroindustrial flexing have been present at every period and turning point in the historical geography of soy in South America and are thus ideal for synthesizing the main mechanisms and trajectories. Yet, these processes have developed in a combined and uneven manner across the continent.

Conclusions

The extensive and high-profile deforestation and ecosystem degradation across South America caused by soy production is not just the most visible indication of ecological degradation through the expansion and intensification of this

Table 9.2 Main mechanisms and trajectories of socioeconomic transformation from soybean production in South America

Scale/location	Transformation	Production of neo-natures	Agroindustrial flexing
Global and continental	Mechanism	- Transformation of soy into a sub-tropical crop - Integration of agricultural heartlands and frontiers areas of South America into global circuits of capital accumulation	- Incorporation of soy into flexible production chains for vegetable oil, livestock feed, biofuel, and industrial products.
	Trajectory	- Competition of South American and US farmers for European and East Asian markets	- Subordination of soy farmers to transnational agroindustrial and trading companies
Established agricultural heartlands	Mechanism	- Commodification of inputs - Homogenization of farming practices - Intensification of technological treadmills - Consolidation of land	- Monopolization of agricultural inputs and trading - Locking in prices for farmers - Financialization of commodity markets
	Trajectory	- Increase of farmer pressure to "get big or get out" - Incorporation of farming activity by agroindustrial and financial investors	- Splintering and proletarianization of farmers - Speculative gains for transnational agroindustrial and trading companies
Agricultural frontier areas	Mechanism	- State support for expansion of commercial agriculture - Land grabbing - Clearing and conversion of vegetation	- Downward pressure on soy prices for farmers in agricultural heartlands
	Trajectory	- Immigration of soy farmers from established agricultural heartlands and farmworkers from marginalized regions - Displacement of indigenous and other local communities	- Intensification of agroindustrial production - Increase of debt and dependence for soy farmers and states

agribusiness. It is more than a mere "environmental impact" of agrarian capitalism and agricultural modernization. This chapter has demonstrated that the political economic distribution of the socio-economic benefits and environmental costs of soy production, processing, and trade are a dialectical outcome of the production of soybean neo-nature. This neo-nature is predicated upon, and further exacerbates, the unequal power dynamics between the transnational and vertically integrated corporations on the one hand and soy farmers on the other hand, not to speak of indigenous communities, poor peasants, and rural workers that are displaced or adversely incorporated into this production system.

The main political implication emerging from this Marxist political ecology of South American soybeans is that by "blaming" soy farmers for this ecological degradation ignores the structural pressures under which they operate. It also overlooks the power dynamics and class conflicts internal to this capitalist agroindustrial production system. Soy farmers are essentially "hostages" to the transnational agroindustrial input suppliers and traders that ultimately drive the homogenization of production practices and the landscapes on which they extend and intensify. Yet, South American governments are also "hostages" to the soybean agribusiness, due to their need for generating dollar-denominated revenue to repay Cold War-era debts incurred by developmentalist misadventures of military dictatorships. Thus, it would be naïve to imagine that ecological sustainability could be prioritized by South American governments or soybean farmers, without first transforming the structural conditions of political economic dependence in which they find themselves, and the discursive power of modernist imaginaries of progress and development that problematize diversified farming systems and glorify the homogeneity of soybean monocultures. The political economic power of soybean agribusiness elites across South America, therefore, rests on deeply seated and globally entrenched practices of agroindustrial flexing, which have generated neo-natures. In this context, the struggles for socio-ecological liberation must be revolutionary.

References

Ahmed, A., Kuusaana, E.D., Gasparatos, A., 2018. "The role of chiefs in large-scale land acquisitions for jatropha production in Ghana: Insights from agrarian political economy." *Land Use Policy* 75: 570–582.

Altieri, M., and W. Pengue. 2006. "GM soybean: Latin America's new colonizer." *Seedling*, January 2006, Barcelona: GRAIN.

Binimelis, R., W. Pengue, and I. Monterroso. 2009. "'Transgenic treadmill': Responses to the emergence and spread of glyphosate-resistant johnsongrass in Argentina." *Geoforum* 40(4): 623–633.

Bernstein, H. 2010. *Class dynamics of agrarian change*. Sterling, VA: Kumarian Press.

Borras, S.M., Franco, J.C., Isakson, S.R., Levidow, L., and Vervest, P. 2016. "The rise of flex crops and commodities: Implications for research." *Journal of Peasant Studies* 43(1): 93–115.

Castree, N. 2000. "Marxism and the production of nature." *Capital & Class* 24(3): 5–36.

Catacora-Vargas, G., Binimelis, R., Myhr, A.I., and Wynne, B. 2018. "Socio-economic research on genetically modified crops: A study of the literature." *Agriculture and Human Values* 35(2): 489–513.

Chase, J. 2003. "Regional prestige: Cooperatives and agroindustrial identity in southwest Goiás, Brazil." *Agriculture and Human Values* 20(1): 37–51.

Correia, J. 2019. "Soy states: Resource politics, violent environments and soybean territorialization in Paraguay." *Journal of Peasant Studies* 46(2): 316–336.

Craviotti, C. 2016. "Which territorial embeddedness? Territorial relationships of recently internationalized firms of the soybean chain." *Journal of Peasant Studies* 43(2): 331–347.

Desconsi, C. 2011. *A Marcha dos 'Pequenos Proprietários Rurais': trajetórias de migrantes do Sul do Brasil para o Mato Grosso.* (The march of the 'small rural landowners': Trajectories of southern Brazilian migrants to Mato Grosso.) Rio de Janeiro: E-papers.

Ducastel, A., and W. Anseew. 2017. "Agriculture as an asset class: reshaping the South African farming sector." *Agriculture and Human Values* 34(1): 199–209.

EcoNexus and Berne Declaration. 2013. *Agropoly: A handful of corporations control world food production.* Available at: econexus.info/publication/agropoly-handful-corporations-control-world-food-production [Accessed August 29, 2014].

Elgert, L., 2016. "'More soy on fewer farms' in Paraguay: Challenging neoliberal agriculture's claims to sustainability." *Journal of Peasant Studies* 43(2): 537–561.

Eloy, L., C. Aubertin, F. Toni, S. Lúcio, and M. Bosgiraud. 2016. "On the margins of soy farms: traditional populations and selective environmental policies in the Brazilian Cerrado." *Journal of Peasant Studies* 43(2): 494–516.

Fair, J. 2019. "Investors warn soy giants of backlash over deforestation in South America." Mongabay, March 18. Available at: news.mongabay.com/2019/03/investors-warn-soy-giants-of-backlash-over-deforestation-in-south-america/ [Accessed on September 3, 2020].

Fairbairn, M. 2014. "'Like gold with yield': Evolving intersections between farmland and finance." *Journal of Peasant Studies* 41(5): 777–795.

Fajardo, S. 2005. "As cooperativas paranaenses e o novo padrão de desenvolvimento agroindustrial." *Formação* 12(1): 165–192.

Galeano, L. 2012. "Paraguay and the expansion of Brazilian and Argentinian agribusiness frontiers." *Canadian Journal of Development Studies* 33(4): 458–470.

Gasparatos, A., G. von Maltitz, F. Johnson, L. Lee, M. Mathai, J. Puppim de Oliveira, and K. Willis. 2015. "Biofuels in sub-Sahara Africa: Drivers, impacts and priority policy areas." *Renewable and Sustainable Energy Review* 45: 879–901.

Goldfarb, L., and G. van der Haar. 2016. "The moving frontiers of genetically modified soy production: Shifts in land control in the Argentinian Chaco." *Journal of Peasant Studies* 43(2): 562–582.

Goldsmith, P., B. Li, J. Fruin, and R. Hirsch. 2004. "Global shifts in agro-industrial capital and the case of soybean crushing: implications for managers and policy makers." *International Food and Agribusiness Management Review* 7(2): 87–115.

Gordillo, G. 2014. *Rubble: The afterlife of destruction.* Durham, NC: Duke University Press.

Harvey, David. 2005. *A brief history of neoliberalism.* Oxford: Oxford University Press.

Hecht, S. 2005. "Soybeans, development and conservation on the Amazon frontier." *Development and Change*, 36(2), 375–404.

Hecht, S., and A. Cockburn. 1989. *The fate of the forest: Developers, destroyers, and defenders of the Amazon.* New York: Verso.

HighQuest Partners, and Soyatech. 2011. *How the global oilseed and grain trade works.* Available at: www.unitedsoybean.org/wp-content/uploads/2013/07/RevisedJan12_GlobalOilSeedGrainTrade_2011.pdf [Accessed on July 30, 2014].

Jepson, W., C. Brannstrom, and A. Filippi. 2010. "Access regimes and regional land change in the Brazilian Cerrado, 1972–2002." *Annals of the Association of American Geographers* 100(1): 87–111.

Kloppenburg, J. 2010. "Impeding dispossession, enabling repossession: Biological open source and the recovery of seed sovereignty." *Journal of Agrarian Change* 10(3): 367–388.

Leguizamón, A. 2016. "Disappearing nature? Agribusiness, biotechnology and distance in Argentine soybean production." *Journal of Peasant Studies* 43(2): 313–330.

Mazoyer, M., and L. Roudart. 2006. *A history of world agriculture: From the neolithic age to the current crisis.* New York: Monthly Review.

McKay, B., and G. Colque. 2016. "Bolivia's soy complex: The development of 'productive exclusion'." *Journal of Peasant Studies* 43(2): 583–610.

McKay, B., S. Sauer, B. Richardson, and R. Herre. 2016. "The political economy of sugarcane flexing: Initial insights from Brazil, Southern Africa and Cambodia." *Journal of Peasant Studies* 43(1): 195–223.

McMichael, P. 2012. "The land grab and corporate food regime restructuring." *Journal of Peasant Studies* 39(3–4): 681–701.

Mier y Terán Giménez Cacho, M., 2016. "Soybean agri-food systems dynamics and the diversity of farming styles on the agricultural frontier in Mato Grosso, Brazil." *Journal of Peasant Studies* 43(2): 419–441.

Morgan, D. 2000 [1979]. *Merchants of grain: The power and profits of the five giant companies at the center of the world's food supply.* Lincoln, NE: iUniverse.

Murphy, S., D. Burch, and J. Clapp. 2012. *Cereal secrets: The world's largest grain traders and global agriculture.* Oxfam Research Report. Oxford: Oxfam.

Nehring, Ryan. 2016. "Yield of dreams: Marching west and the politics of scientific knowledge in the Brazilian Agricultural Research Corporation (Embrapa)." *Geoforum* 77: 206–217.

Oliveira, G. de L.T. 2009. "Uma Descrição Agroecológica da Crise Atual" (An agroecological description of the current crisis). *Revista NERA* 12(15): 66–87.

——— 2016. "The geopolitics of Brazilian soybeans." *Journal of Peasant Studies* 43(2): 348–372.

———2022. "Soy, domestication, and colonialism" in *The Routledge handbook of critical resource geography*, M. Himley, E. Havice, and G. Valdivia (eds.). London and New York: Routledge, pp. 335–344.

Oliveira, G. de L. T., Ben McKay, and C. Plank. 2017. "How biofuel policies backfire: Misguided goals, inefficient mechanisms, and political-ecological blind spots." *Energy Policy* 108: 765–775.

Oliveira, G. de L. T., and M. Schneider. 2016. "The politics of flexing soybeans: China, Brazil, and global agroindustrial restructuring." *Journal of Peasant Studies* 43(1): 167–194.

Oliveira, G. de L. T., and S.B. Hecht. 2016. "Sacred groves, sacrifice zones, and soy production: Globalization, intensification and neo nature in South America." *Journal of Peasant Studies* 43(2): 251–285.

Ouma, S. 2014. "Situating global finance in the Land Rush Debate: A critical review." *Geoforum* 57: 162–166.

Pacheco, P. 2006. "Agricultural expansion and deforestation in lowland Bolivia: The import substitution versus the structural adjustment model." *Land Use Policy* 23(3): 205–225.

Patel, R. 2013. "The long green revolution." *Journal of Peasant Studies* 40(1): 1–63.

Peluso, N.L. 1992. *Rich forests, poor people: Resource control and resistance in Java.* Berkeley, CA: University of California Press.

Peluso, N.L. 2012. "What's nature got to do with it? A situated historical perspective on socio-natural commodities." *Development and Change* 43(1): 79–104.

Pitta, F., and M. Mendonça. 2015. *A empresa Radar S/A e a especulação com terras no Brasil.* Rede Social de Justiça e direitos humanos, GRAIN, Inter Pares, and Solidarity Sweden Latin America, São Paulo.

Rulli, J. 2007. "The refugees of the agroexport model: Impacts of the soy monocultures in Paraguayan peasant communities" in *United Soya Republics: The truth about soya production in South America*, R. Rulli (ed.). Buenos Aires: Grupo de Reflexion Rural, pp. 194–216.

Rumstain, A. 2012. *Peões no trecho: trajetórias e estratégias de mobilidade no Mato Grosso.* (Farmhands on the road: Trajectories and mobility strategies in Mato Grosso.) Rio de Janeiro: E-papers.

Salerno, T. 2017. "Cargill's corporate growth in times of crises: How agrocommodity traders are increasing profits in the midst of volatility." *Agriculture and Human Values* 34(1): 211–222.

Sauer, S. 2018. "Soy expansion into the agricultural frontiers of the Brazilian Amazon: The agribusiness economy and its social and environmental conflicts." *Land Use Policy* 79: 326–338.

Shurtleff, W., and A. Aoyagi. 2009. *History of soybeans and soyfoods in South America (1882–2009) extensively annotated bibliography and sourcebook.* Lafayette, CA: Soyinfo Center.

Silva, M., and L. Costa. 2012. "A indústria de defensivos agrícolas" (The pesticide and herbicide industry). *BNDES Setorial* 35: 233–276.

Stédile, J.P., and B. Mançano Fernandes. 1999. *Brava gente: a trajetória do MST e a luta pela terra no Brasil.* São Paulo: Editora Fundação Perseu Abramo.

Svarstad, H., T.A. Benjaminsen, and R. Overå. 2018. "Power theories in political ecology." *Journal of Political Ecology* 25(1): 350–363.

Thaler, G.M., 2017. "The land sparing complex: Environmental governance, agricultural intensification, and state building in the Brazilian Amazon." *Annals of the American Association of Geographers* 107(6): 1424–1443.

Turzi, M. 2011. "The soybean republic." *Yale Journal of International Affaris* 6(2): 59–68.

Vander Vennet, B., S. Schneider and J. Dessein. 2016. "Different farming styles behind the homogenous soy production in southern Brazil." *Journal of Peasant Studies* 43(2): 396–418.

Watts, M.J. 1983/2013. *Silent violence: Food, famine and peasantry in Northern Nigeria.* Berkeley: University of California Press.

Watts, M., P. Robbins, and R. Peet (eds.). 2010. *Global political ecology.* London: Routledge.

Wesz Jr., V.J. 2016. "Strategies and hybrid dynamics of soy transnational companies in the Southern Cone." *Journal of Peasant Studies* 43(2): 286–312.

10 The political ecology of maize in China

National food security and the reclassification of maize from staple to industrial crop

Li Zhang

Introduction

Maize is the main crop in China in terms of cultivated area and harvested volume (FAOSTAT 2020). It was first introduced from the Americas during the late Ming and early Qing dynasties in the 16th and 17th centuries. However, it was only incorporated into local diets (alongside potatoes) during shortages of other major staple crops such as rice, wheat, and other cereals (Zhang 2010). Until the 1970s, maize only accounted for less than 11% of the total grain harvest in China (Zhang et al. 2019). However, the importance of maize for national food security grew significantly since the 1980s, when surplus maize started being processed into starch for industrial uses (e.g. as input in processed food, paper, and textiles) and used extensively as livestock feed, enabling in the process the industrial concentration of pork production (Schneider and Sharma 2014; Guo 2007).

Currently, China produces approximately 260 Mt of maize annually, which amounts to 23% of the global maize output, making it the second largest producer behind only the US (346 Mt), and well ahead of the third largest producer, Brazil (100 Mt) (USDA 2020). However, unlike the US and Brazil, where maize is undeniably an agroindustrial crop on par with soybeans (Gillon 2016; Oliveira and Schneider 2016; see also Chapter 9), the Chinese government maintains strict regulations on maize, as it is viewed as a strategic food staple alongside rice and wheat. This means that the national government insists on a policy of achieving 95% domestic self-sufficiency for maize and constantly adjusts price protection mechanisms and other regulations that control how much maize can be exported, imported, and processed into ethanol or other industrial products (Zhang et al. 2019).

Nonetheless, political, economic, ecological, and technological transformations in recent decades in the maize system have resulted in less than 5% of maize being consumed directly for food each year in China (Zhang et al. 2019). This creates a mounting pressure for the Chinese government to reclassify maize, not anymore as a strategic food staple, but rather as an industrial crop for use as livestock feed and input for biofuel and hundreds of industrial products such as food preservatives, medicines, cosmetics, glues, plastics, and pesticides (Zhang et al. 2019). Doing so would unleash market forces that could likely set maize on the same trajectory as soy in China, which is now characterized by stagnant domestic production and skyrocketing imports from the US and Brazil to meet the increased demand for agroindustrial processing (Yan et al. 2016; Huang and Yang 2017).

Thus, the future of maize production and use in China is at a major crossroads. On the one hand, the national government is interested in continuing to protect it as a strategic food staple crop through strict self-sufficiency regulations. On the other hand, there are also multiple interests in reclassifying it as an industrial crop to allow its wider use, replacing government regulations with market dynamics. In order to understand this conflict, we must examine the various political, economic, and ecological forces pushing and pulling in each direction and identify where the balance currently lies in this tug-of-war. Doing so enables the proper understanding of recent trends, which can inform the analysis of the political economic and ecological transformations that are expected across various regions of China.

This chapter aims to describe the policy transformations that increasingly promote the reclassification of maize as an industrial crop in China, as well as who stands to benefit and lose through this process. The chapter makes two central arguments. First, the promotion of silage maize has emerged as the main compromise at this crossroads. Second, this reconfiguration of maize as silage, the associated restructuring of the livestock industry, and the accompanying transformations in grain and agroindustrial markets tend to strengthen the position of agribusiness companies that push for the reclassification of maize as an industrial crop, accelerating thus these transformations. This tentative, and relatively short, industrial chain for maize is already reconfiguring local political ecologies, and consequently, the advancement of silage maize is likely to concentrate agribusiness maize production further in the hands of wealthy investors and displace its small-scale agroecological production among peasants.[1]

In order to develop these arguments, this chapter uses a political ecology lens to unravel the debates and processes associated with maize reclassification in China, as well as its main impacts on different stakeholders. This work follows on some of the most classic and well-established frameworks of political ecology that critically examine the links between political economic institutions, environmental policy and management, and socio-ecological outcomes in terms of power relations (Blaikie and Brookfield 1987; Peluso 1992; Rocheleau et al. 1996; Watts and Peet 2004) (Chapter 2). Such work is especially helpful to analyze China's transition to a "hybrid economy", where state controls and market dynamics are constantly in tension with one another, subject to reinterpretation and contestation, with this conflict being central to agrarian and environmental policy and management (Muldavin 1997, 2000). Moreover, this theoretical approach enables a critique of institutional and socio-economic transformations in China that are not necessarily focused superficially on the Chinese government, but rather on the deeper structures of global capitalism and the ideological foundations of classical economics, development, and ecological modernization theory (Muldavin 2007; Jacka 2013).

The chapter draws upon an in-depth analysis of Chinese scientific literature, policy documents, and fieldwork-based observations from two provinces in China, namely Henan in the central region, and Guangxi in the southwestern region. The following section traces the main issues surrounding food security and the classification of staple and industrial crops in China, and the drivers and debates about the reclassification of maize in the past two decades. The third section outlines the methods and study sites, and the following section demonstrates the main actors expected to benefit and lose from the reclassification of maize as an industrial crop in Henan province and Guangxi Zhuang Autonomous Region. The penultimate section discusses how industrial maize is becoming an agent of institutional and socioeconomic transformation.

Reclassification of maize in China

Food security and crop classification

In 1950, the year after the establishment of the People's Republic of China, seven crops were officially classified as strategic staples for domestic food security: wheat, rice, soybean, millet, maize, sorghum, and coarse cereals. A few years later, potatoes were further added to this list of staple food crops. During the socialist period, state-led agricultural production adopted a "grain first" policy, which emphasized on these staples in order to achieve national food security. These crops were subject to strict production quotas and international trade controls, with the agroindustrial processing for non-food use being effectively forbidden (Lardy 1983).

However, following the beginning of market-oriented reforms in 1978, the political and economic importance of these staple food crops has reduced in relation to fruits, vegetables, livestock, and various cash crops. The share of staple crops in national food consumption declined from 75% in 1980 to 29% in 2015, with 50% of the staple crops produced in recent years actually used as livestock feed and the remaining 21% directed toward industrial uses (Zhou et al. 2017). As the relative consumption of staples declined, coarse cereals, millet, sorghum, and potatoes were dropped from the category of staple crops entirely (Yan 2017).[2]

The production of the four remaining staple crops (i.e. wheat, rice, maize, soy) stagnated during the late 1980s and early 1990s, following the easing of government procurement of staples and the increasing profiting of peasants from the production of higher-value cash crops (Song and Ouyang 2012). The Chinese government also began to reassert policies to subsidize, sustain, and control staples for national food security (Song and Ouyang 2012). During the negotiations for China's entrance into the World Trade Organization (WTO), the political and economic architecture around the classification of certain crops as strategic staples was avidly debated. On the one hand, there was a more production-oriented camp organized around the Ministry of Agriculture, which defended strong self-sufficiency in staples (Jiang et al. 2004; Han 2012). On the other hand, there was a more market-oriented camp organized around the Ministry of Commerce, which defended deregulation and reliance upon the imports of land-intensive and low-value crops such as soybean and maize (Huang 2004; Jiang 2005).

The eventual compromise reached with China's transition into the WTO is known as the policy of "two markets, two resources". This essentially meant that China guarantees *food* resources through domestic production and highly controlled markets (95% self-sufficiency of staple crops), but relies upon "free trade" and international markets for *agroindustrial* resources. The central pillar of this reform was the reclassification of soybeans from a strategic staple food crop into an industrial crop (by abandoning government procurement quotas, subsidies, price guarantees, and import tariffs), while maintaining wheat, rice, and maize under strict controls for national food security (Yan et al. 2016). The outcome of this reform was quick and dramatic. Until the mid-1990s, China was not only self-sufficient in soybeans but was even a net exporter. By 2000, however, the Chinese soy sector fell into a crisis, as output stagnated and imported soybeans flooded domestic markets, with China surpassing Europe in becoming the world's largest importer of soy (Yan et al. 2016) (Chapter 9). By 2014, China absorbed about 60% of all international soybean trade, and imports accounted for about 80% of total soybean consumption, mainly for the agroindustrial processing of vegetable oil and livestock feed (Yan et al. 2016).

This reclassification and subsequent transformation of soybean production and trade have been the subject of much dispute, and they feature prominently in current debates about the possible reclassification of maize. On the one hand, some scholars consider this transformation the "inevitable" result of international "comparative advantage" (Chapters 2 and 6) and celebrate the reclassification of soybean as a victory of "rational" allocation of resources, which "frees up" domestic farmland for the production of more essential staple food crops such as wheat and rice, while maximizing economic gains through imports from countries with more abundant natural resources and efficient soybean production systems (Huang 2004; Zhang et al. 2019). Moreover, according to this view, protection policies for staple food crops are inefficient mechanisms for guaranteeing national food security, as market signals are considered to be more effective in generating "farmer enthusiasm" for agricultural production compared to subsidies and price protection policies (Song and Ouyang 2012). Consequently, these scholars interpret the reclassification of soybeans as a successful example, which is suitable for guiding the reform and reclassification of maize (Huang and Yang 2017; Zhang et al. 2019; Song and Ouyang 2012), arguing that "China must rely on the international market to achieve domestic food security" (Mao and Kong 2019: 142).

On the other hand, other scholars lament what they perceive to be as a "crisis" of domestic soybean production, as its reclassification has led to a corporate take-over of soybean processing and production, an increased exposure to genetically modified (GM) foodstuff and the erosion of food sovereignty (Yan et al. 2016) (see Chapter 9 for similar examples from South America). The

Ministry of Agriculture has insisted that "we must firmly control the maize industry in our own hands [and] prevent maize from becoming the 'second soybean'" (Han 2012: 7). The recent US–China trade war has reignited these debates and reinforced the concerns over the "abandonment" of domestic soybean production as the "weak point" of national food security (State Council 2019; Cui and Jiao 2019). Moreover, the COVID-19 pandemic brought further urgency on this topic due to the threat posed by supply chain disruptions not only in China but also in the world's leading exporters of soybeans and maize, namely the US and Brazil (Mai 2020; Cui et al. 2020). This present context will be revisited in the next section, which focuses on the recent drivers and debates about the reclassification of maize in China.

Drivers and debates about the reclassification of maize

Biofuels

Maize production increased dramatically since the late 1990s in response to government policies seeking to reverse the stagnation in staple food crop production (FAOSTAT 2020). However, soon, the government found itself holding an excess of "aged grain" reserves (Cui et al. 2020). In 2000, as soybean imports became the dominant input for livestock feed and cornstarch-based industries could not process the full amount of stockpiled maize, the government established a "grain-to-fuel" program to convert maize into ethanol (Cui et al. 2020). Between 2001 and 2005, maize-based ethanol production increased, accounting for 76% (up from 59%) of all biofuel production in China, exhausting the "aged" maize stocks (Wang 2006). By 2006 China's ethanol blending capacity surpassed 10 million liters, accounting for more than 10% of annual maize consumption (Ma et al. 2006). In 2007, *newly harvested* maize accounted for 86% of all feedstock for ethanol production, causing direct competition between the ethanol and livestock feed industries, and putting pressure on staple food prices (Chen et al. 2013) (Chapter 4).

The expansion of the biofuel industry during the 2000s (not only in China, but especially in the US, Brazil, and the European Union [EU]) certainly contributed to a global rise in agroindustrial commodity prices, which created widespread concerns domestically and internationally about a conflict of "food vs. fuel" (Gasparatos et al. 2011; Oliveira et al. 2017; Ahmed et al. 2019; Martin 2020; see also Chapter 4). He (2007) aptly framed this competition as "the food war between cars and pigs". Concerned about the 2007 spike in the prices of food and other agricultural commodities, and the depletion of maize and other staple food crop stocks, the Chinese government restructured price control mechanisms for maize and reversed its biofuel policy to impose restrictions on the use of maize and other staples for ethanol production (Song et al. 2019). Nonetheless, various agribusiness corporations, including leading state-owned enterprises like COFCO, continued their investments in the sector, with many scholars and policy analysts believing that the development of a biofuel industry in China was "necessary" and "unavoidable" (He 2007; Fang et al. 2004). Thus, the mainstream discussion since the mid-2000s has not been about whether or not China should develop biofuels. Rather, the debate has been about whether the national biofuel industry should continue to rely primarily upon maize, or switch to other "non-food" crops such as sugarcane, cassava, sorghum, jatropha, and cellulose, which can supposedly be grown on "marginal" land and not compete directly with officially designated staples (Ni 2012; Wang 2006; Zhang et al. 2019; Chen et al. 2013; see also Chapter 4).

The argument that only "non-food" crops should be used as biofuel feedstock has received the most support in China, particularly among academics (Ni 2012; Wang 2006; Zhang et al. 2019; Chen et al. 2013) and policymakers associated with the Ministry of Agriculture (Han 2012). Meanwhile, agribusiness and biofuel companies have promoted the argument that maize-based ethanol production should be liberalized, especially those corporate actors that have expanded aggressively during the early 2000s and have since been operating at a loss due to processing overcapacity (Guo 2007; Zhao and Xin 2020). The proponents of liberalization are joined by a minority of scholars (e.g. Cheng 2019) and policymakers associated with the National Energy Commission (NEC). The latter continues to promote ethanol as a means to "optimize the energy structure, improve the environment, regulate the food market, and promote agricultural, rural, and regional economic development" (NEC 2017: online). Although the NEC has not fully reversed restrictions on the use of maize and other staple food crops for biofuel production, it still allows ethanol production as a means of "handling overdue and excessive grain" (NEC 2017: online). For example, the Jilin Fuel Ethanol company, China's first large-scale ethanol enterprise, continues to use as feedstock "excess" maize from northeastern China, maintaining that the "food vs. fuel" controversy is due to the fact that "the public does not understand" agroindustrial development and price dynamics (Zhang and Zhao 2017). They further argue that non-grain ethanol production is not economically viable due to wastewater treatment costs, seeking to reduce popular resistance by promoting maize-based ethanol at the national level (Zhang and Zhao 2017).

However, there is ongoing and high-level resistance to the use of maize for ethanol production, and concerns about national food security aggravated due to the onset of the US–China trade war.[3] Therefore, the Chinese government has continued to adjust price controls for maize and "resolutely maintains" it in the position of a strategic staple for national food security, curtailing its use for ethanol and limiting the development of downstream processing industries based upon cornstarch (State Council 2019; Wang and Yang 2019). Consequently, the China Grain Industry Association (CGIA) now admits that given "China's national conditions, the development of biofuel should take sugarcane, cassava, and cellulose by-products from other plants that are relatively abundant in southern China as the main raw material, and never use grain [including maize] as the main raw material" (CGIA 2019: 24). Yet, Chinese agribusinesses and pro-market scholars and policymakers have not abandoned their efforts to reclassify maize as an industrial crop; rather, they simply shifted their target markets and technological strategies, as discussed in the next section.

Livestock feed, industrial processing, and silage

Following the curtailment of biofuels as the leading non-food market for maize, the proponents of maize reclassification have mainly pursued two other markets: (a) silage for new forms of livestock production intensification through industrial methods (particularly for dairy, beef, and lamb) and (b) "deep" (i.e. downstream) processing of cornstarch into a greater variety of higher-value industrial products.

Their strategy relies on two apparently contradictory arguments. On the one hand, they argue that maize is already an industrial input rather than a staple food crop, so it should no longer be regulated through the national strategic food supply. On the other hand, they also argue that the industrialization of maize can help guarantee national food security. The latter argument goes beyond the rationale used for the past reclassification of soybeans, namely that replacing domestic self-sufficiency in maize with imports will allow Chinese grain farmers to focus on wheat and maize, the two "real" staple crops in China, while simultaneously supplying livestock feed as part of national food security (Huang 2004; Huang and Yang 2017). The CGIA (2019) now argues that the "deep" industrialization of maize can provide for a "higher level of food security" by "storing grains in the industrial chain" itself:

> The view that "deep" [i.e., downstream industrial] processing of maize competes with food for people is a misunderstanding... In essence, "deep" processing of maize creates wealth for society and serves people's needs for a better life... In addition, at present most of China's industrialized maize products are in the food industry... and have not been separated from the food chain. [This] "storage of grain in the industry" increases the flexibility of grain supply and demand... leaving a buffer zone for grain security.
>
> (CGIA 2019: 22)

This argument is strengthened with the critique that the policy of "two markets, two resources" harbors a fundamental contradiction, in that the state-owned agribusiness sector must play two conflicting roles. It has to implement domestic production and price protection policies to protect farmers and safeguard national food security and, at the same time, to liberalize domestic markets and engage in international trade for profit (Jiang 2005; Huang and Yang 2017). This has effectively "weakened the competitiveness of the maize industry downstream", which could generate far more value-added products and strengthen (this new interpretation of) food security through greater agro-industrial capacity and improved rural development (Zhang et al. 2019: 12).

When policymakers are unconvinced about this new approach to food security, the proponents of maize reclassification turn to the opposite argument that "maize has completed its separation from the staple foods and transformed its identity into a feed grain and industrial input," and therefore it "should be viewed as an industry in need of further development, and not managed as part of general food staple control" (Zhang et al. 2019: 14). Indeed, non-feed industrial use of maize has increased from 40 Mt in 2008 to 75 Mt in 2018, representing also an increase in the proportion of total maize consumption from 24% to 31% (Zhang et al. 2019). Given that the use of maize for livestock feed continues to increase as well, only 5% of all maize in China is now used directly for human consumption (Zhang et al. 2019). Consequently, the current limitations on the industrialization of maize create a serious problem for China's agroindustrial sector, which continued to expand with massive government subsidies in the aftermath of the 2007–2008 food price crisis, and now finds itself with "serious overcapacity… industry-wide losses, low-level repetitive construction, insufficient business start-ups, idle equipment, and great waste of maize resources" (CGIA 2019: online).

However, the growing socio-economic inequality sustains the government's need to support grain farmers at the same time as the concerns about self-sufficiency in staple crops have even increased due to the US–China trade war (State Council 2019). Thus, while "deep" maize processing remains the ultimate goal of China's agroindustrial sector (CGIA 2019; Zhang et al. 2019), so far, silage has been the more successful strategy for maize industrialization. This is because silage maize matures more quickly than regular maize and its grains do not need to be harvested and processed by farmers prior to sale to agroindustries. This supposedly leads to higher farmer incomes and lower production costs for livestock producers (Chen et al. 2019). The main downside, from a government and agribusiness perspective, is that silage is only suitable for herbivore livestock such as cattle and lamb, rather than pigs, which have become the predominant meat industry in China during recent decades (Chen et al. 2019). Nonetheless, the massive pandemic of African swine fever in 2018–2019 eliminated at least half of China's pig production, and, combined with the US–China trade war, it has created the conditions for Chinese agribusinesses and policymakers to envision a broad restructuring of the national grain-livestock nexus away from soy and regular maize for the pork and chicken industry, and toward silage for dairy, beef cattle, and lamb (Ministry of Agriculture 2016; Chen et al. 2019; Guo 2019). Therefore, a new compromise has become possible through the Chinese government's grain-to-feed program, which promotes the replacement of regular grain-bearing maize with new varieties grown purely for silage (Ministry of Agriculture 2016; State Council 2019).

This reinvention of maize as silage enables the Chinese government to maintain that "the status of maize as a staple will remain unchanged and price protection [for regular maize] will continue" (Wang and Yang 2019: online), while simultaneously supporting the new agroindustries that drive the expansion of silage and supply downstream agribusinesses that industrialize cornstarch and other byproducts (Han 2018). After all, as an agribusiness executive recently explained, "good silage can be sold to dairy farms [while] inferior quality material can be sold to biofuel refineries for energy generation, providing another source of income, and the state is even providing subsidies" (Wang 2019: online). Indeed, the government is providing direct subsidies for agribusiness companies that harvest and process silage and is also subsidizing the new industrialization of the silage-based livestock industries, particularly in connection with its broader poverty alleviation initiatives (Ministry of Agriculture 2016). Therefore, rather than conceiving of this new emphasis on livestock feed as a turn away from the industrial uses of maize, it should be recognized as a compromise that is in fact laying the foundation for the further industrialization of maize, and its possible reclassification in the future.

Methodology

Data collection and analysis

The primary analysis presented in this chapter draws upon empirical data from 212 semi-structured interviews and participant observation in two rural communities (and their surrounding regions) in Henan province (central China) and Guangxi Zhuang Autonomous Region (southwestern China). The primary data collection was conducted during multiple visits from 2013 to 2017, each lasting between two weeks and two months. Table 10.1 contains the main categories of interviewees, with the numbers of interviews for each category spread relatively evenly between the two field sites (except where noted). Access was facilitated through academic contacts at the China Agricultural University, where the author was affiliated at the time. Data from local government records,

Table 10.1 Respondent breakdown in Henan and Guangxi

Interviewee categories	Number of interviews
Ordinary peasants households	99
Total government officials	47
County-level officials	21
Township-level officials	11
Village-level officials	15
Total agricultural sector professionals	40
Cooperative and corporation leaders	5
Agricultural input dealers	9
Veterinarians (Guangxi only)	3
Butchers (Guangxi only)	4
Food processing and distribution professionals	19
Total additional members of the community	26
Teachers, doctors, and spiritual leaders	16
Elders, disabled, and orphans	10
Total	212

local media, and various company websites was obtained online during June and July 2020 to supplement primary data from fieldwork.

Approximately half of all the respondents were "ordinary peasants". They were identified through a theoretical sampling approach that included households in the entire spectrum found at the villages, from households actively engaged in agricultural production to those "circulating" most of their land (i.e. by renting out their land-use rights), and from households with few family members who migrated out for waged labor in urban areas to households with many family members who out-migrated.

The remaining half of the respondents were key informants at the county, township, and village levels. The first category of key informants including government officials, local government leaders, and lower-level officials stationed in the bureaus of agriculture, animal husbandry, and food and drug administration. The second category includes agricultural sector professionals such as leaders of local cooperatives, agricultural input dealers, and professionals in the food processing and distribution sector, among others. The third category includes other important community members that could provide insights through perspectives and experiences that might be distinct from government officials and agricultural sector professionals.

It should be noted that interviews were originally undertaken to examine food safety politics and agrarian transformations (Zhang 2017, 2020; Zhang and Qi 2019). Thus, field notes were consulted for reference to observations of maize production (including silage maize, and use of maize for livestock feed), and views of the research participants views on policy transformations regarding maize, including not only its classification as a food or industrial crop but also its price controls and subsidies. Once identified through the consultation of field notes, those specific interviews could then examined at greater length for this chapter, utilizing a simple coding of "favorable" or "unfavorable" opinions about silage maize and associated policy transformations.

Study sites

Bian village, Henan province

The first field site is centered upon Bian village (pseudonym) in Lankao county, Henan province. Henan is in China's central plains and is part of the Huanghuaihai region, which is considered to be the main staple food-producing area of the country. In particular, wheat and maize are the predominant crops produced in northern Henan, where Lankao county is located. Henan was among the first provinces to launch biofuel pilot programs in the early 2000s and experience the rapid expansion of maize-based ethanol (from 30,000 tons in 2002 to 1,930,000 tons in 2011) (Wu and Huo 2014). Unlike the other provinces in China's northeast region that are major producers of staple crops and biofuels, Henan also has an extensive livestock sector that includes cattle (for beef and dairy) and lamb. Therefore, Henan is especially well positioned in this new development of agro-pastoral integration, particularly related to combining silage maize with "grass-fed" livestock. For example, in 2015–2016, when the grain-to-feed program was officially established, the area under silage maize in Henan increased by about 37% to 40,000 ha (600,000 mu), further increasing by 80% to 72,000 ha (1.08 million mu) by 2018, surpassing the national government's target at 112% of the policy plan (Henan Daily 2019; Liu 2019). It was expected that by 2020, the area under silage maize in Henan would reach somewhere between 267,000 and 333,000 ha (4 and 5 million mu) (Li 2016; Liu 2019).

Henan is also one of the provinces with the largest population and lowest income per capita in China, thus experiencing one of the largest outmigration of peasants in the country. These migrant workers are primarily engaged in waged employment in the industrial, construction, and service sectors in major cities across China, but some also work for larger-scale "family farms" and agribusinesses in the central region of China such as the ones discussed in the following section (Guan 2019).

Lankao county is located on the banks of the Yellow River, where sandstorms, severe waterlogging, and saline-alkali soils have caused severe damages to local agriculture. These have left the region infamously poor, for example, being one of the areas worst affected by the famine of the early 1960s. Hence, Lankao has also been a prominent site of government-led poverty alleviation interventions, particularly since 2013, when Xi Jinping dedicated personal attention to poverty alleviation in the county.

Bian is a typical village in Lankao county, with about 191 ha (2,860 mu) of arable land and a population of 1,581 people in 392 households. All villagers are ethnically Han Chinese. The main cultivated crops are wheat, maize, and rice, produced almost exclusively for commercialization. Rice production is enabled through the irrigation of about 87 ha (1,300 mu) of paddies, with water from the Yellow River. All maize production utilizes hybrid seed varieties produced by transnational corporations like DuPont Pioneer and Charoen Pokphand. Scholars from the China Agricultural University partnered with local government officials to promote ecological farming in the village since 2005. These ecological farming initiatives are aimed at producing organic rice and high-quality pork, but despite these efforts, all staple crop production utilizes large amounts of chemical fertilizers, pesticides, and herbicides; for more details see (Zhang, 2017; Zhang and Qi, 2019). Still, every household has small backyard plots, producing organic vegetables and free-range chickens for family consumption. By 2014, all households in the village had given up raising pigs due to high production costs, volatile markets for pork meat, and insufficient labor due to outmigration. At that time, seven households raised lambs, scaling up from a handful of sheep to about 30 animals in each herd, and one household started raising cattle, with about 20 beef cows. All lamb and cattle are raised through contract farming with newly established corporations.

Gu village, Guangxi Zhuang Autonomous Region

The second field site is centered on Gu village (a pseudonym), Mashan county, in the Guangxi Zhuang Autonomous Region of southwestern China.

Guangxi is home to China's largest ethnic minority population and has historically been considered as the southern frontier of Chinese imperial dynasties. It is characterized by very mountainous terrain, difficult access from the rest of China, and one of the lowest gross domestic product (GDP) per capita in the country. Rice, maize, and sweet potatoes are the main crops, which are considered to be more significant for local food consumption than national food security. For example, due to cultural preferences and economic necessity, maize is an important local dietary staple (Li 2002; Ely et al. 2016). Therefore, rather than encouraging staple crop production for other provinces, the central government has incentivized the production of sub-tropical cash crops that cannot be so easily grown elsewhere in China, such as cassava, sugarcane, and mulberry (the food source of silkworms) (Zhou et al. 2019).

Thus, Guangxi became prominent in the national biofuels program of the 2000s, with about 70 companies producing ethanol from sugarcane and another 20 from cassava and the ethanol output in Guangxi reaching 500,000 tons by 2008 (Tang 2012). However, sugarcane and cassava prices became too high for economically viable ethanol production, and biofuel companies soon found themselves having to import upwards of 80% of their cassava feedstock (USDA 2019). When Thailand restricted cassava export to China in 2016, much of Guangxi's ethanol industry shut down. This included the leading state-owned company COFCO, which inaugurated a huge facility in 2008 but never reached its capacity of 200,000 tons of cassava-based ethanol per year (USDA 2019). As its biofuel program flopped, the central government included Guangxi among the pilot provinces of the grain-to-feed program as an alternative mechanism for poverty alleviation. Consequently, the area under silage maize increased from 1,333 to 9,307 ha (from 20,000 to 139,600 mu) between 2016 and 2018, which surpassed the government's target at 130% of the policy plan (Zhu 2020).

Mashan county is famous for its black lamb and was one of the 20 counties selected in Guangxi for piloting the grain-to-feed program in 2016. Gu is a typical village in Mashan county, with a total area of 1,425 ha (21,375 mu) but merely 139 ha (2,085 mu) of arable land. The village contains about 3,734 people in 875 households, among whom 75% are ethnically Zhuang, another 20% are Yao, and only 5% are Han Chinese. Maize is the main crop, usually planted in rotation with soy, black, and red beans, and alongside a wide variety of vegetables. About 90% of the harvested crops are consumed locally as staple foods, including maize (which is also used locally as pig and chicken feed). Livestock commercialization is the main source of income for most households. In 2013, there were about 2,000 chickens and 3,000 pigs in Gu village, typically raised in herds of about 20 animals per year. Mashan black lamb fetches higher prices than chicken or pork meat, but it is considered more difficult to raise. As a result, only about 10–15 households raised this type of livestock, mainly in herds ranging between 5 and 20 animals per year.

An important contrast between Gu village in Guangxi and Bian village in Henan regards the development of local agroecological initiatives. The share of maize seeds purchased from transnational corporations is lower in Gu, as collaborations between the village leader and scholars from the Chinese Academy of Sciences enable about 20% of maize cultivation from local seed varieties. In addition to backyard gardens, the village leader established a cooperative to produce organic vegetables for an NGO-led farm-to-table restaurant in Nanning city and farmer markets in Mashan county. In contrast, practices such as maize consumption as human food, organic vegetable production, and cooperative-based production for regional markets are not observed in Bian village. For more details, refer to Ely et al. (2016) and Zhang (2017, 2020).

Results: Who stands to benefit and lose from maize reclassification

Henan province

The government leaders in Lankao county down to the village level in Bian are all actively promoting the grain-to-feed program. This includes the adoption of silage maize, the establishment of new lamb and cattle enterprises, and contract farming with the peasants who are still engaged in agricultural production. Demonstrating advancement in this regard directly favors their political careers. Conversely, lower-level government officials, whether in the fields of agriculture, animal husbandry, or food and drug administration, are far more skeptical about the effectiveness of these poverty alleviation policies. Still, some of them embrace this agro-industrial restructuring due to their own family or business connections with new enterprises investing in the area. Furthermore, they see few other feasible production alternatives due to climate change, which makes conditions increasingly more difficult for rice production that was the main target of the local agroecology-oriented project. Meanwhile, maize has proven to be more drought-resistant, even if it results in lower-quality silage that does not reach the prices usually promoted by local government officials and new corporations.

In recent years, there has been a boom of new agribusiness companies in the region. Their owners and managers have been some of the major boosters of maize industrialization and to an extent the main beneficiaries of the grain-to-feed program. During an interview, the leader of one of these companies said that their main concern about investing in Lankao was the poor environmental conditions and climate change, as major recent droughts that have caused crop loss and poor-quality silage. Nonetheless, he explained that they have dealt with this risk through market diversification, particularly by redirecting the silage contracted from local farmers to biofuel and other industrial purposes when its quality was not adequate for their own livestock. Table 10.2 illustrates the main companies making new investments in Lankao that the author was able to document.

Unlike these agribusiness investors, who are generally able to navigate environmental and market risks to profit from the demand shift toward silage maize (and its industrial processing), very few peasant households are able to

Table 10.2 Agribusinesses investments in silage maize and associated industries in Lankao

Company name	Year	Activity and scale	Notes
Hennan Jinkai Group	2011	Leases 733 ha (11,000 mu) to produce silage maize for 20,000 dairy cows and 7,000 beef cows	Reached an agreement with Lankao County government in 2015 to scale up to 5,333 ha (80,000 mu)
Kaifeng Dasong Food Co.	2012	Leases 800 ha (12,000 mu) for the production of maize, peanut, and soy for industrial processing	NA
Wellhope Agri-tech Joint. Co.	2013	Processes 220,000 tons of silage and livestock feed per year	Listed as a "key project" in Lankao in 2014
Seventeen projects financed by the World Bank	2015	Constructed 9,710m² of cowsheds, 3,920m² of manure storage, and 12,920m³ of silage storage	Coordinated project by the Lankao County Bureau of Finance and the Bureau of Animal Husbandry
Chuying Agropastoral Group	2016	Implements the "grain-to-feed" program as part of poverty alleviation efforts	Halted projects due to corruption investigations in 2018, and financial losses
Zhongyang Animal Husbandry Group	2017	Constructed vertically integrated silage-lamb-fertilizer industry, contracting 300 ha (4,500 mu) of silage maize from 1,200 farmers	In 2018, planned to expand slaughter to 300,000 lambs per year and contract 38,000 tons of silage maize per year
Zhong Chu Cao Ecological Agriculture Technology Co.	2017	Contracted 667 ha (10,000 mu) for the production of silage maize	NA
China Power Construction	2018	Constructed and operates energy plants to produce 50,000 m³ of biogas per day	Signed agreement with Lankao government to process 150,000 tons of silage maize per year
Wanhua Ecological Board Co.	2019	Processes silage maize for plywood production	Expected to consume 5,000 tons of silage maize per year

Source: Author's elaboration from Lankao government and company websites.

consistently benefit. Only the two largest farmers in Bian village, for example, who "circulated" (i.e. leased land use rights) over 12 ha (180 mu) each from other villagers, seemed to be enthusiastic about this new trend of producing silage maize contracted through agribusiness investors. Still, these farmers only dedicated a relatively small portion of their land to silage maize, in order to avoid possible losses due to achieving lower-than-contracted quality.

The vast majority of peasants, on the other hand, farmed much less land and would not risk contracting out their crops for silage maize alone. Instead, they sold their regular maize early as silage when facing financial hardship or environmental stress, particularly in 2014–2015 when most rice crops failed due to a severe drought. Moreover, recent environmental regulations have restricted their ability to burn maize stalks in the fields after harvest, as they had traditionally done. Hence, most peasants now view post-harvest straw purely as a liability and often allow large-scale farmers or agribusiness companies to collect their post-harvest stalks as silage without receiving any pay.

Professionals in the agriculture sector in Bian and the surrounding county tended to be favorable toward "agricultural modernization" initiatives, including the new investments and transformations associated with the grain-to-feed program. However, it was not clear whether they would be able to personally profit from these transformations.

Other community members not directly engaged in local government or farming tended to see these transformations with far more skepticism and concern. In particular, some such respondents suggested that such "agricultural modernization" efforts cause environmental contamination due to the increased use of pesticides and concentrated livestock manure, reinforce the displacement of older peasants and poor households from agriculture, and increase socio-economic inequality.

After all, the expansion of maize production for silage intensifies the use of pesticides and chemical fertilizers. The establishment of large-scale concentrated animal feeding operations (CAFOs) increases the generation of manure and its concentration in ways that prevent traditional recycling into organic fertilizer. Since only the wealthier farmers and investors from outside the village are able to capture the subsidies and profits from this general intensification of agroindustrial production in the region, they become increasingly more capable of renting out the land use rights of poorer peasants. In turn, older and poorer peasants then suffer greater pressure to "circulate" their land to wealthier farmers and agribusiness companies. Such uneven adoption of the grain-to-feed program thus exacerbates the displacement of the most vulnerable from the land and aggravates socio-economic inequality.

Guangxi Zhuang Autonomous Region

Government leaders in Mashan county also actively promote the grain-to-feed program. This includes the adoption of silage maize, the establishment of new lamb and cattle enterprises, and contract farming with peasants who are still engaged in agricultural production. However, the leader in Gu village played a rather different role. While not disregarding the government policies or directives, she emphasized much more the need to strengthen agroecological vegetable production (as an alternative source of income for the poorest households) and the importance of sustaining local varieties of maize and pigs. On the other hand, most lower-level government officials in the bureaus of agriculture, animal husbandry, and food and drug administration shared a similar skepticism to their counterparts in Lankao. However, whenever possible, they also pursued their own family and business partnerships with the local agribusiness sector.

A smaller, but also significant, number of new agribusiness companies started investing in Mashan county, focusing especially on contract farming for silage maize, Mashan black lamb, and cattle (Lu et al. 2017). Their attitude toward the grain-to-feed program seems to reflect their counterparts in Henan. A representative of the China Seed Industry in Guangxi, for example, argued that "we should develop in the direction of feed, industrial inputs, and silage by shifting production towards special-purpose maize" (Shi 2019: 28). In Guangxi, as in Henan province, it seems that agribusiness investors are also the primary beneficiaries of these new paths for the industrialization of maize.

Compared to Bian village, even the wealthiest peasant households in Gu village are not able to "circulate" a large amount of farmland from their neighbors. Consequently, none of the surveyed peasants seemed enthusiastic about contracting with agribusiness investors to produce silage maize. Moreover, a relatively strong local agroecological cooperative focuses on vegetable production, providing thus an economic alternative for the poorest households. Furthermore, the vast majority of peasant households still value grain-bearing maize as essential for producing feed for their own pigs, which is still their main income source. Some recent phone interviews suggest an increase in the number of households raising black lamb between 2017 and 2020 but was not possible to verify the actual numbers through fieldwork. Furthermore, it is not clear whether the expansion of lamb husbandry has occurred through contract farming of independent initiatives. What has been clear in basically all interviews with peasant households is that the middle-aged and older respondents are becoming increasingly concerned about the loss of local varieties of crops and food, especially pigs and maize.

Since Gu village still has a robust livestock husbandry sector, it was possible to interview local butchers and veterinarians. During fieldwork, it became obvious that the prospect of a complete shift away from pigs and chickens toward cattle and lamb was not on their horizon. Yet, most respondents expressed significant concern about the increasingly strict food safety regulations that were being implemented in recent years. These regulations curtailed small-scale operations at the village level and encouraged the large-scale concentration of pigs and more industrialized slaughterhouses at the county level. As it is explained in more detail elsewhere (Zhang 2017; Zhang and Qi 2019), this concentration of livestock did not actually improve food safety but aggravated environmental problems and socio-economic inequality. This occurs basically through the same mechanism described in the section above on Bian village. It is reasonable to conclude that the further industrialization of the regional livestock industry, in conjunction with a shift toward silage maize, would deepen these problems (see also Ely et al. 2016). In particular, local doctors, spiritual leaders, and other community members asserted an increase in public health problems and environmental degradation due to the expansion of agricultural modernization initiatives in the area. Many of them indicated, for example, that elders who continued to practice agroecological farming and rely on traditional diets have remained healthy and strong into old age. In contrast, younger village members that have been more closely engaged in agroindustrial activities and waged labor in urban areas tend to consume more "modern" processed food, with some developing diabetes, heart problems, cancer, and other adverse health conditions at a much younger age. These interviewees attributed this situation directly to farming intensification and the consequent water pollution from high pesticide use, as well as the consumption of unsafe and/or processed food.

Discussion: Industrial maize as an agent of institutional and socioeconomic transformation in China

China's policy framework on national food security and the technological development of the agroindustrial sector has brought maize to a crossroads. Agribusinesses, large-scale farmers, scholars, and government officials associated with the Ministry of Commerce and the National Energy Commission have been arguing that maize should be reclassified, and no longer be considered a staple crop, but rather an industrial input with multiple uses. Following the curtailment of biofuels as the key technological and policy instrument to advance the market-oriented reclassification of maize, this coalition of actors has turned to silage as the main political economic and technological strategy to drive the industrialization of maize.

Since the national government remains unwilling to abandon the classification of maize as one of the three remaining staple food crops due to concerns over peasant incomes and national food security, it has also found a means of responding to the growing demands from agribusinesses and pro-market scholars and policymakers through the promotion of new maize varieties grown specifically for silage. Promoted since 2016 as the grain-to-feed program, this strategy for institutional transformation has become the main compromise that is currently driving the industrialization of maize for emerging livestock industries and indirectly also supplying biomass for ethanol and other "deep" (i.e. downstream) industrial activities.

Although the reconfiguration of maize as silage and the associated restructuring of the livestock industry are still at a relatively initial stage according to the central government grain-to-feed program, there are political, economic, and ecological factors that are likely to entrench this institutional transformation. These include the political pressure from wealthier farmers and agribusiness investors for more government support and lucrative business opportunities, the economic pressures from the US–China trade war (which curtails Chinese access to cheap imports of maize and maize products from the US), and the pressures posed by climate change, which might encourage peasants to plant larger amounts of silage maize as it is a more drought-tolerant crop (relative to

238   *Li Zhang*

rice, for example). Moreover, the continued political, economic, and ecological challenges faced by the pork industry in China in the aftermath of the African swine fever pandemic might also continue to push agribusinesses toward silage-consuming livestock as an investment alternative.

These transformations are already witnessed across various regions of China, from the heartlands of staple food crop production like Henan, where contract farming for the production of silage maize and associated "grass-fed" livestock is expanding, including to mountainous border areas like Guangxi, where these initiatives are still incipient and possibly more limited due to the less favorable socio-ecological conditions. Accompanying this transformation are shifts in grain and agroindustrial markets that tend to strengthen the position of wealthier farmers over poorer peasants who rent out their land-use rights. Similarly, these transformations strengthen agribusiness companies that push for the reclassification of maize as an industrial crop, which would accelerate these shifts even further. Therefore, this early evidence about the development of the grain-to-feed program seems to, so far, indicate that agribusiness investors and large-scale farmers are likely to benefit from the further industrialization of maize, even though a large number of peasants may not reap the benefits of maize reclassification. Indeed many of these peasants may be displaced from the land, while many other smaller-scale actors in agro-food production and distribution chains might be marginalized. The previous sections point to the possible uneven socioeconomic outcomes of maize reclassification in China, having significant ramifications for rural development and rural poverty alleviation.

Conclusions

The emergence of silage as the main compromise in the current debate about the possible reclassification of maize as an industrial crop is clear both nationally and locally. For example, the current trends witnessed in Henan and Guangxi suggest that this process is likely to increase socio-economic inequality. These findings may still be tentative, calling for further research to evaluate in greater detail the transformations due to the recent conversion of maize production for silage and its associated shift from pork and poultry (that feed on grain) toward cattle and lamb (that feed on silage). Such further research is imperative considering the structural conditions in the maize and livestock sectors resulting from the US–China trade war and the African swine fever pandemic, respectively, and the fact that they will likely continue to encourage this shift in the following years. It is also important to further examine the conditions and consequences of this shift toward the greater industrialization of maize, as this process may very well drive an official reclassification of maize as an industrial crop, rather than a food staple. Such a profound institutional transformation could accelerate even further the socio-economic and ecological transformations that are already witnessed across China.

Acknowledgements

I would like to thank the editors for inviting my contribution, Qi Gubo for giving support to my fieldwork, Gustavo Oliveira for assistance in polishing the text, and all participants in this research for their time and contribution.

Notes

1 Agroecology is the scientific practice of applying ecological concepts and principles to the design and management of sustainable agro-ecosystems. It is simultaneously a social movement associated with the maintenance and renovation of the peasantry and the struggles for food sovereignty. Finally, agroecology also refers to the practice of sustainable crop production in diversified farming systems, relying upon practices such as nutrient recycling and integrated pest management rather than synthetic fertilizers, toxic agrochemicals, and other off-farm inputs. For more details on agroecology, see Altieri (2018), Gliessman (2014).
2 The current policy still aims to maintain 50% of potato production for national food consumption, despite the continued pressure to further extend the industrial uses of potatoes as a means of adding value to the crop (Lu 2015).
3 This trade war started in 2018 when US president Donald Trump began imposing tariffs on Chinese imports. In response, the Chinese government increased tariffs on US imports as well, including soybeans and other agricultural commodities. This curtails Chinese imports of soybeans, maize, and other products from the US and increases the cost for China's livestock feed sector that relies primarily on soy and maize. Such uncertainty and higher prices from international markets have increased concern about national food security within China, which in turn raises concerns about the effect of reclassifying maize from a staple crop to an industrial crop. For more information on this topics refer to Zhang (2020).

References

Ahmed, A., Z. Abubakari, and A. Gasparatos. 2019. "Labelling large-scale land acquisitions as land grabs: Procedural and distributional considerations from two cases in Ghana." *Geoforum*, 105: 191–205.

Altieri, M. 1995/2018. *Agroecology: The science of sustainable agriculture*, Second Edition. London and New York: Westview Press.

Blaikie, P. and H. Brookfield. 1987. *Land degradation and society*. New York: Methuen.

Chen, X., F. Zhu, and T. Huang. 2013. "Analysis of the changing characteristics of China's corn market – based on the development background of biofuel ethanol." *Price Theory and Practice*, 9: 88–89.

Cheng, Y. 2019. "The development of China's corn fuel ethanol industry under the China-US trade war." *Chemical Management* 34: 22–25.

Chen, Y., Q. Wang, and Y. Xiang. 2019. "Analysis on the status, superiority and self-sufficiency ratio of maize in China." *Chinese Journal of Agricultural Resources and Regional Planning*, 40(1): 7–16.

China Grain Industry Association (CGIA). 2019. "Vigorously promote the sustainable development of corn deep processing industry in China." *China Grain Economy*, 1: 20–24.

Cui, G. and Y. Jiao. 2019. "China's soybean trade from the perspective of national food security." *Social Sciences* 02: 13–28.

Cui, J., Y. Yang, and Y. Sun. 2020. "Economic welfare and carbon mitigation effect of ethanol policies in China." *China Economic Quarterly* 19(2): 757–776.

Ely, A., S. Geall, and Y. Song. 2016. "Sustainable maize production and consumption in China: practices and politics in transition." *Journal of Cleaner Production* 134: 259–268.

Fang, F., S. Yu, and C. Wang. 2004. "Economic assessment on corn-based fuel ethanol projects in China." *Journal of Agricultural Engineering*, 3: 239–242.

FAOSTAT. 2020. *Food and agriculture organization corporate statistical database*. Rome: Food and Agriculture Organization of the United Nations. www.fao.org/faostat/en/#data

Gasparatos, A., P. Stromberg, and K. Takeuchi. 2011. "Biofuels, ecosystem services and human wellbeing: putting biofuels in the ecosystem services narrative." *Agriculture, Ecosystems & Environment* 142(3–4): 111–128.

Gillon, S. 2016. "Flexible for whom? Flex crops, crises, fixes and the politics of exchanging use values in US corn production." *Journal of Peasant Studies* 43(1): 117–139.

Gliessman, S. 1997/2014. *Agroecology: The ecology of sustainable food systems*, Third Edition. London and New York: CRC Press.

Guan, F. 2019. "Development status and reflections of Henan family farms." *Issues in Agricultural Economy*, 8: 36–37.

Guo, Q. 2007. "Analysis of the development of China's corn processing industry." *Chinese Rural Economy*, 7: 16–22.

Guo, Q. 2019. "A discussion on the quality of farmland in the primary grain producing areas in Northeast China." *Issues in Agricultural Economy* 10: 89–99.

Han, C. 2012. "A brief introduction of corn." *Issues in Agricultural Economy* 33(6): 4–9.

Han, C. 2018. "Grain-to-feed: Corn can also be used in this way." *Farmer's Daily*, November 29. nongye.cnjiwang.com/nyxw/201811/2772749.html. (Accessed July 1, 2020).

He, Q. 2007. "The food war between cars and pigs: In order to avoid competing with pigs and even people, the raw materials of the fuel ethanol industry must start new changes." *IT Manager World*, 18: 54–55.

Huang, J. 2004. "China's food safety issues." *Chinese Rural Economy*, 10: 4–10.

Huang, J. and G. Yang. 2017. "Understanding recent challenges and new food policy in China." *Global Food Security*, 12: 119–126.

Jacka, T. 2013. "Chinese discourses on rurality, gender and development: A feminist critique." *Journal of Peasant Studies*, 40(6): 983–1007.

Jiang, C. 2005. "Some thoughts on China's food security." *Issues in Agricultural Economy*, 2: 44–48.

Jiang, T., X. Liang., X. Wang, and Z. Li. 2004. "Joining WTO and China's food security." *Management World (Monthly)*, 20(3): 82–94.

Lardy, N. 1983. *Agriculture in China's modern economic development*. Cambridge: Cambridge University Press.

Li, H. 2002. "Analysis on the countermeasures of maize development in Guangxi after joining WTO." *Journal of Guangxi Agriculture*, 4: 44–46.

Li, Y. 2016. "The silage corn planting area in Henan reaches 600,000 mu, why is the silage corn popular?" *Henan Daily*, August 17. www.henandaily.cn/content/fzhan/snjjiao/2016/0817/13445.html. (Accessed July 1, 2020).

Liu, H. 2019. "Henan province over-completed the target of Grain-to-Feed in 2018." *Henan Daily*, June 27. m.fx361.com/news/2019/0627/5254565.html. (Accessed July 1, 2020).

Lu, X. 2015. "Strategy of potato as staple food: Significance, bottlenecks and policy suggestions." *Journal of Huazhong Agricultural University (Social Sciences Edition)*, 3: 1–7.

Lu, Z., Y. Fang., Z. Li., R. Luo., H. Chen, X. Bao, and C. Wei. 2017. "Guangxi Grain-to-Feed pilot project effectiveness and existing problems." *Guizhou Animal Husbandry and Veterinary Medicine*, 42(5): 44–45.

Ma, Y., P. Ren, and C. Wang. 2006. "From meat to 'oil,' the Northeast corn is in urgent need of the country." *Xinhua News*, September 29. news.sohu.com/20060929/n245604433.shtml. (Accessed July 1, 2020).

Mai, W. 2020. "China's grain status-staple food is completely self-sufficient, imports account for only 2%." *Xiao Kang*, 14: 12–16.

Mao, X. and X. Kong. 2019. "Reshaping the future of food security concept in China." *Journal of Nanjing Agricultural University (Social Sciences Edition)*, 19(1): 142–168.

Martin, S. 2020. "The political economy of distillers' grains and the frictions of consumption." *Environmental Politics*, 29(2): 297–316.

Ministry of Agriculture (China). 2016. *National Plant Restructuring Plan (2016–2020)*. Beijing: Ministry of Agriculture of the People's Republic of China.

Muldavin, J. 1997. "Environmental degradation in Heilongjiang: Policy reform and agrarian dynamics in China's new hybrid economy." *Annals of the Association of American Geographers*, 87(4): 579–613.

Muldavin, J. 2000. "The paradoxes of environmental policy and resource management in reform-era China." *Economic Geography*, 76(3): 244–271.

Muldavin, J. 2007. "The politics of transition: Critical political ecology, classical economics, and ecological modernization theory in China." In *The political geography handbook*, K. Cox, M. Low, and J. Robinson (eds.). London: Sage, pp. 247–62.

NEC, 2017. "Regarding the implementation plan for expanding the production of biofuel ethanol and promoting the use of automotive ethanol gasoline." National Energy Commission (China). Press release, September 13. www.gov.cn/xinwen/2017-09/13/content_5224735.htm. (Accessed July 1, 2020).

Ni, H. 2012. *Research on biofuel-bioethanol development and its impacts on corn market in China*. PhD dissertation. Nanjing University of Aeronautics and Astronautics.

Oliveira, G., B. McKay, and C. Plank. 2017. "How biofuel policies backfire: Misguided goals, inefficient mechanisms, and political-ecological blind spots." *Energy Policy*, 108: 765–775.

Oliveira, G. and M. Schneider. 2016. "The politics of flexing soybeans: China, Brazil and global agroindustrial restructuring." *Journal of Peasant Studies*, 43(1): 167–194.

Peluso, N. 1992. *Rich forests, poor people: Resource control and resistance in Java*. Berkeley: University of California Press.

Rocheleau, D., B. Thomas-Slayter, and E. Wangari. 1996. *Feminist political ecology: Global issues and local experiences*. New York: Routledge.

Schneider, M. and S. Sharma. 2014. *China's pork miracle? Agribusiness and development in China's pork industry*. Washington, DC: IATP.

Shi, C. 2019. "Development process, existing problems and countermeasures of corn production in Guangxi." *China Seed Industry*, 4: 24–29.

Song, L., B. Cao and Q. Zhu. 2019. "Food security, consumption transformation and policy adjustment." *Xinjiang Social Sciences*, 3: 23–32.

Song, X. and Z. Ouyang. 2012. "Key influencing factors of food security guarantee in China during 1999–2007." *Acta Geographica Sinica*, 67(6): 793–803.

State Council (China). 2019. *China Food Security White Paper*. October 14. Beijing: State Council of the of the People's Republic of China.

Tang, F. 2012. "Guangxi biomass industry development status and countermeasures." *Popular Sciences & Technology*, 14(7): 267–271.

USDA. 2019. "China, Biofuels Annual: China will miss E10by 2020 goal by wide margin." Foreign Agricultural Service, GAIN report number CH19047. Washington, DC: United States Department of Agriculture.

USDA. 2020. "World agricultural production." Foreign Agricultural Service, Circular Series WAP 7-20. Washington, DC: United States Department of Agriculture.

Wang, M. 2019. "'Butterfly effect' caused by 'grain-to-feed': Analysis of the city's innovative 'grain-to-feed' model." *Dongying Daily*, April 16. baijiahao.baidu.com/s?id=1630 931586134019729&wfr=spider&for=pc. (Accessed July 1, 2020).

Wang, C. 2006. "Experts call for careful development of corn fuel ethanol." *China Securities Newspaper*, July 24. finance.sina.com.cn/stock/t/20060724/0855814115. shtml?from=wap. (Accessed July 1, 2020).

Wang, X. and J. Yang. 2019. "Han Changfu: Corn and soybean producers will be subsidized." *China Agricultural Mechanization Herald*, March 11. news.wugu.com.cn/article/1514377.html. (Accessed July 1, 2020).

Watts, M. and R. Peet. 2004. "Liberating political ecology." In *Liberation ecologies: Environment, development, social movements*, R. Peet and M. Watts (eds.), Second edition. London: Routledge, pp. 3–43.

Wu, H. and X. Huo. 2014. "Research on the dynamic correlation of crude oil, corn and fuel ethanol market." *Agricultural Technology and Economy*, 3: 89–96.

Yan, S. 2017. "A study on the strategy of potato staple food in China." *China Science and Technology Information*, 5: 103–104.

Yan, H., Y. Chen, and H. Ku. 2016. "China's soybean crisis: The logic of modernization and its discontents." *Journal of Peasant Studies*, 43(2): 373–395.

Zhang, H. 2020. "The US-China Trade War: Is food China's most powerful weapon?" *Asia Policy*, 27(3): 59–86.

Zhang, L. 2017. *The politics and governance of food and farming sys- tem change in China: Case studies of Bian village in Henan and Gu village in Guangxi*. PhD dissertation, China Agricultural University.

Zhang, L. 2020. "From left behind to leader: Gender, agency, and food sovereignty in China." *Agriculture and Human Values*, 37(4): 1111–1123.

Zhang, L. and G. Qi. 2019. "Bottom-up self-protection responses to China's food safety crisis." *Canadian Journal of Development Studies*, 40(1): 113–130.

Zhang, X. and S. Zhao. 2017. "Assessing the promotion of ethanol gasoline from the development of Jilin Fuel Ethanol Company". *China Petroleum News*, November 7. www.china5e.com/news/news-1008875-1.html. (Accessed July 1, 2020).

Zhang, Z. 2010. "Ecological consequences of corn promotion in the Qing dynasty in the mountainous area of Southwest China." *Journal of Original Ecological National Culture*, 2(3): 40–47.

Zhang, Z., X. Yang, and J. Yang. 2019. "Maize at the crossroads: Staple food grain or feed grain?" *Chinese Rural Economy*, 6: 38–53.

Zhao, D. and Y. Xin. 2020. "Research on the development of corn industry chain in Jilin province." *Journal of Jilin Business and Technology College*, 36(2): 29–31.

Zhou, C., D. Zheng., H. Tan., A. Huang., K. Huang., R. Mo, and R. Zhai. 2019. "Research on the status and development strategies of corn production in Guangxi." *Southern Agriculture*, 13(8): 139–141.

Zhou, D., P. Zhang, H. Sun, R. Zhong, Y. Huang, Y. Fang, Q. Li, and T. Wang. 2017. "Regional difference of grain production and its consumed fraction in China." *Soils and Crops*, 6(3): 161–173.

Zhu, Z. 2020. "Investigation report on the implementation of Guangxi Grain-to-Feed project." *Guangxi Animal Husbandry and Veterinary Medicine*, 36(2): 53–56.

11 Institutional and socioeconomic transformation from sugarcane expansion in northern Eswatini

Alexandros Gasparatos, Graham von Maltitz,
Nikole Roland, Abubakari Ahmed, Shakespear Mudombi
and Marcin Pawel Jarzebski

Introduction

Sugarcane is one of the major industrial crops in Sub-Saharan Africa (SSA), with its production expanding rapidly across the region (Figure 11.1) (Chapter 1). With many parts of the continent having very favourable conditions to achieve very high yields (Cutz and Nogueira, 2018), sugarcane is expected to expand even further over the next decades in countries such as Ethiopia, Mozambique, Zambia, and Tanzania among others (Hess et al., 2016; IRENA, 2019).

One of the major strengths of sugarcane as an industrial crop is its great flexibility, as it can be used for the production of sugar, ethanol, biomaterials, and even electricity (Chapter 1). As a result, sugarcane has been seen promoted in many SSA countries to boost economic growth and rural development (Chapter 3), eventually dominating the economies of some small countries such as Mauritius and Eswatini (Gasparatos et al., 2015). Furthermore, the great potential of sugarcane to contribute to national energy security through the production of ethanol fuel for transport and electricity from bagasse has driven in the past its production in countries such as Malawi, Ethiopia, and Zimbabwe (Gasparatos et al., 2015; Johnson and Silveira, 2014).

Due to the perishability of harvested cane, the need to achieve economies of scale, and political pressure, sugarcane is usually produced in hybrid systems that combine large core plantations (mostly irrigated) and mills, surrounded by hundreds or even thousands of irrigated and/or rainfed smallholders (von Maltitz et al., 2019) (Chapter 1). However, there are also some examples of sugarcane production solely in large plantations (Palliere and Cochet, 2018) or smallholder settings (Ahmed and Gasparatos, 2020). The former is usually for commercial purposes targeting international markets, while the latter is usually associated with rudimentary production techniques for household use or local food markets. Similar to other industrial crops, sugarcane production increasingly becomes an agent of ecological, agrarian, socioeconomic, and institutional transformation in SSA (Chapter 1, 3).

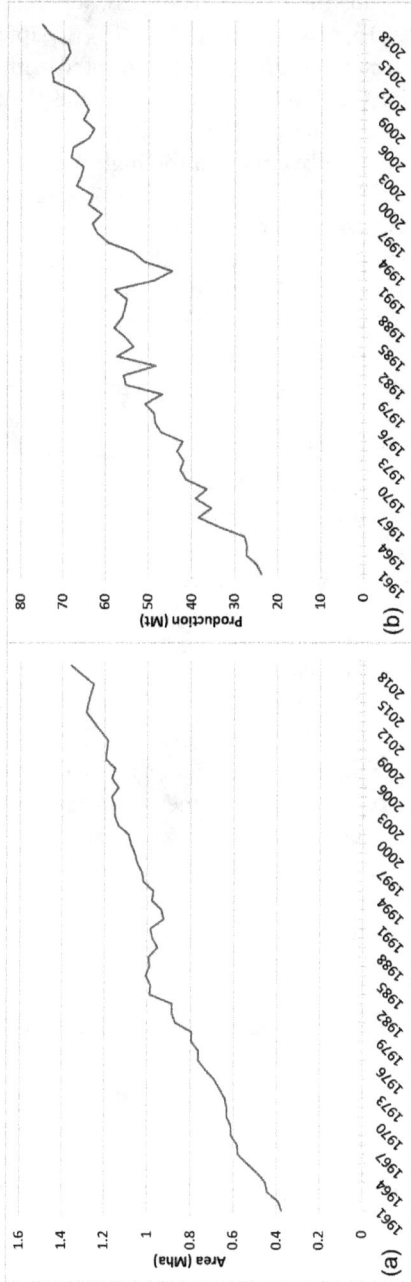

Figure 11.1 Sugarcane output and area under cultivation in Sub-Saharan Africa.

In terms of ecological transformation, the generally high land requirement of (and intensified production practices in) large-scale systems in SSA has been linked to landscape degradation, biodiversity loss, water depletion, and water pollution (Hess et al., 2016; Gasparatos et al., 2015) (Chapters 1, 3). Landscape degradation and biodiversity loss are mainly associated with the extensive land use change caused during the development of sugarcane plantations, smallholder schemes, and ancillary infrastructure (Romeu-Dalmau et al., 2018; Semie et al., 2019). Water depletion is usually linked to the high water demand of the dense biomass in sugarcane fields, which is often met through irrigation, even in areas characterized by chronic water scarcity (Ngcobo and Jewitt, 2017; Hess et al., 2016). Water pollution is usually caused by the extensive use of fertilizers and agrochemicals (Lwimbo et al., 2019; van der Laan et al., 2012).

In terms of agrarian and socioeconomic transformation, large-scale sugarcane production in SSA has been linked to the (a) generation of off-farm employment and income opportunities in large-scale plantations (e.g. Matenga and Hichaambwa, 2017); (b) promotion of commercial smallholder-based production models and linkage to markets (von Maltitz et al., 2019); and (c) broader regional development through secondary job and income generation in sugarcane areas, and other wider benefits to the national economy (Schuenemann et al., 2017). Such mechanisms are associated with changes in both the agrarian structure and the socioeconomic status of local communities (Chapters 3 and 8).

In more detail, sugarcane production has been linked to rural poverty alleviation, including infrastructure development and maintenance (Mudombi et al., 2018; Herrmann et al., 2018). However, the actual livelihood outcomes can be rather context-specific and depend between groups. For example, plantation employees (though generally fewer) are better off compared to outgrowers and non-involved community members in some areas, and worse off in other areas (Mudombi et al., 2021; Matenga and Hichaambwa, 2017). Some studies have even pointed to the counterproductive livelihood outcomes of sugarcane production through, for example, government coercion to force smallholders to engage in sugarcane production, loss of smallholders' access to land, unclear and unbalanced rules of smallholder engagement in value chains, and predatory/unethical corporate practices (e.g. Wendimu et al., 2016; Manda et al., 2020). Furthermore, it has been shown that sugarcane production can intersect with food security through different and context-specific mechanisms including the (a) loss of agricultural land or food crops, (b) improved ability to purchase food due to income generation, and (c) improved access to irrigation and agricultural inputs to increase food crop yields (Jarzebski et al., 2020; Herrmann et al., 2018) (Chapter 3). Many of these socioeconomic outcomes are gender-differentiated, with women often being less well integrated into sugarcane value chains and, as a result, reaping fewer benefits (Sulle and Dancer, 2020) (Chapter 3).

In terms of institutional transformation, sugarcane production in SSA has required the development of comprehensive policy frameworks to enable large-scale production by domestic and international investors but has also reconfigured access to natural resources when such systems are operationalized or expand in local contexts (Maconachie, 2019; Dubb et al., 2017; Chinsinga, 2017). Some of the more prevalent examples include (a) informal processes during large-scale acquisitions (LSLAs), (b) coerced transfer of lands from rural communities to corporate players, (c) privatization of common-pool resources and loss of access for local communities, and (d) development of new markets and integration of smallholders and rural areas to global agricultural value chains (Maconachie, 2019; Dubb et al., 2017; Chinsinga, 2017; FAO, 2013). Many of these institutional changes have been linked to changes in land/water access and grabbing, with institutional change being both a driver and outcome of the sugarcane expansion and reconfiguration of access.

Eswatini (formerly known as Swaziland) is one of the SSA countries with a very dominant sugarcane sector. The sector is essential for the national economy, accounting for roughly 13% of the GDP, 16% of national workforce (directly or indirectly), and 25% of exports (FAO, 2017). The sector is also pivotal for rural development through the economic integration of thousands of smallholders through the development of irrigation and capacity for sugarcane cultivation (Terry and Ogg, 2017).

Large-scale sugarcane production started in the late 1950s and has grown considerably since then. Currently, there are two main sugarcane production zones located at the northern and central lowveld[1]. These areas contain large sugarcane plantations operated by private companies, namely Illovo and the Royal Swaziland Sugar Corporation (RSSC), which are surrounded by smaller independent producers and thousands of outgrowers organized in smaller irrigated community plantations (von Maltitz et al., 2019; Terry and Ogg, 2017). However, the development and operation of the sugarcane sector in these two areas has had both positive and negative outcomes for local communities and ecosystems. On the one hand, sugarcane production has improved the livelihoods and access to infrastructure for large segments of the rural population (Mudombi et al., 2018, 2021) but on the other hand, it has caused extensive land use change, water over-abstraction, and ecosystem degradation (Romeu-Dalmau et al., 2018; FAO, 2017). More importantly, sugarcane production has catalyzed major reconfigurations in access to land and water resources (Terry and Ogg, 2017).

It is worth mentioning that Eswatini experiences periodic droughts, some of which are very severe such as the 2015–2016 drought associated with the El Nino event that impacted much of southern Africa (FAO, 2016; Siderius et al., 2018). This particular drought affected significantly the agricultural sector and many rural areas (SEPARC, 2017). It has been suggested that the sheer size of the sugarcane sector in parts of the country and the generally problematic water management system (both nationally and internationally) exacerbate the effects of water scarcity in drought-affected areas (Hess et al., 2016). Arguably, with sugarcane production being the backbone of the national economy and ongoing rural development efforts, and with climate change increasing the unpredictability and severity of droughts in the region, it is important to understand better the intersection of sugarcane production and water access and management.

This chapter uses an analytical lens that draws from actor-oriented political ecology and the theory of access to unravel the history, processes, and impacts of sugarcane production in Eswatini. We focus on the major sugarcane zone located in the northern lowveld and combine primary and secondary information from policy documents, expert interviews, focus group discussions (FGDs), and household surveys. The section "Methodology" outlines the study site and the main data collection and analysis methods. The section " Evolution of sugarcane production and water management" highlights the history and major changes that the sugarcane sector has caused in the study area (see " Evolution of sugarcane production and water management") and how it has become an agent of institutional (see " Institutional change: From policies of expansion to reconfiguration of access to land and water") and socioeconomic change (see " Socioeconomic change: Differentiated livelihood outcomes"). The section "Distributional aspects of sugarcane production" synthesizes the main findings to show the distributional effects of sugarcane production.

Methodology

Research approach

This chapter combines concepts from actor-oriented political ecology (Bury, 2008; Svarstad et al., 2018) and the theory of access (Ribot and Peluso, 2003). These rather complementary lenses have often been used jointly to articulate how power relations intersect with environmental change, and how they affect access to natural resources for different social groups and actors (Svarstad et al., 2018).

In more detail, actor-oriented political ecology rests on the axiom that multiple actors, and their subjective perspectives and entrenched interests, are involved in (and inherently affect) processes of environmental change (Svarstad et al., 2018). In the context of LSLAs and agrarian transformation in SSA, actor-oriented political ecology assumes that the processes underpinning land acquisition, land use change, water use, and agrarian transformation are not politically neutral (Ahmed and Gasparatos, 2020). Instead, they are directly and indirectly shaped by the entrenched interests of the different actors involved in these processes, which will essentially dictate which actors will benefit or lose (and to what extent) (Bury, 2008).

Actor-oriented political ecology can thus help disentangle interactions between actors in the context of environmental change and resource use, and rationalize how power differentials permeate agrarian social-ecological systems and give rise to socioeconomic impacts that may be distributed unequally (Bury, 2008; Svarstad et al., 2018). Several studies have used actor-oriented political ecology in rural SSA contexts to explore power interplay among diverse actors such as corporations, local communities, chiefs, and subsistence farmers

(Ahmed et al., 2019; Bergius et al., 2018; Svarstad et al., 2018). In most of these studies, power interplays are often the outcomes of struggles in accessing land-based livelihoods and natural resources, and the restrictions that some actors place (Ahmed and Gasparatos, 2020).

The theory of access can help disentangle how the processes unfolding during institutional change, agrarian transformation, and/or environmental change (a) permeate and reconfigure access to natural resources such as land and water and (b) mediate impacts and their (often unequal) distributions among actors (Ribot and Peluso, 2003). The theory of access differentiates between legal and illegal rights-based access to natural resources and identifies various structural and relational mechanisms that underlie it, including access to technology, capital, markets, labor markets/opportunities, knowledge, authority, identity, or social relations (Ribot and Peluso, 2003).

This methodological combination of actor-oriented political ecology and the theory of access is ideal for interrogating the trajectories and outcomes of industrial crop production in SSA. On the one hand, there is clear evidence that the interests and influence of multiple actors intersect during industrial crop production, whether through LSLAs or outgrower schemes (Chapters 3 and 5). On the other hand, industrial crop production is often undertaken in areas with amenable environmental conditions (e.g. good water availability), causing coupled changes in the access to land and water (Mehta et al., 2012; Dell'Angelo et al., 2018). Many studies have pointed that changes in access to land and water are often the basis of contestations during LSLAs and agrarian transformation in SSA (Adams et al., 2019; Ayelazuno, 2019) and that local institutional and power dynamics often underpin competition and contestations in access to land and water during commercial crop production (Allan et al., 2014; Dell'Angelo et al., 2017; Adams et al., 2019).

Study site

The sugarcane areas of the northern lowveld are located around the town of Tshaneni, close to the borders with South Africa. Sugarcane production started in the late 1950s (see "Evolution of sugarcane production and water management"), and the broader area now contains (a) a large sugarcane estate, two mills (in Mhulme and Shimunye), and an ethanol distillery operated by the RSSC; (b) multiple smaller plantations operated by relatively large independent producers; and (c) community plantations operated by outgrowers.

The large-scale commercial sugarcane production undertaken by RSSC and the smaller independent producers has a long history of land transfers and consolidation, which started in the 1950s (see "Evolution of sugarcane production and water management"). The smallholder-based irrigated sugarcane production started in the 1960s, but expanded significantly since the 1990s due to a series of domestic policies and international circumstances (see "Evolution of sugarcane production and water management"). This was facilitated through an extensive community development programme promoted by the Swazi government, which sought to develop capacity for irrigated sugarcane production in local communities located in the "catchment area" of the two RSSC mills (see "Evolution of sugarcane production and water management"). Smallholders from communities adjacent to the Komati River were essentially incentivized to pool their dryland plots used mainly for subsistence food crop production into large consolidated plots for sugarcane production (see "Institutional change: From policies of expansion to reconfiguration of access to land and water"). Through this process of land consolidation, irrigation provision, and capacity-building, the selected members from local communities along the Komati river formed 29 associations that operate as independent companies[2]. The smallholders involved in these associations are equal partners and receive annual dividends by selling sugarcane to the RSSC mills (Terry and Ogg, 2017).

For this chapter, we focus on four associations located in four different communities along the Komati River. These communities are from downstream to upstream: (a) Sihhoye; (b) Mafucula; (c) Malibeni; and (d) Ekuvinjelweni (Figure 11.2).

Data collection and analysis

This chapter synthesizes selected findings from multiple fieldwork campaigns in Eswatini conducted between 2015 and 2018. During this period we conducted 35 expert interviews with local- and national-level stakeholders, 12 FGDs in the four study communities, and 397 household surveys in the wider sugarcane production area with households with different types of involvement in the sugarcane sector.

The expert interviews targeted experts affiliated with the major organizations involved in the sugarcane and water management sectors, namely government agencies (N = 10), parastatal agencies (N = 6), civil society (N = 4), private sector (N = 3), and traditional authorities and community leaders (N = 12). The expert interviews elicited information about the structure and characteristics of the sugarcane sector, as well as its impacts especially in relation to land and water access and use. The interviews were semi-structured and open-ended to allow respondents to expand according to their expertise. All interviews were conducted in English, apart from those with experts from traditional authorities that were conducted in Swazi and then translated to English. Interviews lasted between 45 and 60 minutes and were transcribed and manually coded for further analysis (see below).

The FGDs were undertaken in three of the four communities alongside the Komati River, namely Sihhoye, Mafucula, and Malibeni (see "Research approach")[3]. In each of the three communities, we conducted four FGDs divided across gender (male vs. female) and engagement in sugarcane production (sugarcane growers vs. non-growers). This division sought to facilitate the elicitation of group-differentiated perceptions about the modalities and outcomes of smallholder-based irrigated sugarcane production in each community. This included questions about access to water and water tensions before

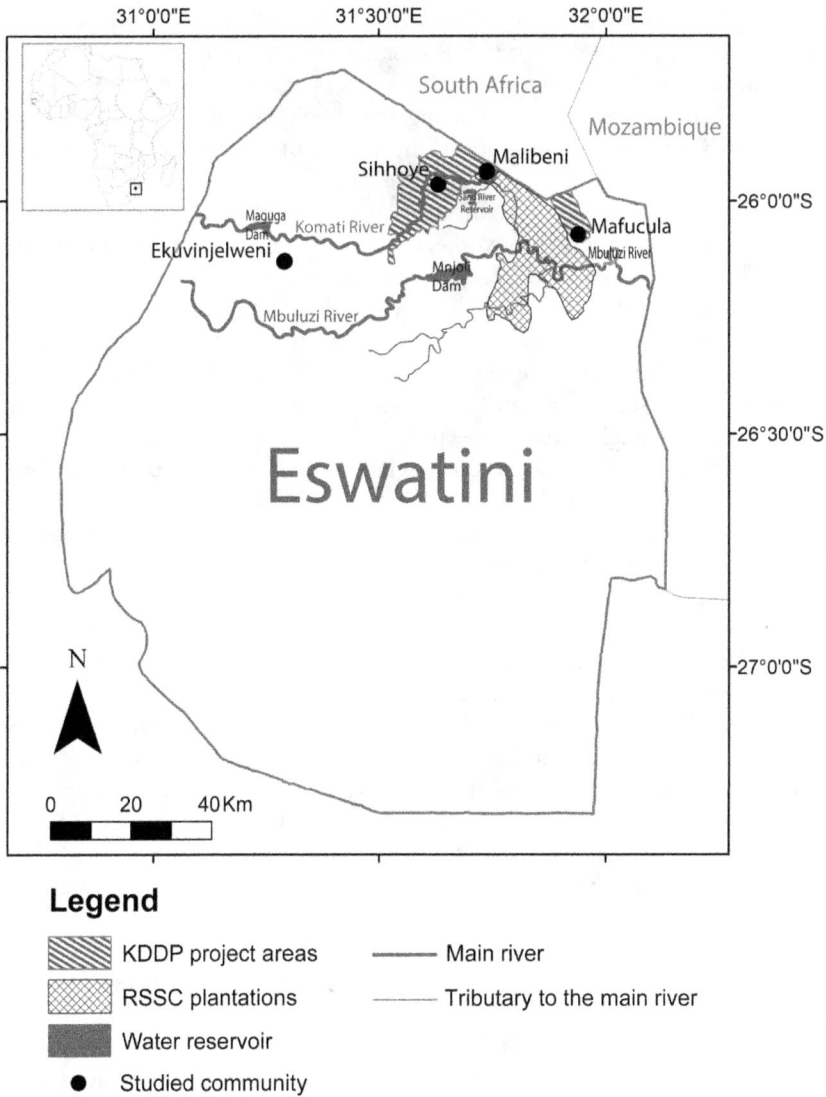

Figure 11.2 Study area.

and during the 2015–2016 drought that affected the country severely, including the sugarcane areas (SEPARC, 2017). Each FGD lasted about 60–90 minutes and was conducted in Swazi. The FGDs were translated to and transcribed in English for further analysis (see below).

The transcribed interviews and FGDs were manually coded using concepts and approaches from grounded theory such as line-by-line coding (Charmaz,

2014) to expand. Codes were informed from prior literature reviews about the broad sugarcane impacts in SSA (Gasparatos et al., 2015; Richardson-Ngwenya and Richardson, 2014; Terry and Ryder, 2007) and processes in Eswatini's sugarcane sector (Terry and Ogg, 2017). When patterns were identified in the coded data, we re-coded following the typologies of access mechanisms and rights-based classifications according to the theory of access (Ribot and Peluso, 2003).

We surveyed households with different types of engagement with the sugarcane sector, namely (a) workers at the large-scale RSSC plantation, (b) workers at the irrigated community plantations, (c) irrigated sugarcane smallholders, and (d) households not involved in sugarcane production (i.e. control group). We randomly selected RSSC plantation workers below mid-level management using the full list of employees obtained from RSSC's human resources office. Sugarcane smallholders and community plantation workers were selected through a two-stage process. Firstly, we randomly selected 14 from the 27 smallholder associations operating in the study area and obtained a list of all smallholders and community plantation workers belonging to each of the 14 associations. Secondly, we randomly selected respondents based on the overall number of farmers and workers belonging to each of the 14 associations. To prevent oversampling, we used weights proportionate to the number of workers and smallholders in each of the 14 associations. Overall, this two-stage sampling process allowed for a good spread of respondents within the sugarcane production landscape, without diluting sample sizes by surveying all associations. Control groups were selected in the same local communities as the study associations through transect walks to randomize respondent selection.

The household surveys were structured and were conducted in the Swazi language. They contained questions about household demographics, livelihoods, food security, and farming practices. In this chapter, we report a sub-set of the study variables, namely (a) income, (b) access to land, (c) access to agricultural inputs, (d) asset ownership, (e) livestock ownership, and (f) multi-dimensional poverty (see "Socioeconomic change: Differentiated livelihood outcomes"). Gasparatos et al. (2018) contains a detailed description of the study protocol, including the survey design and sampling.

Income is estimated through the summation of all major and minor on- and off-farm income sources; see Mudombi et al. (2021) for the calculation procedure. Similarly, access to land and agricultural inputs is calculated through the summation of the size of all land plots and fertilizer amounts respectively. When estimating livestock ownership, we use an "Exchange Ratio" that allows for the conversion of the different livestock species to a common unit, called the Tropical Livestock Unit (TLU) (Chilonda and Otte, 2006). For asset ownership, we estimate the Household Domestic Asset Index (HDAI) that considers all movable assets. Apart from the mono-dimensional measures of income and wealth mentioned above, we also estimate the multi-dimensional poverty index (MPI), which is a standardized measure of poverty consisting of health, education, and living standards dimensions; see Mudombi et al. (2018) for the calculation procedure.

Finally, we conduct a land use change analysis to understand the overall landscape conversion processes during sugarcane expansion from RSSC and the smallholder associations, as well as of the individual converted land uses. This entails the geospatial analysis of remote sensing data for the years 1976 (the first year with available satellite data) and 2014. A detailed description of the methodological approach can be found in Romeu-Dalmau et al. (2018).

Results and discussion

Evolution of sugarcane production and water management

Table 11.1 briefly summarizes the history of sugarcane production in the northern lowveld. It juxtaposes it alongside major changes in land and water rights

Table 11.1 Evolution of the sugarcane sector and water management in Eswatini and the northern lowveld

Year	Event
1907	Land partition to clarify title deed land (TDL) (see Table 11.2)
1907–1959	First Swaziland Water Law (1907) and the Proclamation No.11 of 1910 to protect the rights of riparian owners to water in normal flows, by granting only unused and surplus flows to non-riparian users
1944	Distribution and use of water in Swazi areas is given to the King-in-Council in trust for the Nation
1950s	Normal flows were claimed as surplus flows to be used for non-riparian, irrigative purposes in the decline of the mining industry. One estate was awarded most of the available dry-season water flow along the Komati river
1958	Commencement of sugarcane production in the Mhulme estate through foreign capital
1959	Swaziland Act No. 73 prioritizes (unclearly) potential Swazi users and leads to the protection of existing water abstractors rights, limits allocable water, and establishes Irrigation Boards to exercise, unsuccessfully, control over water users
1962	Commencement of the first irrigated sugarcane smallholder project (Vuvulane Irrigated Farms)
1967–1970	Water Act of 1967 and Bill No.11 of 1970 protects the previously awarded water rights of riparian owners and awards permits of the current water use by the Swazi Nation instead of reserving potential water needs as originally intended
1967	Commencement of Swaziland Sugar Association control over sugar sales from all national producers
1968	Independence from the UK and establishment of Tibiyo (Royal Swazi investment fund)
1978	Introduction of the tinkhundla system (form of limited democracy under the king)

(continued)

Table 11.1 Cont.

Year	Event
1979	Commencement of the Simunye estate and forced movement of approximately 5,000 people
1979	Creation of the Royal Swaziland Sugar Corporation (RSSC)
1983–1984	4th National Development Plan and strategic switch to sugarcane expansion through smallholders
1986	Control of Vuvulane Irrigated Farms smallholder land from Tibiyo
1991	Establishment of sugar quota for smallholder growers (10,000 t, schedule D quota)
1992	The Treaty on the Development and Utilization of the Water Resources of the Komati River Basin portions the water flowing in the Komati River Basin between South Africa and Eswatini (Swaziland at the time)
1993–1996	Gradual expansion of the Schedule D quota and smallholders
1998	Construction of the Maguga dam
1999	National Development Strategy and strategic shift of the smallholder sector from rain-fed maize subsistence farming to irrigated sugarcane production
1999	Development of Swaziland Water and Agricultural Development Enterprise (SWADE; currently ESWADE) to undertake the large-scale development of irrigated sugarcane production
2000	Establishment of the Komati Downstream Development Project (KDDP) through SWADE to provide irrigation to 7,400 ha and 14,500 smallholders
2000	Revised Protocol on Shared Watercourses in the Southern African Development Community promotes Integrated Water Resource (IWR) management, monitoring of water extractions, and establishes an economic value for water as a means of increasing water efficiency
2002	Merger of the RSSC and Mhlume Sugar Company and listing of the new company on Swazi Stock Exchange
2002	Tripartite Interim Agreement expands the 1992 treaty to include Mozambique in the portioning of water from the Komati River Basin
2003	Water Act of 2002 establishes a decentralized approach to water management and requires water permits for existing and new water users
2005	Changes in the EU sugar protocol and sugar prices
2006	Implementation of EU Accompanying Measures for Sugar Protocol (AMSP) to support certain exporting countries (incl. Eswatini)
2013	Expansion of the Mhulme mill

Source: (Terry and Ogg, 2017; Heilbronn, 1981; GoS, 2003; SADC, 2000; KOBWA, 2008).

in the broader region, which are central to understanding the political ecology of sugarcane production. According to Terry and Ogg (2017), a major element underlying the evolution and impact of the sugarcane sector are the shifts in the land tenure system (Table 11.2) and the gradual consolidation of sugarcane land

Table 11.2 Major land types and tenure systems in Eswatini

Type	Explanation
Title Deed Land (TDL)	Land held by colonists, foreigners, and companies. This land was clarified in 1907 to avoid land overlapping
Swazi National Land (SNL)	Land controlled by the king and administered through the local chiefs or the national government. This includes: • Land always under customary law that cannot be bought, mortgaged, leased, or sold, with the chiefs having land allocation powers. Individuals can acquire this land through inheritance or pledging allegiance to the chief • Land previously under TDL that is leased out to commercial enterprises. This land is managed by the national government or controlled by Tibiyo (Table 11.1).
Crown Land	Land owned by the Swazi government. This land is not relevant for the sugarcane sector.

Source: (Terry and Ogg, 2017; FAO, 1998).

under the Royal Swazi investment fund (Tibiyo), which is essentially directly controlled by the king.

Sugarcane production in the northern lowveld started in 1958 in the Mhulme estate. This production occurred on title deed land (TDL) through foreign capital that was provided by the Colonial Development Corporation (CDC) and JL Hulett. The first smallholder scheme was established on this TDL a few years later (in 1962), rather than customary land. This was a major departure from the usual approach to commercial agriculture that was prevalent in the country up to that point (Table 11.1); for more details refer to Terry and Ogg (2017). In 1973, the second mill/estate was established in Simunye, but this time on customary Swazi National Land (SNL). This caused the forced eviction of 5,000 residents and had various negative ramifications for local communities (Terry and Ogg, 2017).

In 1979, the RSSC was established through a Joint Venture Agreement between seven partners and the national Government and Tibiyo. This illustrates the gradual emergence of the king as a major player in the sugar industry through the Tibiyo. The role of the king is further consolidated in 1986 through the transfer of the land of the Vuvulane Irrigated Farms smallholder scheme from the TDL. Eventually, by the mid-1990s, Tibiyo controlled 34% of sugarcane growing and 50% of milling operations in the northern lowveld, mostly under direct royal control that was outside the remit of the national government (Terry and Ogg, 2017).

However, the shifting political priorities in the early 1980s gradually changed the focus of sugarcane expansion from large-scale estates/mills on SNL to smallholder-based schemes on SNL. This was largely due to lower employment generation in large plantations compared to the initial investment (Terry and Ogg, 2017). This gradual shift started in 1991 through the allocation of a sugarcane quota to smallholders (10,000 t), which increased gradually over the next few years (Terry and Ogg, 2017). This National Development Strategy of 1999 pushed more aggressively this shift towards smallholder-based sugarcane expansion, by further advocating the strategic shift of the smallholder sector from rain-fed maize subsistence farming to irrigated sugarcane production (Table 11.1). In a sense, by the late 1990s,

> the expansion of smallholder production of sugar cane was clearly established as the only option if the industry was to expand. The lack of suitable TDL for sugar cane production and the availability of suitable soils on SNL near the mills, coupled with the negative publicity associated with the expulsion of communities during the creation of Simunye, meant that expansion via large-scale estate production was no longer politically acceptable.
>
> (Terry and Ogg, 2017: 11)

The Swaziland Water and Agricultural Enterprise (SWADE, currently ESWADE), a para-statal agency, was instrumental in "operationalizing" this shift in priorities. SWADE was mandated to promote irrigation adoption and capacity development for sugarcane smallholders (ESWADE, 2019). SWADE essentially provided the social, technical, and financial support that was identified during the earlier smallholder schemes as critical for a viable smallholder sector. These mainly included (a) the adoption of a development philosophy centred around community empowerment and knowledge transfer; (b) the provision of technical and financial advice to catalyze a shift from subsistence rain-fed agriculture to co-operatively managed irrigated commercial agriculture, and (c) the development of an enabling environment to catalyze the creation of independent businesses (Terry and Ogg, 2017). This clearly indicates that the driving philosophy of smallholder-based sugarcane expansion essentially reflects a market-led approach to rural development (Terry and Ogg, 2017).

The Komati Downstream Development Project (KDDP) enabled the smallholder-based sugarcane expansion through a three-pronged approach of (a) land consolidation, (b) irrigation provision, and (c) capacity-building for commercial sugarcane production. The land consolidation process entailed the pooling of the small landholdings of individual households (that were under customary tenure arrangements) into much larger block plots that could receive irrigation to accommodate sugarcane production[4]. This process essentially required the customary landholders to "return" their individual land to the king, who then reallocated it to a company that was held in equal share by all farmers that contributed land. However, in reality, the KDDP only targeted farmers with land holdings in suitable locations close to the Komati River that was deemed to be cost-effective for irrigation expansion. Following the land pooling and consolidation process, the sugarcane fields essentially engulfed the suitable dryland agricultural areas, with houses and small home gardens practically becoming small islands within the sugarcane landscape (Figure 11.3). Another aspect of the KDDP was the provision of potable water to all

(a)   (b)

Figure 11.3 Stylisized representation of land consolidation process from dryland farms
    (a) to sugarcane fields (b).

households in each target community, regardless of engagement in irrigated sugarcane production.

The targeted households were arranged in 29 farmers associations, each containing 37–220 members. These associations operate as individual companies (or, more precisely, commercial cooperatives; see Footnote 2) that produce sugarcane under their own management system and sell it to RSSC mills, which is practically the only market in the region. The constituent farmers are essentially equal partners within these private entities and receive equal annual dividends from the profits accruing through sugarcane sales, regardless of the amount of land they contributed during the consolidation process. SWADE and the KDDP "enforced" equal partnership as a prerequisite for offering support, which has, however, remained a contentious issue among some farmers, especially those who contributed large landholding.

Eventually, this consolidation and irrigation development project reached a total of 5,206 ha of dryland farms from 2,360 homesteads. Of these, 4,616 ha were transformed into irrigated sugarcane fields and 590 ha into food cropland and home gardens to ensure some level of food security (ESWADE, 2019). The smallholders have become a critical component of the sugarcane sector, currently supplying 52% of the sugarcane processed in the Mhlume mill and 25% in the Simunye mill (Terry and Ogg, 2017).

Since the 1950s (and especially since the 1970s), sugarcane areas have expanded significantly in the northern lowveld (Figure 11.4), mainly converting agricultural land, low-density forest and high-density forest (Figure 11.5) (Romeu-Dalmau et al., 2018). Most of these woodlands have been traditionally used by local communities for the extraction of timber, fuelwood, and other forest products, which contribute significantly to the local livelihoods (see also Chapter 8).

The sugarcane sector consumes 96.6% of irrigation water in Eswatini (FAO, 2017). Thus, water management policies and institutions influence significantly

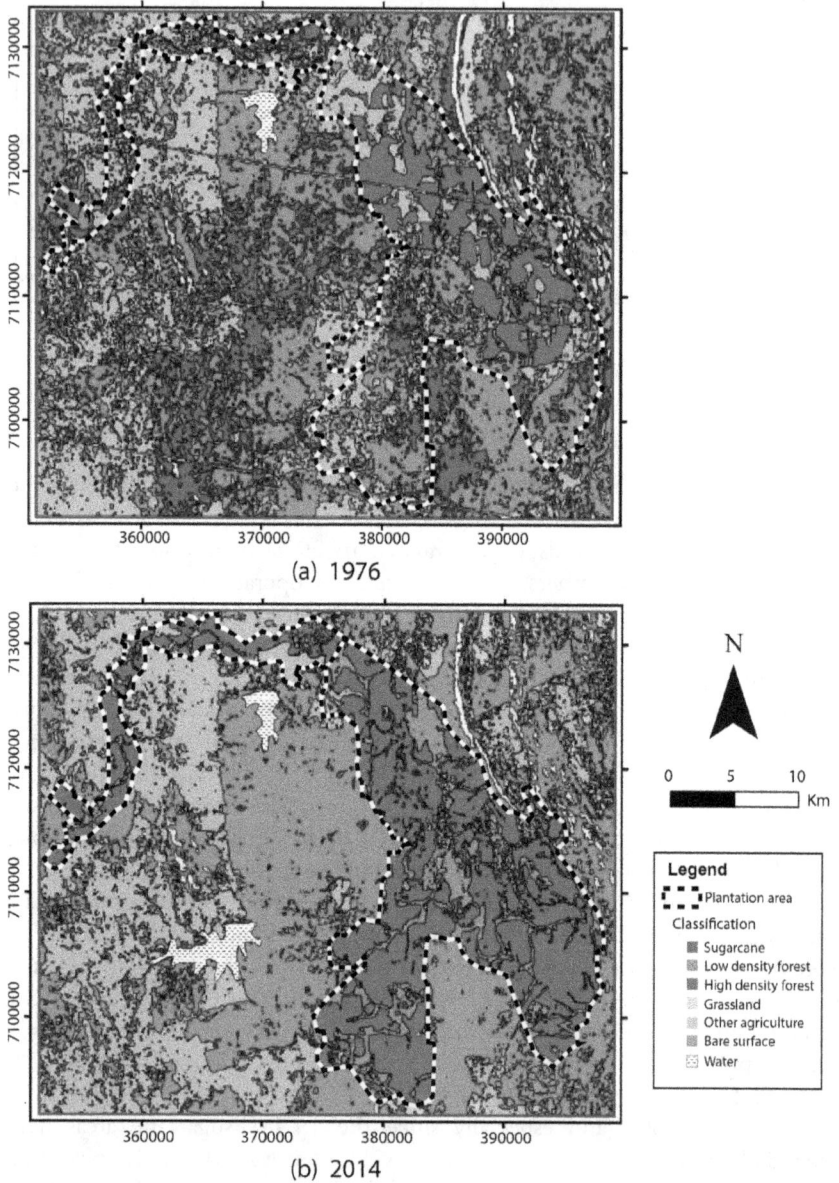

Figure 11.4 Land use maps of the sugarcane production areas in the northern lowveld for the years 1976 (Panel A) and 2014 (Panel B).

Source: (Romeu-Dalmau et al., 2018)

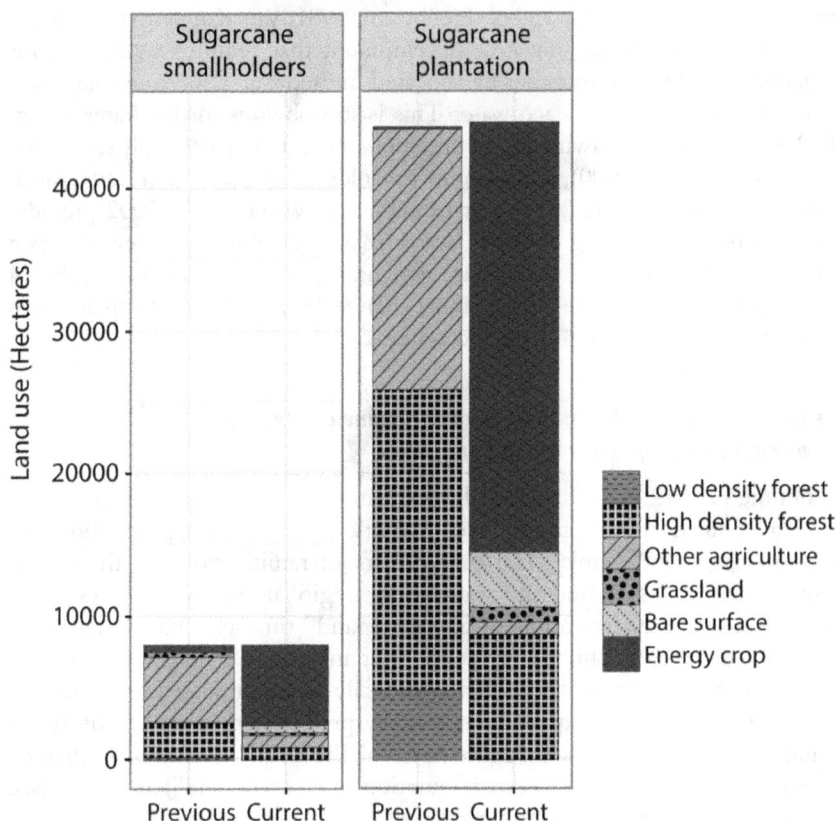

Figure 11.5 Land use change due to sugarcane production in the northern lowveld.
Source: Adapted from (Romeu-Dalmau et al., 2018)

the trajectory and outcomes of the sugarcane sector. However, water management has a long and turbulent history, spanning to the early entanglement of the Swazi nation with the Boer and British colonies in the mid-to-late 1800s. These early experiences have influenced until the present day the eligibility of acquiring water rights and whose rights to water are to be protected. In more detail, in order to keep its independence, the Swazi nation agreed to various forms of concessions with the Boers and the British, but due to various circumstances, it became a British Protectorate from 1906 until its independence in 1967. During this period, Eswatini was ruled through an amalgamation of Roman-Dutch and British laws (Pain, 1978), which held water resources tied to the land adjacent to it (and essentially to the economic use of the land). Under these laws, legal rights to water were predominantly given to colonist farmers (Pain, 1978; Heilbronn, 1981). This mindset permeated both the early water legislation and the post-independence water laws, which have strongly articulated the connections between water and land and have been protecting previously rewarded water rights (Heilbronn, 1981; van Koppen et al., 2014) (Table 11.1). As a result, any new development that requires water (e.g. for irrigation) would need to either be allocated unused water, increase water efficiency, or harvest new surface water. This is still obvious in the Water Act of 2002, which, though drawing from contemporary and internationally supported legislation (SADC, 2000), still mimics the older Roman-Dutch and British laws (van Koppen et al., 2014). For example, the Water Act of 2002 provides water permits to existing and new water users, and although it establishes a decentralized approach to water management, it protects the existing rights of the former (Table 11.1, see "Institutional change: From policies of expansion to reconfiguration of access to land and water").

Institutional change: From policies of sugarcane expansion to reconfiguration of access to land and water

As outlined above, the actual modalities of land and water management in the context of sugarcane expansion (see "Evolution of sugarcane production and water management", Table 11.2) have had major ramifications for the reconfiguration of access to land and water at the regional and the local level (see sub-sections "Reconfiguration of access to land" through "Reconfiguration of access to water"). This clearly shows that institutional change was necessary to enable sugarcane expansion (nationally through a series of policies), but also that sugarcane expansion eventually precipitated further institutional changes (locally). Below we discuss some of the major institutional changes linked to access to land (see "Reconfiguration of access to land") and water (see "Reconfiguration of access to water").

Reconfiguration of access to land

At the regional level, the reconfiguration of land access is mostly linked to the large-scale acquisition of land for sugarcane production in the core estates. In particular, the most important aspect of this reconfiguration relates to the acquisition of private land under TDL by the Royal Swazi investment fund (Tibiyo), often through the contribution of foreign capital (Table 11.2). Terry and Ogg (2017) estimate that by 1997, this non-customary type of SNL under Tibiyo expanded from 2% to 19% of the total Swazi area, much of which was for sugarcane production in association with large agri-businesses. This suggests the major consolidation of land under the direct power of the king, especially following the introduction of the Tinkhundla system in 1978 (see "Evolution of sugarcane production and water management", Table 11.1). Following the typology of Ribot and Peluso (2003), this constitutes major shifts in the rights-based legal access to land, with the underlying mechanisms being access to authority (the king himself in this case) and to capital, considering the centrality of foreign capital in many of these land acquisitions (see "Evolution of sugarcane production and water management").

At the local level, the reconfiguration of land access is mostly linked to the consolidation of individual dryland farms to large land blocks able to receive irrigation and accommodate sugarcane production. This meant that the households engaged in irrigated sugarcane production essentially surrendered their access to individual land for dryland agriculture for shares in a form of communal access to land (and the benefits provided by sugarcane production on this land). This signifies a major shift in rights-based legal access to land. However, in this case (and in contrast to the regional level above), the initial land used by individual families and the land consolidated for sugarcane production controlled by the sugarcane associations was and remains SNL under customary law. The initial access to this land was mediated through mechanisms linked to access to authority and access through social identity (Ribot and Peluso, 2003), considering how the customary land tenure system operates in Eswatini (see Table 11.2 and Footnote 1). Land consolidation and the subsequent production of sugarcane through irrigation were facilitated via access to knowledge (i.e. capacity for sugarcane production built from SWADE), access to technology (i.e. irrigation equipment provided by SWADE), and access to finance (Ribot and Peluso, 2003). As discussed below, the access to irrigation technology has become the major mechanism permeating the observed change in access to water.

Reconfiguration of access to water

At the international level, the treaties of the Southern African Development Community (SADC) seek to, among others, rationalize water resource management and facilitate more equal water sharing along transboundary river basins and within national borders through decentralized water management (SADC, 2000, 2006). This has major implications for water management at different levels, considering that most major rivers and catchments in the country (including the Komati River) are shared between Eswatini, South Africa, and Mozambique. This has influenced the development of tripartite agreements to manage water resources (Table 11.1) and regulate better water flows between countries such as the Maguga dam at the Komati River (see below). As discussed below, these international initiatives have had major implications for water management at the national, regional, and local levels.

At the national level, unlike past national water laws (Table 11.1), the Swazi Water Act of 2002 adopts the SADC guidelines to rationalize water allocations in river basins. This "requires" individuals and companies to go through the same process when requesting water, with priority given to household consumptive use[5]. However, much like the 1959 and 1967 water legislations, the previously awarded permits and rights are protected. Thus, for rivers such as the Komati whose water resources have already been fully allocated, there is a need to find ways to increase water supply in order to cater to population increase and new economic activities. Furthermore, the Water Act of 2002 and SADC guidelines may contain another roadblock for smaller and new developments (whether agricultural or not), in that water receives a monetary value to supposedly improve the management of this limited and increasingly scarce resource (see "Introduction"). Many expert interviews pointed that strong domestic players or international investments (e.g. large sugarcane plantations or individual producers with better access to financial resources) may also obtain better access to water resources when compared to smaller players. In this sense, while there seems to be an extension of rights-based legal access to water resources (see "Evolution of sugarcane production and water management", Table 11.1), increasingly mechanisms such as access to capital may improve the ability of different players to obtain access to water through the existing legal instruments, and thus consolidate their control and subsequent benefits.

At the regional level (i.e. Komati river level), the Maguga Dam has provided this opportunity for new economic developments by harvesting water while meeting the pre-determined water allocations to existing users and downstream nations. Following the tripartite agreements (Table 11.1) and some bilateral agreements with South Africa, the national apportionment of guaranteed water rights is provided to national users following the process set in the Water Act of 2002 (see above). During the dam's inception, the KDDP was specifically reserved water portions. This was because the KDDP was perceived to be a cornerstone of the regional poverty alleviation efforts. Through SWADE, the sugarcane associations along the Komati River received and were guaranteed water permits. Following Ribot and Peluso (2003), this constitutes a clear example of legal, state-granted rights-based access to water. However, as explained below, various mechanisms shape how the benefits are gained, controlled, and maintained.

At the local level (i.e. community level), the smallholder sugarcane associations receive access to irrigation infrastructure as part of SWADE's support package, while the individual households generate income through dividends from ownership of these associations (see "Evolution of sugarcane production and water management" and "Socioeconomic change: Differentiated livelihood outcomes"). In the context of the Water Act of 2002, the sugarcane associations are able to obtain and maintain water permits due to their economic ability, secured financial support, and their streamlined and secured market access, which has major ramifications about water access at the community level, especially during the 2015–2016 drought (see "Distributional aspects of sugarcane production"). When considering that the proximity to the Komati River was a major factor affecting the inclusion of households in sugarcane associations (see "Institutional change: From policies of expansion to reconfiguration of access to land and water"), the effects of mechanisms related to access to water are amplified, regardless of the legal state-granted rights protecting the household consumption of water for all (Ribot and Peluso, 2003). In particular, association members actualize their rights-based access to water through different mechanisms, including infrastructure, newly forged social identities as investors and shareholders, and new financial capabilities from dividends (Ribot and Peluso, 2003). Conversely, non-sugarcane growers relying on nearby springs, small water bodies, rainfall, as well as farmer associations had to share their water apportioned for sugarcane production. When looked through the lens of the theory of access, both cases demonstrate the significance of access mechanisms beyond means specified by governmentally recognized legal rights (Ribot and Peluso, 2003).

It should be noted that apart from the legal rights-based access to water outlined above, there is extensive illegal access, both at the regional and the local level. Illegal is defined as water access that does not follow the legal means of surface water abstraction indicated in the Water Act of 2002, namely through payments to a registered macro-rural water system that provide monitored amounts of water or direct physical abstraction[6] from natural water bodies. Illegal extraction encompassed the local population's means to livelihood and sustenance[7], both for association shareholders and the non-shareholders alike. This illegal access is partly due to limitations and ambiguities in the Water Act of 2002[8] and may be partly due to the non-recognition of livelihoods or needs when calculating water use. These collectively contribute to water deficits, especially during periods of water scarcity. Illegal water extraction from the local to international level is a rather contentious issue often featuring in agitated discussions between South Africa, Eswatini, and Mozambique both during periods of drought or relative water abundance.

Socioeconomic change: Differentiated livelihood outcomes of sugarcane expansion

According to Table 11.3, the RSSC plantation workers have the highest total income, followed by irrigated sugarcane farmers and community plantation workers. Groups not involved in sugarcane production through waged planta-tion employment or sugarcane cultivation have the lowest income. The results indicate that RSSC plantation workers tend to specialize in waged employ-ment as they hold little land and livestock, but they hold many other assets (Table 11.3) (see Chapter 8 for similar patterns). Indeed, from the groups

Table 11.3 Income, wealth, and access to land and agricultural inputs across the study groups

Respondent type	Land (ha)	Fertilizer (kg)	Livestock ownership (TLU)	Asset ownership (HDAI)	Total income (USD)
Plantation workers (RSSC)	0.1	1	0.6	118.8	11,988
Plantation workers (Community)	2.6	21	3.3	116.7	5,998
Irrigated sugarcane farmers	1.6[a]	46	4.0	174.3	6,728
Control group	1.6	10	2.1	89.4	4,415

Note: [a] does not include sugarcane land.

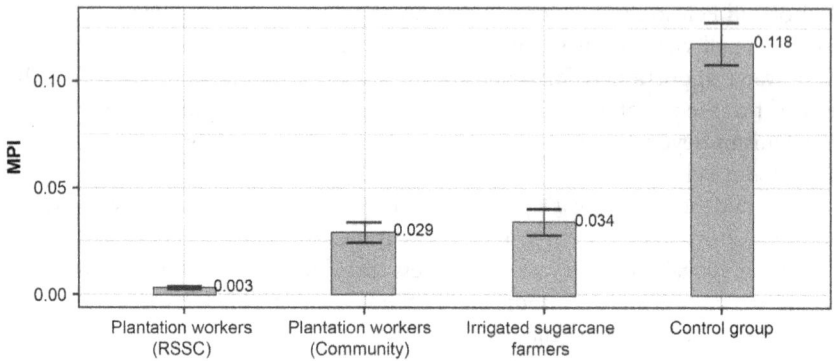

Figure 11.6 Multi-dimensional poverty profiles of the study groups.
Source: Adapted from (Mudombi et al., 2018)

engaged in agricultural production, irrigated sugarcane farmers own consistently more assets and livestock and have better access to fertilizer, followed by community plantation workers and control groups. Even though we cannot preclude some historic accumulation of assets and livestock (possibly due to better access to water from proximity to the Komati River, although farms were still under dryland agriculture), when considering the project engagement process (see "Evolution of sugarcane production and water management"), we believe that this an outcome of engagement in irrigated sugarcane production.

Similarly, groups involved in sugarcane production have much lower levels of multi-dimensional poverty when compared to the control group. This is evident both for the aggregate multidimensional poverty index (MPI) (Figure 11.6) and the deprivation scores for each individual indicator (Table 11.4). Indeed, with the exception of child school attendance, RSSC workers have by far the lowest deprivation across all indicators. This is because many of these workers live in company-owned properties, which explains their generally better access to sanitation, clean fuel, and electricity. Conversely, the control group has the highest deprivation for most indicators (Table 11.4). One very interesting finding is the much lower deprivation of irrigated smallholders and community plantation workers for drinking water when compared to the control group. This is a clear indication of the differentiated access to water through the engagement with irrigated sugarcane production (see "Institutional change: From policies of expansion to reconfiguration of access to land and water").

Distributional aspects of sugarcane production

Formation of the sugarcane elite

As discussed, sugarcane expansion in the northern lowveld has entailed massive shifts in land control, from private interests (TDL) to the monarchy under the

Table 11.4 Deprivation scores for the individual MPI indicators across the study groups

	Years of schooling	Child school attendance	Nutrition	Child mortality	Drinking water	Improved sanitation	Clean cooking fuel	Electricity	Flooring material
Plantation workers (RSSC)	4%	15%	2%	4%	0%	5%	3%	57%	0%
Plantation workers (Community)	19%	10%	3%	4%	5%	6%	97%	88%	2%
Irrigated sugarcane farmers	15%	2%	7%	4%	2%	15%	99%	71%	6%
Control group	23%	3%	18%	6%	31%	32%	99%	78%	17%

Tibiyo system (see "Evolution of sugarcane production and water management" through "Institutional change: From policies of expansion to reconfiguration of access to land and water"). Essentially, this has made Tibiyo the *"central mechanism of domestic capital formation"* (Terry and Ogg, 2017: 5), which, combined with the tinkhundla system introduced in 1978, gives unprecedented land control powers to the monarchy considering that the national government has practically no overseeing mechanisms (see "Evolution of sugarcane production and water management"). This situation has essentially formed a new elite that found in sugarcane a new way to benefit and accumulate wealth. According to Terry and Ogg (2017: 17), citing (Levin, 2001: 148) this

> *provided the material basis for the transformation of the entire social structure of Swaziland by providing the aristocracy with an independent basis for capital accumulation. Through involvement in Tibiyo, Swazis with both royal and non-royal origin found themselves part of a new elite club.*

This elite has often benefited at the expense of local communities, as exemplified by the eviction and relocation of 5,000 people from local communities during the sugarcane expansion in the 1970s (see "Evolution of sugarcane production and water management"). According to Terry and Ogg (2017: 6), this reflected

> *the sense of powerlessness of people residing on SNL in the face of coercive power exercised by the king through his control the chiefs. This has taken the form of forced labour, forced financial contributions and forced removal of groups considered to be either a threat to royal power or a barrier to Tibiyo-sponsored development.*

The development of smallholder-based sugarcane production on customary SNL through the KDDP further reinforces the benefit of this elite in a rather interesting manner. By influencing or arguably mandating sugarcane production (over other crops) in large community areas, this elite ensures the stable flow of sugarcane in the RSSC mills, consolidating thus its economic benefits due to its de facto control of the sugarcane sector. In this sense, regardless of the rural development and poverty alleviation benefits of sugarcane, at least for some community members, (see "Socioeconomic change: Differentiated livelihood outcomes" and " Growing inequalities within local communities"), the sugarcane elite reaps huge benefits.

It should be noted that apart from this sugarcane elite, certain types of estate workers have also benefited from the expansion of the sugarcane sector. Although poverty alleviation was not a goal of the sugarcane sector (at least in its early stages), the stable wages, housing, and other social services offered by the large sugarcane estates afforded some of their workers a better standard of living compared to other members of the local communities (Terry and Ogg, 2017). This is still evident considering that RSSC plantation workers have by far the highest incomes and lowest multi-dimensional poverty compared to other study groups (Tables 11.3–11.4, Figure 11.6) (Mudombi et al., 2018,

2021). Such observations have been made and for other sugarcane plantations in SSA (Matenga and Hichaambwa, 2017).

While it would be an exaggeration to designate estate workers as an elite, it is reasonable to argue that they have morphed into a rather distinct social group within the wider lowveld. Plantation workers seem to divert all their labour and specialize in waged employment, having very low levels of land owner-ship (Table 11.3) (see also Chapter 8). This is in stark contrast with most local residents that depend directly on land for their livelihoods (e.g. through the pro-duction of sugarcane, food crops, livestock), or migrant labour and remittances. This seems to suggest the gradual consolidation of plantation workers in large estates into a distinct social class (Cousins, 2010).

Growing inequalities within local communities

Sugarcane expansion, and the associated reconfiguration of access to land and water, have catalyzed growing inequalities within local communities. As outlined in the sub-sections "Evolution of sugarcane production and water management" through "Institutional change: From policies of expansion to reconfiguration of access to land and water", only those community members with land close to the Komati River were beneficiaries of the KDDP project, receiving access to irrigation and technical training for sugarcane cultivation. Their engagement in irrigated sugarcane cultivation increased substantially their incomes and reduced their multi-dimensional poverty (see "Socioeconomic change: Differentiated livelihood outcomes") (Mudombi et al., 2018, 2021). Furthermore, engagement in irrigated sugarcane production has possibly had multiplier effects, as the stable access to credit, fertilizers, and a mature sugarcane market (von Maltitz et al., 2019) might further increase the income and food security of sugarcane smallholders.

At the same time, the KDDP and the Water Act of 2002 (see "Institutional change: From policies of expansion to reconfiguration of access to land and water") have mediated the re-configuration and formalization of access to water in the smallholder sugarcane areas, leaving most of the non-sugarcane farmers in the local communities with precarious access to water. This manifested aptly during the 2015–2016 drought, when the escalating water shortages required the rationing of water originally allocated through permits. In more detail, the combined outcome of water rationing during this period and the supported claims over the economic importance of water for sugarcane production (as a major contributor to the national GDP) meant that both sugarcane and non-sugarcane growers lost the water rights for their subsistence farms and gardens. Furthermore, due to this water rationing and the stricter water use monitoring (from governmental water management agencies), sugarcane associations prioritized their allocated water for sugarcane production (i.e. the intended use of the allocated water). As a result, both sugarcane growers and non-growers in many local communities had to resort to some illegal avenues to access water for their needs. For example, some of the sugarcane growers covertly diverted portions of their (or the associations') allocated water to other family members or their home gardens, while non-growers were no longer able to access water through social relations (see "Institutional change: From policies of expansion to reconfiguration of access to land and water"). Some sugarcane growers and non-growers alike resorted to stealing water, purchasing from unregistered vendors (who had in turn accessed it illegally), or accessing it through labour-intensive means from ever-scarcer surface water bodies.

Overall, while members of sugarcane associations coped better with the drought, this created some local contestations, as non-growers saw their access, the actualization of their rights, to water decline. These phenomena are aptly captured in the following quotes from FGDs and exert interviews:

> Those who are still on the other side of the road, those who are not part of the project, …they are surviving because they have got some water. Where do they get their water? From the river.
>
> (pers. comm. Malibeni F-Shareholders FDG)

> They were denied access. But before the drought, they were given water
>
> (pers. comm. Malibeni M-non-Shareholders FGD)

> The water is no longer enough for the community due to the population increase
>
> (pers. comm. Sihhoye Water Development Committee Chairperson)

> They are not allowed to farm their gardens with this water…[but] they wish to … irrigate their home gardens.
>
> (pers. comm. Sihhoye F-non-Shareholders FGD)

> So now the communities brought up the verbal agreement issue, [lightly quoting community members] "[It's] blo[cking] our way to the river", "he promised to that you will supply us with water". So then they were fighting.
>
> (pers. comm. Emandla Ekuphila Water User District)

Furthermore, the consolidated sugarcane plantations are located between the communities and the Komati River, obstructing favourable paths for watering livestock, especially when surface water is limited, as during the 2015–2017 drought. With fewer paths, cattle have to traverse longer distances to reach available surface water[9]. Even though the expert interviews suggest that the dearth of food was the most important factor for cattle loss during the drought, these longer walks for watering contributed to cattle mortality by increasing exhaustion (pers. comm. Sihhoye traditional authority).

The above are a testament to the truly transformative, yet unevenly distributed, livelihood outcomes of irrigated sugarcane production in the northern lowveld. Arguably, and despite its many benefits, the KDDP has created communities divided into two segments; one that benefits extensively from sugarcane production and one that does not benefit at all (or even experiences negative effects). This alludes to the exclusion of segments of the local communities from the direct benefits of agricultural modernization, which is has been discussed extensively in the political ecology literature (Hall, 2011; Svarstad et al., 2018) (Chapter 8). With entry to the irrigated sugarcane growers' associations now almost impossible due to multiple technical and institutional barriers (Terry and Ogg, 2017), low-skilled employment in the RSSC plantation or community plantations is the only major income generation option for non-sugarcane growers who cannot leave the area (see "Emergence of new social classes"). This might lead to the long-term accumulation of the benefits from engagement in sugarcane production by only a segment of the local community, creating a sense of unfairness as was pointed in most FGDs with non-growers.

While tempting, it is still not warranted to view the above as the emergence of a local sugarcane elite at the community level. Considering that incomes are still low and poverty high among irrigated sugarcane farmers, it is not easy to predict whether such a community elite will emerge. It is possible that this elite would be formed through the constant accumulation of capital and other benefits (see also " Emergence of new social classes"), but it is equally possible that these smallholders will be stuck in a form of poverty trap considering that many associations have been oversubscribed during their formation (see also similar findings in Matenga and Hichaambwa, 2017; Wendimu et al., 2016; Hall et al., 2017). Actually, in many expert interviews and FGDs, it was alluded to that in many associations, the dividends are quite diluted, and essentially not enough to ensure the initially expected poverty alleviation benefits. For example, as was pointed during an expert interview:

> Upon re-location they were told that the sugarcane production will give them good dividends to buy food as they have lost their productive land. However, they are currently not getting the dividends and there are high food insecurity cases.
>
> (pers. comm. Ekuvinjewlni Water Development Committee Chair)

Emergence of new social classes

The large estates and the community plantations have brought major changes in the local labour system through the generation of waged off-farm employment opportunities. Even though community plantations create on average more jobs per unit area than the large estates (Terry and Ogg, 2017), employment in both requires the substantial diversion of workers' labour from other agricultural activities (Chapter 3, 8).

RSSC workers tend to specialize in waged employment considering their very low land ownership, while community plantation workers continue farming their own plots, though possibly less intensively (Chapter 8). Community plantation workers have higher incomes and lower poverty than community members not involved in sugarcane activities (control group), but these incomes are much lower compared to estate workers (Table 11.2; Figure 11.6). However, they can vary substantially between types of employment, with the better-paying jobs offered to well-connected individuals or even the family members of the irrigated sugarcane growers (Terry and Ogg, 2017) (Chapter 8).

In this context of rural employment generation, it is interesting to note that by virtue of being the owners of the sugarcane associations, the irrigated sugarcane smallholders essentially become the employers of other community members not owning sugarcane land. Considering the very evident social differentiation between study groups (see sub-sections "Institutional change: From policies of expansion to reconfiguration of access to land and water" through "Socioeconomic change: Differentiated livelihood outcomes"), it can be argued that sugarcane production has possibly led to the gradual emergence of new social classes in the northern lowveld, though this is rather difficult to ascertain at this stage.

Following a class-analytic typology of small-scale agricultural producers, it can be argued that the irrigated sugarcane smallholders share the characteristics of "small-scale capitalist farmers" that "*rely substantially on hired labour and can begin to engage in expanded reproduction and capital accumulation*" (Cousins, 2010: 10). Conversely, considering their different engagement in agricultural activities in their own farms, RSSC estate workers share the characteristics of "allotment holding wage workers" and community plantation workers of "worker-peasants" (Cousins, 2010: 10).

Such shifts from engagement in waged plantation employment are rather common in SSA (Chapter 8). However, such a clear-cut example of an "agro-capitalist" class is rather uncommon in agrarian contexts of SSA, as this role is usually associated with external investors operating large plantations (Smalley, 2013; Gibbon, 2011) (Chapter 8). As community plantations were established on SNL under customary law (Table 11.2), the eventual tying of land to sugarcane production and the inability of new farmers to join the associations creates a form of accumulation by dispossession (Harvey, 2003), which is "from below" (Cousins, 2010). However, such a reading can be challenged by the fact that the sugarcane smallholders received the land that ended up for irrigated sugarcane production through customary avenues (FAO, 1998) much before the commencement of the KDDP, which was in turn largely beyond the agency and influence of local communities and chiefs.

Conclusions

This chapter has used an analytical lens drawing from actor-oriented political ecology and the theory of access to unravel the history, processes, and impacts of sugarcane production in Eswatini. The results clearly show that sugarcane production in large estates and irrigated community plantations have had major transformative effects on the national economy, agrarian systems, and the socio-economic status of local communities. On the one hand, the sugarcane industry has contributed substantially to economic growth and has created a relatively thriving local economy in an area that is rather marginal, with low suitability even for dryland agriculture. However, this happened concurrently with major institutional transformations that have reconfigured access to land and water and has created a sugarcane elite. This has permeated the uneven distribution of the benefits of sugarcane production with the major beneficiaries being the sugarcane elite, workers in large estates, and irrigated smallholders. Conversely, community plantation workers and community members not involved in the sugarcane sector reap little-to-no direct benefits and arguably are losing access to water. Arguably, this social differentiation could catalyze the emergence of new social classes, though this is rather difficult to ascertain at this stage.

Acknowledgements

This development of this chapter was supported through grants offered by the Ecosystem Services for Poverty Alleviation Program (ESPA; grant number: NE/L001373/1) and the Japan Science and Technology Agency (JST) for the Belmont Forum project FICESSA.

Notes

1 The term lowveld refers to two low-lying areas of southern Africa with an elevation ranging from 150 to 600 m.
2 Legally, smallholder associations are considered to be individual companies or more precisely "commercial cooperatives" (Terry and Ogg, 2017). The main goal of these cooperatives extends beyond economic profitability to also include the social and cultural needs of the farmer group and is viewed as an alternative means of conducting business. For reasons of simplicity, in this chapter, we use the term associations rather than "social cooperatives" or "companies" in order to avoid confusion with large sugarcane companies such as RSSC.
3 In Ekuvinjelweni, we only conducted expert interviews with the traditional authorities, as we were not granted access to interview non-sugarcane growers.
4 According to the FAO (1998), in Swaziland, customary land "*is held by the King in trust for the nation and is allocated by chiefs to homestead heads, who under Swazi law and custom are men. Although many women are de facto heads of homestead, land is allocated to them through male proxies. Membership in a local community is the condition for the right to receive or to be allocated land. Land can also be acquired through inheritance.*" The households "*receive usufruct rights to land, but the administrative rights remain with the king*" (Terry and Ogg, 2017: 5).
5 Household consumptive use is defined slightly differently in the SADC guidelines and the Water Act of 2002, with the latter being more inclusive. However, both designate the highest priority to these household consumptive uses.
6 The Water Act of 2002 protects the use of wheelbarrows and containers for water abstraction but requires permits for mechanical pumps even if the water is used for household consumptive uses.
7 According to legal rights, allocated water should only be used for the respective purpose of the permit, e.g. either for agriculture or household consumption. The use of agricultural water permits to supply household water consumption or subsistence farms is considered to be a breach of the permit-holder's contract. Thus, the government views many of the mechanisms described throughout this section as illegal extractions.
8 Differences in the definition of seemingly similar, yet legally different, terms (i.e. domestic, primary, and cultural purposes) further complicate the protection and

administration of water rights in transboundary rivers. For example, primary water extraction for sustenance (e.g. household consumptive use, subsistence farming) is protected by the Water Act of 2002, though with land-size and location limitations. However, the "*use for primary purpose… does not include the use of water by a local authority for distribution to the inhabitants of the area*" (GoS, 2003). Also, while primary water is highly protected at the national level, there is no accurate formula to estimate it and deduct it from estimates of water availability (pers. comm., Department of Water Affairs, 2017). Thus, subsistence farming may not be protected as *primary water* when managed by decentralized water management bodies. At the international level, subsistence farming falls within agricultural water use. Domestic water is supplied for household consumptive needs and does not include home gardens (pers. comm., KOBWA, 2017). Realistically, these parameters may complicate access to water for households located far from natural water bodies or when there is no adequate rainfall for rain-fed agriculture during droughts.

9   The water legislation considers cattle under the primary water stipulation (see "Reconfiguration of access to water"). Thus, they cannot be denied water from water bodies, with the exception of the RSSC canal, which is considered as already-allocated water. However, the water legislation does not touch on how easily the cattle are to obtain their water, nor the water indirectly required to feed them.

References

Adams, E.A., Kuusaana, E.D., Ahmed, A., Campion, B.B., 2019. Land dispossessions and water appropriations: Political ecology of land and water grabs in Ghana. Land Use Policy, 87, 104068

Ahmed, A., Abubakari, Z., Gasparatos, A., 2019. Labelling large-scale land acquisitions as land grabbing: Procedural and distributional considerations from two cases in Ghana. Geoforum, 105, 191–205.

Ahmed, A., Gasparatos, A., 2020. Reconfiguration of land politics in Community Resource Management Areas in Ghana: insights from the Avu Lagoon CREMA. Land Use Policy, 97, 104786

Allan, J.A., Keulertz, M., Sojamo, S., Warner, J., (Eds.), 2014. Handbook of Land and Water Grabs in Africa: Foreign Direct Investment and Food and Water Security. Routledge, Oxford. p. 513

Ayelazuno, J.A., 2019. Water and land investment in the "overseas" of Northern Ghana: The land question, agrarian change, and development implications. Land Use Policy, 81, 915–928,

Bergius, M., Benjaminsen, T.A., Widgren. M., 2018. Green economy, Scandinavian investments and agricultural modernization in Tanzania. Journal of Peasant Studies, 45, 825–852.

Bury, J., 2008. Transnational corporations and livelihood transformations in the Peruvian Andes: An actor-oriented political ecology. Human Organization, 67, 307–321.

Charmaz, K., 2014. Constructing Grounded Theory. SAGE Publications, London.

Chilonda, P, Otte, J., 2006. Indicators to monitor trends in livestock production at national, regional and international levels. Livestock Research for Rural Development, 8, 117.

Chinsinga, B., 2017. The Green Belt initiative, politics and sugar production in Malawi. Journal of Southern African Studies, 43, 501–515.

Cousins, B., 2010. What is a "smallholder"?: Class-analytic perspectives on small-scale farming and agrarian reform in South Africa, in: Hebinck, P., Shackleton, C. (Eds.), Reforming Land and Resource Use in South Africa: Impact on Livelihoods. Routledge, London, pp. 86–109. doi.org/10.4324/9780203839645

Cutz, L., Nogueira, L.A.H., 2018. The potential of bioenergy from sugarcane in Latin America, the Caribbean and Sothern Africa, In Cortez, L.A.B., Leal, M.R.L.V., Nogueira, L.A.H. (Eds.) Sugarcane Bioenergy for sustainable development. Expanding production in Latin America, and Africa. Routledge, Oxford, pp. 141–153.

Dell'Angelo, J., D'Odorico, P., Rulli, M.C., 2017. Threats to sustainable development posed by land and water grabbing. Current Opinion in Environmental Sustainability, 26–27, 120–128.

Dell'Angelo, J., Rulli, M.C., D'Odorico, P., 2018. The global water grabbing syndrome. Ecological Economics, 143, 276–285.

Dubb, A., Scoones, I., Woodhouse, P., 2017. The political economy of sugar in Southern Africa – introduction. Journal of Southern African Studies, 43, 447–470.

ESWADE, 2019. Annual Report 2018/2019. Eswatini Water and Agricultural Development Enterprise (ESWADE), Mbabane.

FAO, 1998. Land Reform: Land Settlement and Cooperatives. Food and Agriculture Organisation (FAO), Rome.

FAO, 2013. Structural Changes in the Sugar Market and Implications for Sugarcane Smallholders in Developing Countries: Country Case Studies for Ethiopia and the United Republic of Tanzania. Food and Agriculture Organisation (FAO), Rome.

FAO, 2016. Southern Africa El Niño Response Plan (2016/17). Food and Agriculture Organisation (FAO), Rome.

FAO, 2017. The State of Eswatini's Biodiversity for Food and Agriculture. Food and Agriculture Organisation (FAO), Rome.

Gasparatos, A., von Maltitz, G., Johnson, F.X., Lee, L., Mathai, M., Puppim de Oliveira, J., Willis, K., 2015. Biofuels in Africa: Drivers, impacts and priority policy areas. Renewable and Sustainable Energy Reviews, 45, 879–901.

Gasparatos, A., von Maltitz, G., Johnson, F.X., Romeu-Dalmau, C., Jumbe, C., Ochieng, C., Mudombi, S., Balde, B., Luhanga, D., Nyambane, A., Lopes, P., Jarzebski, M., Willis, K.J., 2018. Survey of local impacts of biofuel crop production and adoption of ethanol stoves in southern Africa. Nature: Scientific Data, 5, 180186

Gibbon, P., 2011. Experiences of Plantation and Large-Scale Farming in 20th Century Africa: DIIS Working Paper. The Danish Institute for International Studies (DIIS), Copenhagen.

GoS, 2003. Water Act of 2002. Government of Swaziland, Mbabane

Hall, R., 2011. Land grabbing in Southern Africa: The many faces of the investor rush. Review of African Political Economy, 38, 193–214.

Hall, R., Scoones I., Tsikata, D., 2017. Plantations, outgrowers and commercial farming in Africa: agricultural commercialisation and implications for agrarian change. Journal of Peasant Studies, 44, 515–537

Harvey, D., 2003. The New Imperialism. Oxford University Press, Oxford.

Heilbronn, S., 1981. Water laws, prior rights and government apportionment of water in Swaziland, Southern Africa. Journal of African Law, 25, 136–149

Herrmann, R., Jumbe, C., Bruentrup, M., Osabuohien, E., 2018. Competition between biofuel feedstock and food production: Empirical evidence from sugarcane outgrower settings in Malawi. Biomass Bioenergy, 114, 100–111

Hess, T.M., Sumberg, J., Biggs, T., Georgescu, M., Haro-Monteagudo, D., Jewitt, G., Ozdogan, M. Marshall, M., Thenkabail, P., Daccache, A., Marin, F., Knox, J.W., 2016. A sweet deal? Sugarcane, water and agricultural transformation in Sub-Saharan Africa. Global Environmental Change, 39, 181–194

IRENA, 2019. Sugarcane bioenergy in southern Africa: Economic potential for sustainable scale-up. International Renewable Energy Agency (IRENA), Abu Dhabi.

Jarzebski, M.P., Ahmed, A., Boafo, Y.A., Balde, B.S., Chinangwa, L., Saito, O., von Maltitz, G., Gasparatos, A., 2020. Food security impacts of industrial crop production in sub-Saharan Africa: A systematic review of the impact mechanisms. Food Security, 12, 105–135.

Johnson, F.X., Silveira, S., 2014. Pioneer countries in the transition to alternative transport fuels: Comparison of ethanol programmes and policies in Brazil, Malawi and Sweden. Environmental Innovation and Societal Transitions, 11, 1–24

KOBWA, 2008. Dams and Development: The KOBWA-Experience Practices for Balancing Social, Environmental and Economic Aspects in Water Resources Management. Komati Basin Water Authority (KOBWA), Mbabane.

Levin, R., 2001. When the Sleeping Grass Awakens: Land and Power in Swaziland. Witwatersrand University Press, Johannesburg.

Lwimbo, Z.D., Komakech, H.C., Muzuka, A.N., 2019. Impacts of emerging agricultural practices on groundwater quality in Kahe catchment, Tanzania. Water, 11, 2263.

Maconachie, R., 2019. Green grabs and rural development: How sustainable is biofuel production in post-war Sierra Leone? Land Use Policy, 81, 871–877.

Manda, S., Tallontire, A., Dougill, A.J., 2020. Outgrower schemes and sugar value-chains in Zambia: Rethinking determinants of rural inclusion and exclusion. World Development, 129, 104877

Matenga, C.R., Hichaambwa, M., 2017. Impacts of land and agricultural commercialisation on local livelihoods in Zambia: Evidence from three models. The Journal of Peasant Studies, 44, 574–593,

Mehta, S., Veldwisch, G.J., Franco, J., (Eds.), 2012. Water grabbing? Focus on the (re) appropriation of finite water resources. Water Alternatives, 5, 193–468.

Mudombi, S., Ochieng, C., Johnson, F.X., von Maltitz, G., Luhanga, D., Dompreh, E.B., Romeu-Dalmau, C., Gasparatos, A., 2021. Fuelling rural development? The impact of biofuel feedstock production in southern Africa on household income and expenditures. Energy Research and Social Science, 76, 102053

Mudombi, S., von Maltitz, G.P., Gasparatos, A., Romeu-Dalmau, C., Johnson, F.X., Jumbe, C., Ochieng, C., Luhanga, D., Lopes, P., Balde, B.S., Willis, K.J., 2018. Multi-dimensional poverty effects around operational biofuelprojects in Malawi, Mozambique and Swaziland. Biomass and Bioenergy, 114, 41–54.

Ngcobo, S., Jewitt, G. 2017. Multiscale drivers of sugarcane expansion and impacts on water resources in Southern Africa. Environmental Development, 24, 63–76.

Pain, J., 1978. The reception of English and Roman-Dutch law in Africa with reference to Botswana, Lesotho and Swaziland. Comparative and International Law Journal of Southern Africa, 11, 137–167.

Palliere, A., Cochet, H., 2018. Large private agricultural projects and job creation: From discourse to reality. Case study in Sella Limba, Sierra Leone. Land Use Policy, 76, 422–431.

Ribot, J.C., Peluso, N.L., 2003. A Theory of Access★. Rural Sociol, 68, 153–181.

Richardson-Ngwenya, P., Richardson, B., 2014. Aid for trade and African agriculture: The bittersweet case of Swazi sugar. Review of African Political Economy, 41, 201–215.

Romeu-Dalmau, C., Gasparatos, A., von Maltitz, G., Graham, A., Almagro-Garcia, J., Wilebore, B., Willis, K.J., 2018. Impacts of land use change due to biofuel crops on climate regulation services: Five case studies in A Theory of Access★. Rural Sociol, 68, 153–181.Malawi, Mozambique and Swaziland. Biomass and Bioenergy, 114, 30–40.

SADC, 2000. Revised Protocol on Shared Watercourses, South African Development Community (SADC), Gaborone.

SADC, 2006. Regional Water Strategy. South African Development Community (SADC), Gaborone.

Schuenemann, F., Thurlow, J., Zeller, M. 2017. Leveling the field for biofuels: Comparing the economic and environmental impacts of biofuel and other export crops in Malawi. Agricultural Economics, 48, 301–315.

Semie, T.K., Silalertruksa, T., Gheewala, S.H., 2019. The impact of sugarcane production on biodiversity related to land use change in Ethiopia. Global Ecology and Conservation, 18, e00650

SEPARC, 2017. The socio-economic impacts of the 2015/16 EL Niño induced drought in Swaziland. Swaziland Economic Policy Analysis and Research Centre (SEPARC), Mbabane.

Siderius, C., Gannon, K.E., Ndiyoi, M., Opere, A., Batisani, N., Olago, D., Pardoe, J., Conway, D., 2018. Hydrological response and complex impact pathways of the 2015/2016 El Niño in Eastern and Southern Africa. Earth's Future, 6, 2–22

Smalley, R., 2013. Farming and Commercial Farming Areas in Africa: A Comparative Review. Institute for Poverty, Land and Agrarian Studies (PLAAS), Cape Town.

Sulle, E., Dancer, H., 2020. Gender, politics and sugarcane commercialisation in Tanzania. The Journal of Peasant Studies, 47(5), 973–992

Svarstad, H., Benjaminsen T.A., Overå, R. (Eds.), 2018. Power theories in political ecology. Journal of Political Ecology, 25, 350–425.

Terry A., Ogg M., 2017. Restructuring the Swazi sugar industry: The changing role and political significance of smallholders. Journal of Southern African Studies, 43, 585–603.

Terry, A., Ryder, M., 2007. Improving food security in Swaziland: The transition from subsistence to communally managed cash cropping. Natural Resources Forum, 31, 263–272.

van der Laan, M., van Antwerpen, R., Bristow, K.L., 2012. River water quality in the northern sugarcane-producing regions of South Africa and implications for irrigation: A scoping study. Water SA, 38, 87–96

van Koppen, B., van der Zaag, P., Manzungu, E., Tapela, B., 2014. Roman water law in rural Africa: The unfinished business of colonial dispossession. Water International, 39, 49–62.

von Maltitz, G.P., Henley, G., Ogg, M., Samboko, P.C., Gasparatos, A., Ahmed, A., Read, M., Engelbrecht, F., 2019. Institutional arrangements of outgrower sugarcane production in southern Africa. Development Southern Africa, 36, 175–197.

Wendimu, M.A., Henningsen, A. Gibbon, P., 2016. Sugarcane Outgrowers in Ethiopia: "Forced" to remain poor? World Development, 83, 84–97.

Part V

Synthesis

12 Political ecology of industrial crops

Towards a synthesis and systematization

Alexandros Gasparatos and Abubakari Ahmed

Towards a political ecology of industrial crops

The chapters included in this volume have highlighted how industrial crops can become agents of ecological, agrarian, socioeconomic, and institutional transformation. The actual production of industrial crops, whether large-scale or smallholder-based (Ahmed et al., 2021), often causes substantial ecological transformation through the extensive conversion of agricultural land and natural vegetation and intensified production using genetically modified seeds and agrochemicals, among others (e.g. Urteaga-Crovetto and Segura-Urrunaga, 2021; Gasparatos and Ahmed, 2021; Oliveira, 2021; Flachs, 2021; Ahmed and Gasparatos, 2021a, 2021b) (see "Local level"). At the same time, through the combined effects of labour diversion, adoption of new production models, linkage to markets, and social differentiation, industrial crops tend to transform agrarian systems and socioeconomic conditions (Ahmed and Gasparatos, 2021a, 2021b; Flachs, 2021; Zhang, 2021; Gasparatos and Ahmed, 2021; Oliveira, 2021) (see "Local level"). Interestingly industrial, crops have a dual role related to institutional transformation. On the one hand, the promotion, expansion, and trade of industrial crops require the reconfiguration and transformation of institutions across different levels in order to facilitate industrial crop expansion (e.g. Ahmed and Gasparatos, 2021a; Lundy, 2021; Zhang, 2021; Urteaga-Crovetto and Segura-Urrunaga, 2021) (see sub-sections "International level" through "National level"). On the other hand, the production of industrial crops often reconfigures access to natural resources (i.e. changes in land rights) at the local level (Gasparatos and Ahmed, 2021; Ahmed and Gasparatos, 2021b).

Arguably, there are many different and context-specific factors mediating the transformative power of industrial crops, as discussed throughout this volume. Political ecology can help delineate the type and role of these mediating factors through its critical and interdisciplinary lens. See sub-sections "Multi-scalar dynamics" through "Social differentiation", which synthesize and systematize the main findings of this edited volume using some of the main conceptual tools and perspectives of political ecology, namely multi-scalar analysis and attention to marginalization, social difference, and discourses and narratives (Moseley, 2021).

Multi-scalar dynamics

Many chapters have made abundantly clear that institutions with different remits, and actors with different interests and perspectives, interact at multiple spatial scales (Ahmed and Gasparatos, 2021a). Thus a multi-scalar lens is a prerequisite to better understand the drivers, impacts, and dynamics of industrial crops' promotion, production, trade, and use. The most pertinent scales for this volume include the international, national, and local scales (see sub-sections "International level" through "Local level").

International level

Historically, international demand has perhaps been the most important and prevalent driver of industrial crop expansion in many parts of the global South. The most obvious recent example has been the growing biofuel demand in the European Union, which caused a global interest in bioenergy crop production and eventually catalyzed to a large degree the land rush of the mid-2000s and early 2010s (Ahmed and Gasparatos, 2021a; Shortall and Helliwell, 2021). Approaches to meet the growing international demand have manifested in rather different manners across time periods and geographic contexts, ranging from economic incentives to attract international land-based investments for industrial crops (e.g. Urteaga-Crovetto and Segura-Urrunaga, 2021) to the almost forced expansion and transformation of industrial crop systems to increase productivity (e.g. Flachs, 2021). This volume has discussed many examples of how international demand has affected industrial crop production, including the cases of smallholder-based cotton production in India (Flachs, 2021) and cashew in Guinea-Bissau (Lundy, 2021), and the large-scale production of soybean in South America (Oliveira, 2021), sugarcane in Eswatini (Gasparatos and Ahmed, 2021), and jatropha and oil palm in Ghana (Ahmed and Gasparatos, 2021b). Often, this expansion to meet international demand has been justified through the concept of comparative advantage (Lundy, 2021; Moseley, 2021) (see "Social differentiation").

However, international dynamics sometimes take a much more tangible form beyond simple demand for industrial crops. For example, current global debates over technological innovation for agricultural sustainability drive the promotion and adoption of mutually exclusive technologies in the Indian cotton sector (i.e. genetically modified vs. organic cotton) (Flachs, 2021). Multinational and vertically integrated companies that cater to international financial interests dominate the production, marketing, and trade of soybeans in South America, as well as its inputs (i.e. seeds, agrochemicals) (Oliveira, 2021). Furthermore, the growing focus of many international investors on marginal land has been sparked not only by the growing demand for industrial crops but also by highly visible international criticisms regarding the competition between bioenergy and food crops (Shortall and Helliwell, 2021). In some extreme cases, this has led to the speculation and trading of this land in international land markets (often through the "support" of international funding institutions such as the World Bank) (Urteaga-Crovetto and Segura-Urrunaga, 2021). International trade wars can also influence industrial crop systems, as it has become evident during the recent trade war between the US and China and the apparent inability of China to access cheap maize, creating concerns over national food security in the face of a possible re-classification of maize from a staple food crop to an industrial crop (Zhang, 2021).

National level

At the national level in many countries, the development of national laws, regulations, and policy frameworks have created institutional environments that are amenable (or even welcoming) to large-scale agro-industrial development. Considering the multiple elements of industrial crop systems, ranging from land acquisition to agricultural production, processing, and trade, these laws, regulations, and policy frameworks span many different policy dimensions. These can be as diverse as land use and zoning, land compensation, access to land/water, investment regulation, infrastructure development, and biofuel mandates, as seen, for example, in many parts of South America (Oliveira, 2021; Urteaga-Crovetto and Segura-Urrunaga, 2021) and Sub-Saharan Africa (Gasparatos and Ahmed, 2021; Lundy, 2021; Ahmed and Gasparatos, 2021b). National governments often further assist this enabling policy environment by developing and disseminating new crop varieties, facilitating certification through national land registries, and providing economic incentives such as subsidies for agrochemicals or removal of export taxes (Flachs, 2021; Oliveira, 2021). Discourses and narratives centred around economic growth, rural development, and agricultural modernization usually frame such national-level processes (see "Narratives and discourses").

However, such national-level processes create the pre-conditions for some of the negative outcomes of industrial crop production that manifest at the local level (see "Local level"). Such examples include national policies, regulations, and policy frameworks that (a) accelerate ecological transformation by facilitating the targeting of marginal and ecologically important areas for industrial crop production (Urteaga-Crovetto and Segura-Urrunaga, 2021; Oliveira, 2021), (b) increase the vulnerability of national economies through an opportunistic mindset that decreases national economic diversification and value-addition (Lundy, 2021; Gasparatos et al., 2021), (c) facilitate the consolidation of power to just a few actors (Gasparatos et al., 2021; Oliveira, 2021), or (d) repress peasant demands for agrarian reform and indigenous struggles for autonomy and territorial control (Oliveira, 2021). Many of these outcomes are linked to discourses and narratives centred around marginal land, land grabbing, food security/competition with food, and reconfiguration of access to resources (see "Narratives and discourses").

It is worth noting the, sometimes, great polarization among interests at the national level, with such an example being the possible re-classification of maize from a staple food crop to an industrial crop in China (Zhang, 2021). On the one hand, agribusinesses, large-scale farmers, and government officials associated with the Ministry of Commerce and the National Energy Commission argue that maize should be re-classified, originally for biofuel purposes and gradually as silage, to boost economic diversification and agricultural modernization (see "Narratives and discourses"). The "grain-to-fuel" and "grain-to-feed" programme is the main policy instrument advancing this position. Conversely, scholars and policy-makers mainly associated with the Ministry of Agriculture have opposed this re-classification, on the grounds of national food security and protecting the peasantry (see "Narratives and discourses").

Local level

Many of the major ecological, agrarian, and socioeconomic transformations associated with industrial crops unfold at the local level. This is often linked to the manifestation of different impacts related to landscape conversion and the reconfiguration of institutions and local livelihoods.

When it comes to ecological transformation, as discussed throughout this edited volume, the large-scale and smallholder-based industrial crop systems convert extensive land areas (often natural vegetation), to create large monocultures (Ahmed and Gasparatos, 2021a; Ahmed et al., 2021). These processes degrade natural ecosystems, with some of the examples discussed in this volume including cashew nuts in Guinea-Bissau (Lundy, 2021), cocoa and sugarcane in Peru (Urteaga-Crovetto and Segura-Urrunaga, 2021), sugarcane in Eswatini (Gasparatos and Ahmed, 2021), jatropha and oil palm in Ghana (Ahmed and Gasparatos, 2021b) and soybeans in South America (Oliveira, 2021). Often industrial crop production is intensified through environmentally detrimental production practices such as extensive fertilizer and agrochemical use (Flachs, 2021; Zhang, 2021) or extensive irrigation, sometimes in water-scarce environments (Gasparatos and Ahmed, 2021; Urteaga-Crovetto and Segura-Urrunaga, 2021). However, some production practices seek to reduce the overuse of agrochemicals or increase agro-diversity, e.g. organic cotton (Flachs, 2021). It is worth noting that the shift to monocultural production models could increase the vulnerability of these cropping systems to environmental conditions, pests, and diseases, with significant ramifications for rural livelihoods and national economies dependent on industrial crops (Lundy, 2021).

Industrial crops are major agents of agrarian transformation in the global South, as they are usually promoted in rural areas containing agrarian systems mainly geared towards food production, and especially subsistence agriculture. When promoted in such agrarian contexts, industrial crops can catalyze shifts from traditional farming practices and markets as witnesses, for example, in the (a) shift from dryland food crop production to the commercial production of sugarcane in Eswatini (Gasparatos and Ahmed, 2021) and oil palm in Ghana (Ahmed and Gasparatos, 2021b), and (b) the adoption of new technologies for smallholders-based production in India (Flachs, 2021) and soybeans in

South America (Oliveira, 2021). Furthermore, industrial crops can alter radically power dynamics in some rural areas as witnessed through the consolidation of power to few actors during soybean production in South America (Oliveira, 2021), sugarcane production in Eswatini (Gasparatos et al., 2021), cocoa and sugarcane production in Peru (Urteaga-Crovetto and Segura-Urrunaga, 2021), or even the re-classification of maize in China (Zhang, 2021). In some extreme cases, the agrarian transformation caused by industrial crops can lead to deep and lasting changes in agrarian systems such as the emergence of new social classes (Ahmed and Gasparatos, 2021b; Gasparatos et al., 2021) or major shifts in gender roles (Ahmed and Gasparatos, 2021b).

Socioeconomic transformation is usually associated with the changes that industrial crops bring on rural livelihood strategies, both for those directly engaged in industrial crop production, as well as those that are not (Ahmed and Gasparatos, 2021a). For example, smallholder-based cashew production is the main income source for a large fraction of the rural population in Guinea-Bissau, with many rural households specializing completely in cashew (Lundy, 2021). Similar examples include smallholder-based irrigated sugarcane production in Eswatini (Gasparatos and Ahmed, 2021), cotton production in India (Flachs, 2021), and oil palm production in Ghana (Ahmed and Gasparatos, 2021b). Sometimes, large plantations can provide wider rural development benefits (Ahmed and Gasparatos, 2021a), influencing the generation of new off-farm income opportunities such as petty trading (Ahmed and Gasparatos, 2021b) and a more diversified local economy (Gasparatos et al., 2021).

However, more often than not, engagement in waged plantation employment offers low salaries and entails precarious working conditions (Ahmed and Gasparatos, 2021a, 2021b). Similarly, industrial crop production (especially large-scale) can have negative effects on surrounding communities (including indigenous communities) and local community members not involved in industrial crop production, especially when land displacement and loss occur (Ahmed and Gasparatos, 2021a; Oliveira, 2021; Urteaga-Crovetto and Segura-Urrunaga, 2021). This can manifest in, among others, the loss of livelihood options linked to forest resources such as fuelwood, timber, and non-timber forest products following the extensive ecological transformation around large-scale industrial crop systems (Gasparatos and Ahmed, 2021; Ahmed and Gasparatos, 2021b).

Some additional socioeconomic ramifications include food insecurity, gender inequality, and social conflicts (Ahmed and Gasparatos, 2021a). For example, the risk of food insecurity might increase due to the (a) extensive conversion of food cropland for cashew production in Guinea-Bissau (Lundy, 2021) and oil palm in Ghana (Ahmed and Gasparatos, 2021b) and (b) complete or partial diversion of labour for waged plantation employment in Ghana (Ahmed and Gasparatos, 2021b) and Eswatini (Gasparatos and Ahmed, 2021). Many of these socioeconomic outcomes are gender-differentiated with women having fewer opportunities to engage meaningfully in industrial crop production or experiencing disproportionally negative impacts (see "Social differentiation" for more details from Ghana). Finally, some of these uneven socioeconomic outcomes can give rise to local agitations and social conflicts, as has been witnessed in the context of large-scale land acquisitions (LSLAs) in Peru (Urteaga-Crovetto and Segura-Urrunaga) and Ghana (Ahmed and Gasparatos, 2021b).

Marginalization

Marginalization and industrial crops intersect in multiple ways as observed throughout the edited volume (Ahmed and Gasparatos, 2021a). Some of the more prevalent mechanisms include (a) industrial crop production (or specific production approaches/models) marginalizing (some) producers, (b) industrial crop production marginalizing entire local communities or community members not involved in it, (c) some marginalized groups being more likely to engage in industrial crop production through specific roles, which subsequently reinforces their marginalization, (d) some modes and approaches of industrial crop production being promoted as an intervention to reduce marginalization.

Related to (a), small-scale farmers are increasingly becoming marginalized and playing a secondary role in the prevailing consolidated and integrated system of soy production in South America, often getting proletarianized in the process through the *"pools de siembra"* system (Oliveira, 2021). Similarly, there are concerns that a large proportion of small-scale farmers will not be able to benefit from maize re-classification in China, thus becoming marginalized in the process by being excluded from the possible benefits, and possibly being displaced from their land (Zhang, 2021).

Related to (b), the prevailing soybean production systems in South America have caused the marginalization of many indigenous communities, poor peasants, and rural workers due to displacement, land loss, and unfair incorporation into the highly consolidated and integrated production system (Oliveira, 2021). Similarly the conversion of marginal land for cocoa and sugarcane production in Peru "appreciated" its value, marginalizing in the process the local communities that lost access to this land, and essentially some of their main livelihood options (Urteaga-Crovetto and Segura-Urrunaga, 2021). The selective inclusion of some community members in irrigated sugarcane smallholder schemes in Eswatini has "created" local communities of "haves|" and "have nots", not only due to access to the more lucrative livelihood options associated with sugarcane production but also due to better access to water resources especially during periods of water scarcity (Gasparatos et al., 2021).

Related to (c), marginalized groups such as landless, poor, and migrants often seek employment in industrial crop plantations due to the lack of other livelihood alternatives. Considering that waged plantation employment is often low skilled, badly paid, precarious, and lacking labour standards, this can further reinforce their marginalization over time (Ahmed and Gasparatos, 2021a). For example, many of the worst-paying and more precarious jobs in oil palm and jatropha plantations in Ghana have been taken up by landless community members, women, and migrants, largely due to their lack of access to land that prevents their engagement in the production of oil palm or other crops

(Ahmed and Gasparatos, 2021b). Similarly, the workers in community sugarcane plantations in Eswatini often come from households on marginal land that was excluded from involvement in the more lucrative irrigated sugarcane production. Thus they gain much lower income over time, which might reinforce their marginalization (Gasparatos et al., 2021).

Related to (d), industrial crops have, on many occasions, been promoted with the aim of reducing or avoiding marginalization. For example, cotton cooperatives in India seek to improve access to knowledge and build social capital among farmers as a means of enhancing their ability to benefit from GM and organic cotton production (Flachs, 2021). This includes farmers who are highly marginalized, poor, and at a high risk of suicide due to lack of access to resources, and especially reliable electricity and irrigation (Flachs, 2021). Furthermore, the notion of inclusivity in value chains and social benefits usually frames the efforts of NGOs and other actors in the organic cotton sector, which can help reduce marginalization (Flachs, 2021). Similarly, despite some counter-productive outcomes, the initial development of the irrigated sugarcane smallholder schemes in Eswatini (and the support offered by the national government) was in an effort to both bring marginal lands to commercial agriculture and reduce the marginalization of these truly poor and vulnerable agrarian communities (Gasparatos and Ahmed, 2021).

Social differentiation

Social differentiation manifests in many forms in industrial crop settings; for example, across lines of gender, age, education, wealth, or social class, among others (Ahmed and Gasparatos, 2021a). Usually, this reflects (a) the differentiated ability of social groups to engage in (and essentially benefit from) industrial crop production; (b) the differentiated ability of some industrial crop producers to access or benefit from prevailing/novel industrial crop production practices and markets; and (c) the differentiated experience of the negative impacts of industrial crop production.

Related to (a), a major example is the lower ability of youth and women to engage in cashew production in Guinea-Bissau due to their differentiated (lower in this case) access to land (Lundy, 2021). Another example from Ghana is the lower ability of women to both engage in smallholder-based oil palm production due to prevailing land tenure systems, and be employed in skilled and better-paying tasks in oil palm and jatropha plantations due to perceptions of high work requirement and/or low capacity (Ahmed and Gasparatos, 2021b). The third example comes from Eswatini and relates to how only some community members were able to be involved in irrigated sugarcane production and reap both long-term economic benefits and better access to water (Gasparatos and Ahmed, 2021). Even though in this case the differentiation was rather arbitrary based on proximity to the Komati River, it has caused profound and lasting changes in the agrarian and socioeconomic make-up of the local communities (Gasparatos and Ahmed, 2021).

Related to (b), an example is the increasingly consolidated and integrated nature of some industrial crop production systems differentiates the ability of some stakeholders to obtain benefits. For example, some transnational and vertically integrated corporations and independent soy farmers in South America reap disproportional benefits compared to smaller family farmers (Oliveira, 2021). The former form essentially an elite that dominates the soybean production system, causing most of the ecological transformation (see "Local level") and gaining disproportional benefits from the increased flexing in the sector (see "Narratives and discourses"), while the latter gain less and are increasingly squeezed out during the transition of the soybean production system (Oliveira, 2021).

Another example is the visible social differentiation in who gains access to the benefits of maize re-classification in China (Zhang, 2021). In this case, the main differentiation relates to wealth and political/business connections. In terms of wealth, wealthier farmers are more able to circulate land, and thus produce maize for silage, increasing their economic benefits from this new market. In terms of connections, some government leaders (from the county to village levels) or farmers with political or business connections seem to be more able to navigate the investment opportunities and institutions created by the grain-to-feed programme. The former benefit in terms of career advancement while the latter in monetary terms by producing silage maize and/or being engaged in new livestock business opportunities.

A third example discussed in this volume is how caste and educational differences seem to underpin the success of cotton cooperatives in India (Flachs, 2021). In this case, high-caste farmers tend to invest more money in their cooperatives compared to poorer and lower-caste farmers, thus having a greater say in how money is to be spent from the cooperatives (Flachs, 2021). This affords them also better access to information and extension services, benefiting them significantly in the long run.

Related to (c), some of the negative impacts of industrial crop systems (see "Local level") are experienced differently among social groups, with some groups experiencing the negative impacts disproportionally. For example, there is gender differentiation in the negative impacts of GM cotton cultivation in India, as women are usually responsible for agrochemical application and are thus more exposed to them (Flachs, 2021). Similarly, women-headed households around oil palm and jatropha plantations in Ghana experience more the loss of ecosystem services due to landscape conversion (Ahmed and Gasparatos, 2021b).

Narratives and discourses

Various narratives and discourses have been mobilized to justify the actions of proponents and critics of industrial crops alike, across the different stages of industrial crop adoption, production, processing, trade, and use (Ahmed and Gasparatos, 2021a). Narratives and discourses are usually mobilized to (a) justify or oppose industrial crop promotion and expansion, (b) assist or hinder the operationalization of industrial crop systems, and (c) criticize or praise the impacts and outcomes of industrial crop systems. Sometimes, these narratives and discourses are more academic, sometimes they originate from (or target) policy and practice communities, and sometimes these lines are rather blurred.

Related to (a), the notion of the "comparative advantage" has been one of the most popular narratives and discourses mobilized to justify the promotion of industrial crops (Moseley, 2021; Ahmed and Gasparatos, 2021a). One of the more visible uses of this discourse discussed in this edited volume has been in the context of cashew production in Guinea–Bissau. In particular, the comparative advantage has been used over decades and successive governments, leading to the current over-reliance of the national economy on the crop (see "National level"), with all the associated vulnerabilities and negative impacts this entails (see "Local level") (Lundy, 2021).

The notion of comparative advantage is often linked to narratives and discourses of economic growth, rural development, and foreign exchange generation. For example, in India, the cotton sector has been viewed as an important engine of rural economic development that needs to be strengthened through appropriate economic incentives such as subsidies (see "National level") (Flachs, 2021). Similarly, soybean expansion in South America has been justified on grounds of its foreign exchange generation potential (Oliveira, 2021), agro–industrial projects in Peru based on their perceived ability to attract international investments (Urteaga-Crovetto and Segura-Urrunaga, 2021), sugarcane in Eswatini based on its employment and poverty alleviation potential (Gasparatos and Ahmed, 2021), and the re-classification of maize in China based on expectations of market diversification and the generation of higher-value products (e.g. biofuels, biomaterials, livestock feed) (Zhang, 2021).

"Technological optimism" and "agricultural modernization" have been some allied discourses and narratives mobilized to justify the promotion of industrial crops. For example, in India, some technological options such as GM and organic cotton are seen as uncomplicated interventions to increase yields in the former case or enhance sustainable cotton production in the latter case (Flachs, 2021). Furthermore, implicit economic signals and incentives influence many private companies in South America to improve technological and logistical aspects throughout the soy value chain to increase production efficiency (Oliveira, 2021). Local-level agricultural modernization initiatives have been also used to justify the re-classification of maize in China, including new investments and transformations associated with the "grain-to-feed" programme (Zhang, 2021).

On the contrary, discourses and narratives linked to food security and the competition between bioenergy and food crop production have been used to oppose industrial crop production. For example, "food vs. fuel" has been one of the more critical such narratives used to oppose the production of first-generation biofuels from food crops (Shortall and Helliwell, 2021). Similar, concerns over national food security have posed one of the main obstacles for the re-classification of maize and soy in China from staple food crops to industrial crops (Zhang, 2021). This has become very noticeable during the recent trade wars that complicated the access to cheap maize in China and soybeans from the US, reinforcing these concerns over food insecurity (Zhang, 2021).

"Land grabbing" has been another prominent narrative and discourse mobilized to oppose industrial crop expansion in the global South. A key element of this discourse is that LSLAs often result in major changes in land access and tenure rights, which usually benefit large investors and disadvantage local communities (Ahmed and Gasparatos, 2021a). This can happen through different mechanisms, ranging from complicated processes during LSLAs that are difficult for some actors to follow to misinformation and intimidation (Urteaga-Crovetto and Segura-Urrunaga, 2021; Ahmed and Gasparatos, 2021a, 2021b).

Related to (b), the notion of "marginal land" is one of the more common discourses and narratives used to facilitate the operationalization of industrial crop production. Marginal land is essentially a discursive construct used to denote land of lower ability to grow food crops or host cost-effective agricultural production, which is as a result fit to grow industrial crops (Shortall and Helliwell, 2021; Ahmed et al., 2021). Often, there is a concerted effort to establish land marginality through narratives that craft land irrelevance, which is followed by narratives of value crafting in order to appreciate the land value and transform it into a lucrative commodity that can be traded internationally (Urteaga-Crovetto and Segura-Urrunaga, 2021).

"Flexing" is another discursive construct used to facilitate the operationalization of industrial crop production. In particular, flexing is mobilized to convey the increasingly dynamic production approach for certain industrial crops in terms of agricultural inputs and outputs for diverse markets (e.g. food, livestock feed, fuel, industrial products). The concept of flexing has been implied or used directly in the context of the transformation of soybean production in South America (Oliveira, 2021) and the re-classification of maize in China (Zhang, 2021).

Related to (c), "accumulation by dispossession" and "class formation" have been two rather common and complementary discourses and narratives used to critically appraise industrial crop production in the global South. The entry point for the former is that the consolidation of family and common land for LSLAs creates the pre-conditions for widespread land dispossession and the elite capture of benefits (Ahmed and Gasparatos, 2021a; Gasparatos et al., 2021). In some cases, this has led to the accumulation of benefits by international investors through the dispossession of local communities (Urteaga-Crovetto and Segura-Urrunaga, 2021). In other cases, this dispossession has created landless households that have no other option than to become labourers in large-scale plantations (Ahmed and Gasparatos, 2021b). Over time, this can deepen social differentiation and give rise to new social classes, as it seems to become the case around large-scale production sites in Ghana and Eswatini (Ahmed and Gasparatos, 2021b; Gasparatos and Ahmed, 2021).

"Neo-natures" is a rather powerful narrative to critically discuss the ecological transformation caused by industrial crops. For example, this narrative has been employed in the context of soybean expansion in South America to highlight how ecological transformations are linked to the development of (a) biotechnological innovations that produce novel soybean varieties; (b) anthropogenic landscapes dominated by soybean monocultures emerging from the conversion of rainforest and grassland; (c) new types of socio-natures emerging through the *commodification* of "nature's products, places, and processes" (Oliveira, 2021).

Finally, the reconfiguration of "access to resources" has been also mobilized to critically discuss the ecological and institutional transformation in the context of industrial crop production. For example, the reconfiguration of access to land and water were major features of irrigated sugarcane production in Eswatini (Gasparatos and Ahmed, 2021) and agro-industrial development in Peru (Urteaga-Crovetto and Segura-Urrunaga, 2021).

Outro: Political ecology as a critical lens for industrial crop systems

Industrial crops are a very diverse class of crops that are promoted in different parts of the global South for various reasons, through different processes and discourses, and by actors with varying vested interests. Throughout this volume, we acknowledge that industrial crops are agents of ecological, agrarian, socio-economic, and institutional transformations. Yet, we argue that industrial crop systems are not inherently good or bad for ecosystems, local communities, and national economies, but we acknowledge that the processes harnessing their production and outcomes are indeed very differentiated depending on a host of context-specific factors. This makes it extremely difficult to distill an all-encompassing guideline on how to ensure their production in an environmentally, economically, and socially sustainable manner. This is largely due to the fact that their production always creates winners and losers at different spatial and temporal levels, which are very socially differentiated.

Considering the above, we argue that it is essential to critically understand the main elements of industrial crop promotion, justification, production, trade, and use on a case-by-case basis through robust and interdisciplinary lenses. The different contributions collected in this volume have made a strong case that the interdisciplinary field of political ecology can offer such a lens with its inherent focus on multi-scalar analysis, marginalization, social differentiation, and underlying narratives and discourses. Such a lens can critically deconstruct how industrial crop systems act as agents of ecological, agrarian, socioeconomic, and institutional transformations, and thus identify pressure points and opportunities to improve their performance and reduce, to the extent possible, their negative outcomes. This would be critical in ensuring that industrial crop systems can fulfil their potential for economic growth and rural development in the global South while preventing or minimizing destructive ecological, agrarian, socioeconomic, and institutional transformations.

References

Ahmed, A., Gasparatos, A., 2021a. Political ecology of large-scale land acquisitions and land grabbing for industrial crops, in Ahmed, A., Gasparatos, A. (Eds.), Political Ecology of Industrial Crops. Earthscan/Routledge, London.

Ahmed, A., Gasparatos, A., 2021b. Changing agrarian dynamics in oil palm and jatropha production areas of Ghana: A feminist political ecology perspective, in Ahmed, A., Gasparatos, A. (Eds.), Political Ecology of Industrial Crops. Routledge, London.

Ahmed, A., Jarzebski, M., Gasparatos, A., 2021. Industrial crops as agents of transformation: Justifying a political ecology lens, in Ahmed, A., Gasparatos, A. (Eds.), Political Ecology of Industrial Crops. Routledge, London.

Flachs, A., 2021. The political ecology of genetically modified and organic cotton in India as agents of agrarian transformation, in Ahmed, A., Gasparatos, A. (Eds.), Political Ecology of Industrial Crops. Routledge, London.

Gasparatos, A., Ahmed, A., 2021. Political ecology of industrial crops: Towards a synthesis and systematization, in Ahmed, A., Gasparatos, A. (Eds.), Political Ecology of Industrial Crops. Earthscan/Routledge, London.

Lundy, B.D., 2021. Cashews in conflict: The political ecology of cashew pomiculture in Guinea-Bissau, in Ahmed, A., Gasparatos, A. (Eds.), Political Ecology of Industrial Crops. Routledge, London.

Moseley, W.G., 2021. Political agronomy 101: An introduction to the political ecology of industrial cropping systems, in Ahmed, A., Gasparatos, A. (Eds.), Political Ecology of Industrial Crops. Routledge, London.

Oliveira, G.L.T., 2021. Political ecology of soybeans in South America, in Ahmed, A., Gasparatos, A. (Eds.), Political Ecology of Industrial Crops. Routledge, London.

Shortall, O., Helliwell, R., 2021. Marginal land for bioenergy crop production: Ambiguities, contradictions and cultural significance in policy and farmer discourses, in Ahmed, A., Gasparatos, A. (Eds.), Political Ecology of Industrial Crops. Routledge, London.

Urteaga-Crovetto, P., Segura-Urrunaga, F., 2021. Transforming nature, crafting irrelevance: The commodification of marginal land for sugarcane and cocoa agroindustry in Peru, in Ahmed, A., Gasparatos, A. (Eds.), Political Ecology of Industrial Crops. Routledge, London.

Zhang, L., 2021. The political ecology of maize in China: National food security and the reclassification of maize from staple to industrial crop, in Ahmed, A., Gasparatos, A. (Eds.), Political Ecology of Industrial Crops. Routledge, London.

Index

For Product Safety Concerns and Information please contact our EU
representative GPSR@taylorandfrancis.com
Taylor & Francis Verlag GmbH, Kaufingerstraße 24, 80331 München, Germany

www.ingramcontent.com/pod-product-compliance
Lightning Source LLC
Chambersburg PA
CBHW060447240326
41598CB00088B/3909

9 781032 062136